The Body, the Dance and the Text

The Body, the Dance and the Text

Essays on Performance and the Margins of History

EDITED BY BRYNN WEIN SHIOVITZ

McFarland & Company, Inc., Publishers
Jefferson, North Carolina

ISBN (print) 978-1-4766-7189-5
ISBN (ebook) 978-1-4766-3485-2

LIBRARY OF CONGRESS CATALOGUING DATA ARE AVAILABLE

British Library cataloguing data are available

© 2019 Brynn Wein Shiovitz. All rights reserved

No part of this book may be reproduced or transmitted in any form or by any means, electronic or mechanical, including photocopying or recording, or by any information storage and retrieval system, without permission in writing from the publisher.

Front cover image by Linda Carreiro

Printed in the United States of America

*McFarland & Company, Inc., Publishers
Box 611, Jefferson, North Carolina 28640
www.mcfarlandpub.com*

For Josh

Acknowledgments

This collection would not have come to fruition without my aunt Toni Wein's zealous encouragement to take on a project of this magnitude. I want to especially thank my colleagues Gwyneth Shanks and Melissa Melpignano who convened with me at the Nineteenth Century Studies conference in Charleston to explore this volume in its earliest stages and provided feedback on my introduction. I am fortunate for Margaret Morrison's support of my research on Billy Kersands and other forgotten tap dancers. It is easy to get discouraged in the archive and her enthusiasm for my work kept me going. An enormous amount of gratitude goes out to my mentor Susan Leigh Foster who really taught me how to write about the body. A big thank you to Josh, Zoe, Mom, Dad and the rest of my inner circle for tolerating me throughout this process. Finally, I thank all of the authors who trusted me with their ideas and contributed to this volume.

Table of Contents

Acknowledgments vi

Introduction: Writing the Body, Staging the Other
 Brynn Wein Shiovitz 1

Part 1: Writing the Body

Mouth Over Matter: Writing Early Tap Dance in the Margins of the Black Body
 Brynn Wein Shiovitz 18

Spain in the Basement: Dancing Race and Nation at the Paris Exposition, 1900
 Kiko Mora *and* K. Meira Goldberg 41

White Dreadlocks: Black Aesthetics in the Work of Louise Lecavalier and La La La Human Steps
 MJ Thompson 68

Flowers of Menace: Stephen Petronio's Rites of Spring
 Constance Valis Hill 85

Part 2: Transmissions and Traces

Othering the Religious Right: Ameritude, Whiteness and the USA Freedom Kids
 Michelle T. Summers 104

Escape Routes and Roots: Rewriting the Narrative of the Vulgar Body
 A'Keitha Carey 117

Lo Que Queda/That Which Remains: Dancing Bodies, Historical Erasure and Cultural Transmission
    MICHELLE HEFFNER HAYES   134

Screaming Soundscapes: The Sounds of Puerto Rican Contemporary Performance in the Work of Teresa Hernández and Ivette Román
    LYDIA PLATÓN LÁZARO   152

Part 3: Staging the Other

Always Already: The Jewish Body as Victim and Victimizer
    REBECCA K. PAPPAS   172

Israel Galván's Aesthetic Anarchism: An Ethics Instantiated in Motion
    NINOTCHKA D. BENNAHUM   190

Brown and Black: Performing Transmission in Trisha Brown's *Locus* and Hosoe Eikoh and Hijikata Tatsumi's *Kamaitachi*
    MICHAEL SAKAMOTO *and* CHRISTOPHER-RASHEEM MCMILLAN   202

The Bustle, the Body and Stillness: Re-Centering Modernities through the Broadway Musical
    GWYNETH SHANKS   219

Gradations of Presence: Armida in Nineteenth-Century Italian Dance Librettos
    MELISSA MELPIGNANO   235

Choreogrammatics: About the Cover of This Book
    LINDA CARREIRO   253

About the Contributors   263

Index   265

Introduction

Writing the Body, Staging the Other

BRYNN WEIN SHIOVITZ

I learned rather quickly that trying to trace the origins of dance steps popularized on the minstrel stage is an almost futile endeavor. Even after a full dissertation, I had made little progress in the way of tracing century-old tap dances back to particular "authors." Instead arose a frustrating puzzle of trying to figure out which writers were in fact telling the truth. That is, depending on the race of any given journalist, dances like the Virginia essence or the buck-and-wing took on entirely different genealogies. White journalists, for example, tended to write about the black performers of these dances as curiosities, whereas these writers' black contemporaries would focus their attention on these tap dancers' physical talents. Perhaps there was an important lesson to be had here: vernacular dance has no *one* author, but the process by which we arrive at a particular step speaks to a gamut of historical processes worthwhile of further scholastic inquiry. The corpus of literature which remains documents, as Susan Foster noted in the mid–1990s, an "encounter between bodies and some of the discursive and institutional frameworks that touch them, operated on and through them, in different ways."[1] The archive—and the bodies it embalms—has thus been "choreographed" to a certain degree.

The Billy Rose Theater Division—a subset of the New York Public Library for the Performing Arts—has yet to digitize most of its materials and still uses the outdated card catalog; the kind of dance research I do tends to fall within the limits of this tangible index. While at first enchanted by the prospect of digging with my hands, I became increasingly frustrated by my findings, or lack thereof. My searches often yielded directions rather than book titles or call numbers: In a search for "Billy Kersands," for example, the card said, "see Georgia Minstrels"; the card for Georgia Minstrels instructed me to "see Brooker and Clayton's Georgia Minstrels," and the card for Brooker and Clayton's Georgia Minstrels advised me to "see Charles Hicks/see also Ethiopian minstrels/see also blackface minstrel performers." The card for blackface minstrel performers was rich with names but lacking in call numbers so I randomly selected a few names and searched for those individual performers. If present these cards told me to "see blackface minstrelsy," or, better yet, "see [the name of someone for whom I had already searched]." With aching quads and fatigued eyes (a "bodily researching"), I had danced in a small circle around the Billy Rose Theater Division with nothing to show for it other than half-finished call slips, hopeful scribbles, and many questions: *What do all of these names have in common? Why are these names listed if the library has no record of them? Have I been looking in the wrong places? Are these people real?* It was then that I realized that these people *were* real, but that they did not *really* exist in the archive. These were the names of people who librarians must have come across in random clippings while trying to categorize the histories of other, more famous minstrel performers. George Primrose, for example, takes up a good quarter of a drawer.

It was not until I came across a tab labeled "blackface minstrelsy" a couple years ago that I had any material to request from special collections. Ironically my "discovery" marked one of the more discouraging moments in my process. Since "blackface minstrelsy" was allegedly the only common denominator among these findings—and of course there was no distinction made between black-on-white minstrelsy and black-on-black minstrelsy, companies, individuals, or the "art of blacking up"—I sat for the next year reading newspapers from the 1850s up through the early twentieth century, word for word, name for name, trying to make sense of the relations between these performers. Every day I would leave the library with a long list of names and dance steps, but no closer to any sort of conclusion about what these steps looked like or where I might find more information about each of these particular performers. Unable to locate even 70 percent of the materials I requested, the lead research librarian commented on how I was "really striking out this week." After exhausting the "blackface minstrelsy" category, the only other place where I found anything remotely helpful was in a small section labeled "Negroes in the Theater," a title which insinuated that nobody had updated the indexical titles since the early 1940s.

While it would have been easier to have requests retrieved without complications and with no confusion as to why articles spanning two centuries were to be found within the same box, my frustrations are what in fact inspired this edited collection. History is not only shaped by how people write the body, nor is it solely the way information is transmitted through the body or the traces those bodies leave behind. It is also not merely a matter of how playwrights, directors, and choreographers stage the body. History is shaped by a combination of all of these things, and, in particular, the ways in which the archive writes (or writes the body out of) History. What my findings—or lack of findings—say about some of the most famous black performers of the minstrel stage is not just that they have been invisibilized, but that their legacy has been completely erased.

In many ways this book takes a cue from the dialogue set in motion by Thomas DeFrantz in his 2002 edited collection, *Dancing Many Drums*.[2] DeFrantz's book convenes around issues of historiography, identity, and categorization of African American dance. *The Body, the Dance and the Text* explores a certain ordering of the body, but extends its scope beyond the African American tradition, instead interrogating a more generalized "Other." That is, my main goal in editing this volume is to explore the process by and through which bodies come to be marginalized and how History is thus mediated by certain taxonomical approaches. The lack of specialized collections chronicling a history of the Other (and others) and the supposed disappearance of credible source material extends far beyond the Billy Rose Theater Division and my quest to find three black tap dancers. This volume seeks to complicate the archive and the repertoire,[3] dislocated bodies and the writing of them, their traces and how they have come to be transmitted, the visible text of History and its interstices.

Writing the Body

The idea of "writing the body" is nothing new: Susan Foster has introduced to dance studies both the "bodily writing"[4] and the idea of viewing the historian as a sort of choreographer. Foster's theories surrounding the relationship between the body and writing, or the dance and text act as one of this collection's primary frameworks: the authors of this volume unanimously accept that bodies, whether dancing or still, are thrown into a "discursive framework" that continually "sustain[s] a conversation" and "lucidly enunciates their specific corporeal identities."[5] Each of this book's fourteen essays examines the ways in which distinct institutional frameworks operate upon and through bodies, the writing of these bodies, and these bodies' writing, in distinct ways.[6] Some of the essays in this collection will focus on bodies writing while others concentrate more on the writers and thinkers who have written the body, literally and figuratively. Each author explores both the way that history has shaped certain bodies and the way that the construction of History informs bodily theories. Together we ask: *How can we trouble the certainty of narratives which have existed thus? What bodies are recuperated during this excavation? What are the relations—rather than the differences— between these bodies we recover? How can we learn from and re-write these narratives?*

The writing of bodies is directly related to the ways in which bodies are oriented. As Sara Ahmed has analyzed, because different bodies are differently "oriented"—gendered, sexualized, raced, in time, in space—the different "orientations that one takes toward the objects with which it comes into contact" determines how something might appear.[7] Different orientations can create a set of frictions, frictions which the written word tends to exacerbate in virtue of its permanence. I view the combination of dance's ephemerality and writing's permanence as being a major contributing factor in identity formation and reinforcing notions of otherness.

This book's second thread is in dialogue with what André Lepecki identifies as the "photological project," or the process wherein dancers are subjected to the "archival structure of command."[8] This is not to say that the authors of this text are photologists, but instead to note how our projects are related: the documentation of bodies has historically fixed bodies and the narratives that accompany them. Bodies which do not fit into a particular set of social values have been "cast off" or abjected. As Lepecki has acknowledged, taking a Derridian stance on writing as *différance*[9] offers dance studies a space where "both writing and dancing plunge into ephemerality. With Derrida, dance finally finds the form of writing that is in harmony with dance's current ontological status ... the return to symmetry derives from the acknowledgment that both writing and dancing participate in the same motion of the trace: that which will always be already behind at the moment of its appearance."[10] The authors of this text *read* dance as photologists and offer readers fourteen new narratives of writing the body as *différance*.

Staging the Other

I have organized this book in a manner that seeks to decolonize the photological project as well as the reification that occurs during this historical process of trying to hold on to bodies which colonial approaches deem "different" or "other." It became evident in my researching of Kersands and other famed black tap dancers of the nineteenth century that white newspapers across the United States preserved an Antebellum-era mindset well into the twentieth century by freezing antiquated (and false) descriptions of the black body through writing. This is but one facet of the colonial fantasy where the writing, to use Homi Bhabha's words, dramatizes "the separation—*between* races, cultures, history, within histories—a separation between *before* and *after* that repeats obsessively the mystical moment or disjunction."[11] The question of this volume becomes how to bring the dislocated, the dejected, and the abject together.

The other key thematic of this book is "staging the other," here defined by the viewing of the performance space as a safe zone to scrutinize the dejected, as Judith Butler[12] has theorized, along with an uncountable number of performance studies scholars who look at specific others and groups of others on the stage. This phrase also encompasses the ways in which directors, choreographers, and performers might use the stage as a way of inverting shame into something productive. This is a theory which writers like Ralph Ellison or Eric Lott have explored in relation to the blackface mask, but some of the authors in this volume examine this psychological inversion on a broader plane. For example, Lydia Platón Lázaro's essay "Screaming Soundscapes" does this through an examination of the body's aural register and how the voice is capable of obscuring silences of oppression onstage. Finally, I define "staging the other" as a space where performers might consciously resist reification, classification, or even challenge standards as broad as those that define a white Western postmodernity. Thus, if identification and "misidentification" are implied by my first understanding of staging the other, authors taking up this second definition of the phrase explore the moving/writing body as exemplifying a kind of "dis-identification," or one that resists dominant ideologies.[13] Accordingly, several of the essays in this book interrogate the location of agency through bodily performances that contest prevailing social systems and corporeal stigmas. Together these definitions seek to challenge previously conceived notions of difference and ask, as Karen Shimakawa does, "What is the productive potential of abjection?"[14] "Staging the other" uses "writing the body" within a performance space to transmit the body in new lights and to leave behind unique traces.

While one of the goals of this book is to try and draw parallels between different types and tokens of Otherness,[15] we must take each type of other as its own entity and try to understand its distinctness as coming out of a particular time and place. In other words, understanding the context out of which each other stems allows us to recognize each's subjectivity and unique experience while at the same time acknowledging the ways in which the psychological, linguistic, and corporeal process of othering may transcend time and place. Thus, studies of time and space figure into the constructions of otherness in each of these essays.

Time and Space

We see the role of time unfold as both a specific temporality—i.e., as a Western chronology—and as a way of thinking about particular historical moments or periods—e.g., modernity or postmodernity—which organize time in terms of values or scientific ideas. The essays in this volume are not organized chronologically but they do cover a period of about three centuries.

Melissa Melpignano's contribution, "Gradations of Presence," takes us back to the early nineteenth century as she uncovers the corporeal presence of dancing bodies in Italian dance manuals. Michelle T. Summers, on the other hand, offers the book a very contemporary analysis as she explores the othering of the religious right which has transpired in the wake of Donald Trump's presidency.

Questions of modernity and postmodernity arise for many of these authors, and often these inquiries overlap with transatlantic circulation which in turn seems to necessitate hybridity of form. In "The Bustle, the Body and Stillness," Gwyneth Shanks examines Dot's desire to be in step with modernity, to be *en vogue* as a means of connecting her to affluent women across Europe and the United States. Kiko Mora and K. Meira Goldberg's essay "Spain in the Basement" analyzes the way Spain absorbed the American cakewalk into flamenco repertoire as well as the modernist imagery of jazz into the figure of a nativized *Gitano*. This Spanish substitute for negritude in Paris might have gestured at a sort of avant-garde modernism, but as these two authors go on to unpack, the adoption of such an aesthetic reinforced a preestablished racial hierarchy between Spain and the rest of Europe and was ultimately viewed as an "uncultured spectacle." Ninotchka D. Bennahum revisits flamenco's ties to modernism in "Israel Galván's Aesthetic Anarchism," where she offers readers a glimpse into the very anti-modernist choreography of Galván. Bennahum reads Galván's break from his predecessors less as rebellion and more as a "spiritual reassessment of flamenco's historical undercurrents." In escaping flamenco's modernist undercurrents, Galván draws from a number of sources including the feminist pedestrianism of radical dance artists Anna Halprin, Simone Forti, and Yvonne Rainer as well as from Pina Bausch's neorealism, just to name a few. Again, this transatlantic dialogue sparks new traditions and forms within a given time and place. Constance Valis Hill's essay "Flowers of Menace" also tracks a lineage of modernism and postmodernism by following Petronio's earliest manifestation of modernist transformations through his *Rite* series. While not the focus of her essay, Petronio's most recent work, BLOODLINES, has been an ongoing project and tribute to a lineage of pioneering American postmodernists such as Merce Cunningham and Trisha Brown. Such play on a postmodern mindset surfaces in Christopher-Rasheem McMillan's reinterpretation of Trisha Brown's 1975 dance inside a cube, *Locus*, and Michael Sakamoto's butoh/hip hop hybrid.

In addition to considering how the body may underwrite a rupturing of time, this volume includes many essays which look closely at writing which contributes to a fracturing of the body. Several analyze the othering that occurs when one takes a synechdochic approach to the dancing body. To understand what lay at the heart of this splitting of the body into its parts requires a certain grasp of abjection's process. While the term may carry a few different meanings, I understand the process of abjection as aligned with Julia Kristeva's psychological definition. Karen Shimakawa sums it up perfectly when she writes of the process as "an attempt to circumscribe and radically differentiate something that, although deemed repulsively *other* is, paradoxically, at some fundamental level, an undifferentiable part of the whole."[16]

A handful of the essays in this book study the fracturing of the body as an attempt to "radically differentiate" the other from self, or whole human form. We see this in my essay, "Mouth Over Matter," where I explore nineteenth-century journalists' oral fixation on Billy Kersands who History has remembered as "the man with the biggest mouth in the United States," rather than as a superb tap dancer or the highest paid black performer of that century. In "White Dreadlocks" MJ Thompson also examines the ruptured body as a form of synecdoche when she looks closely at the implications of La La La Human Steps' principal dancer Louise Lecavalier's appropriation of dreadlocks not as theft, but instead as a way to challenge notions of purity and authenticity. Similarly, A'Keitha Carey in "Escape Routes and Roots" works to dismantle the negative connotations surrounding the pelvis as overly sexual and vulgar, showing that this body part and the dance movements confined to its region can, in fact, serve as a form of redemption and empowerment for many. Kiko Mora and K. Meira Goldberg discuss flamenco's exploitation at the 1900 Paris Exhibition and the way that the French used this dance as a synecdoche of Andalucía and Andalucía as a synecdoche of Spain in order to continually remind Spaniards of their subordinate status in the eyes of Europe. Finally, Gwyneth Shanks zooms in on the female lead Dot and her bustle in Sondheim's *Sunday in the Park with George*. Here we might think of Dot as standing in for one part—or literal point—of Seurat's famous *Sunday Afternoon on the Island of La Grande Jatte*, and Dot's bustle as metonym for femininity within both Seurat's pointillism and Sondheim's Broadway musical.

Methodology

Before the book even begins, I invite our readers to engage with the collection's cover, a still image from Linda Carreiro's *Shadow Boxed* (2017). While a description of this image and its theoretical underpinnings can easily be found at the end of this volume, it might be fun for you to see this work first as the artist intended: without explanation and as a "bodily viewing." As Carreiro will explain in the book's back matter, *Shadow Boxed* opens up the possibilities of bodily movement while reading. Her work as a whole emphasizes the relationship between kinesthetics and reading to reveal how the meaning of words "can be opened and even intensified through a bodily performance of text." It is thus our hope that "Choreogrammatics" will frame your experience of the text writ large and that you may continue *to move and be moved* while reading.

I have organized this book into three parts: "Writing the Body"; "Transmissions and Traces"; and "Staging the Other." While I could envision numerous configurations for these essays, this particular tripartite structure moves the reader away from obvious distinctions based on discipline, performance genre, or writing style and confronts the reader with a series of thematics which bring different kinds of "Others" into dialogue. Here Others who have historically been defined by their differences must be recognized according to those attributes which they share.

According to Julia Kristeva, the subject comes into being and maintains self-consciousness by creating borders around itself and casting off that which it deems abhorrent. This process defines individual subject formation and, according to Karen Shimakawa, is "necessary to and mutually constitutive of *national* subject formation."[17] In this light, abjection functions at the individual, national, and cultural levels and in order for any sort of coherent identity to be maintained, the Other must continually be both "made and jettisoned."[18] By making "the jettisoned" a thematic, patterns emerge between how Others are made and how identities are shaped.

The first section, "Writing the Body," takes as its central theme the way in which the written word has shaped the Other's body. The four essays in this first part confront the press and examine *who* has written the body and *how* the act of writing has both created and reinforced certain narratives. The first two essays look at dance in a more vernacular context at the *fin de siècle*. These authors understand particular "writings" of the body as ignoring national diversity by at once accentuating difference on the stage *and* archiving difference in a way that undervalues the aesthetic contributions of "Others" in the nation-building process.

In my essay "Mouth Over Matter," I aim to recover the lost dancing bodies of three black American tap dancers by highlighting the way that white journalists elided tap dance's aural nature in favor of sensationalist language. Such writing privileged the dancers' physical attributes over their artistic innovations. Uncovering a Bourdieusian "logic of transmission" I locate some of the conceivable motivations behind such writing and some probable reconstructions of Kersands, Sam Lucas, and Tom McIntosh's dancing.

Mora and Goldberg similarly recover the identity and dancing of circus equestrian, bullfighter, and flamenco singer/dancer Jacinto Padilla, "El Negro Meri," in the Lumière Brothers' two short films of Spanish dance at the 1900 Paris Exhibition. Mora and Goldberg reveal a semiotic ambivalence in the Spanish Pavillion's architecture and a recurring mislabeling of Spanish identity, including the way in which Padilla's dancing has been attributed to the renowned Sevillan maestro José Otero Aranda. Such archival mishaps they argue are "perhaps indicative of French audience's lack of discernment regarding Spanish dance."

The next two authors move this conversation forward 80 years and analyze a different kind of dance criticism on the concert stage. MJ Thompson exposes a semiotic ambivalence in descriptions of Louise Lecavalier's style. Thompson questions the way in which dance writers have, for decades, focused on Lecavalier's punk, queer, aesthetic and ignored the way that black aesthetics might be central to the company's identity. Thompson, like the previous two essays in this part, believes that eliding the diasporic threads of the dance flatten the national image by excluding major pieces of the identity puzzle. As a product of Quebec society, La La La Human Steps has come to stand in for the image of Quebec or "New France" as "culturally unique, boldly innovative, and internationally recognized." Considering the ethnic complexity embedded in Lecavalier's dreadlocks, Thompson re-writes the blackness into Lecavalier's roots and uncovers the diasporic routes of Édourard Lock's choreography.

We move from the gender/ethnic bending performances of Lecavalier to Stephen Petronio's gender play in Constance Valis Hill's "Flowers of Menace." Hill's essay takes us back in time to Nijinsky's 1913 *Le Sacre du Printemps* (*The Rite of Spring*) and its subsequent failure in the eyes of Parisians and forward in time to Petronio's radical reconstruction of *Rite*. While this essay would have been equally fitting in the book's final part on "Staging the Other," Hill's focus on Petronio's text (lay text), interviews, and the popular press to bolster many of her claims, adds a layer of "writing the body" this section has yet to explore: Hill looks at Petronio's *Rite* works—a series of dances which re-envision Stravinsky's original 1913 *Sacre* score made between 1991 and 2006—within the context of broader cultural currents of the 1980s and 1990s. Here Petronio uses several forms of body writing to "accommodate contemporary ideas about brutality, violence, and sexuality." Keen on provoking and enlightening, Petronio's choreography often writes the body in such a way that simultaneously disturbs and excites his audiences. As choreography, Petronio writes sex into his narrative: "It doesn't look like sex, but you get a very visceral feeling.... Sex is here. It belongs here and is a part of who we are." As voice, Petronio writes the gay community into the public narrative, continually confronting readers with the AIDS crisis, the complexity of gender, and abjection. Hill's essay makes clear the way in which dance, as a form of writing the body, can say something which sometimes words alone cannot.

Understanding the broader cultural currents of the *present* is an important element in this collection's second part: the essays in "Transmissions and Traces" convene around questions of location and purpose. *Where* is the body transmitted? *What* traces does it leave behind? *Why now*? The first two essays study dance in an outdoor setting while the last two look at dance on the concert stage. All four of these essays consider an Africanist aesthetic from a contemporary perspective but explore it from a variety of angles, spaces, and styles. In "Othering the Religious Right" Michelle T. Summers asks how a Trump brand fervor—as exemplified by the USA Freedom Kids' performance—mobilizes the "Christian right." By examining a few different transmissions of the girls' single "Freedom's Call," including its viral backlash, Summers explores the values these children enact as well as how liberal discourses have framed these girls as "Other." By pointing out the traces of precision drill team dance that reside in "Freedom's Call," she revisits this idea of whitewashing black dance forms for the sake of homogenization (and simplicity) that MJ Thompson previously addressed. It is through such choreographic traces and the symbolic link these girls carry between Trump and the Christian right's family values platform that Summers argues might be at the heart of why the Christian right was able to vote for Trump.

A'Keitha Carey locates a trace of the precision drill team style at the Miami Broward Carnival where she is reminded of the drum major of an HBCU (Historically Black College University) marching band drum line. The call of the drum roll and the singer's response act as both trace and command: "These lyrics served as an offering of freedom and liberation from societal constraints and oppressive occurrences." Carey, like Summers, explores the rhetoric of freedom carried by dancing bodies and the possible narratives produced through their transmission. However, unlike the whitewashing that many of these authors record, Carey's essay attempts purely to reclaim the bodies of women of color. As a major practice of Caribbean identity, the "hip wine," she theorizes, is a "subversive practice rooted in the spiritual and sublime"; it signifies redemption and empowerment.

Part practice and part research, Carey's essay segues its readers into Michelle Heffner Hayes' practice as research (PaR) conducted with her students at the University of Kansas. In "*Lo Que Queda/That Which Remains*," Hayes closely examines musician and anthropologist Raúl Rodríguez's "imaginary folklore," a style that incorporates the varied intercultural dialogue that eventually became flamenco. Through Rodríguez Hayes unpacks flamenco's diasporic underpinnings, complementing Mora and Goldberg's discoveries in "Spain in the Basement": transatlantic migration connects many of the case studies in this volume. Flamenco's influences come from Andalucía (itself a rich mix of cultures: African, Arab, Sephardic, Roma, and regional folklore), the Caribbean and the American South. By way of demonstration, Hayes shows her readers how her attempt to reinvent the "phantom dances" of Rodríguez's imaginary folklore confirmed that training shapes systems of organizing the body and its vocabulary and furthermore, that dance is "reflective of and constitutive of the conditions that produce/are produced by it."

Lydia Platón Lázaro continues this thread of cultural transmission as enacted on/through the body in "Screaming Soundscapes," where she probes the question of how to articulate freedom through the voice. Concentrating on artists Teresa Hernández and Ivette Román, Platón Lázaro studies these performers' vocal "disidentification" as attempts to escape the fixed regimes of cultural representation which have negatively marked Caribbean subjects. These "voices of migration," as she calls them resist the double colonialism in Puerto Rico by rehearsing the colonial trace inherent in the voice.

We might understand the body in many of these essays as *realizing* a discourse of difference. The essays in part three, "Staging the Other," recognize the body as capable of both realizing *and* challenging such a discourse. Together these five essays offer ways that the body might resist certain accepted narratives, either by staging ambivalent identities or through counter readings of the body.

In "Always Already," Rebecca K. Pappas explores questions of identity through her own body. Her dance "Monster" asks that both she and her audience examine how Judaism has "marked" her body: *What is the inheritance of being Jewish? What are the burdens and gains of carrying such legacies? Can we choose to own some markings and not others?* These questions of transmission and the traces—physical and political—which ethnic and cultural identity carry segue readers from the essays of the last section to those that grapple with the Other as "always already" ambivalent. "To be a body is to be many things at once," she writes. Looking at what she labels a victim/victimizer dichotomy, Pappas stages some of the many complexities of identity and uncovers her own unique relationship to Judaism in the process.

Israel Galván, the subject of Ninotchka D. Bennahum's essay on his "Aesthetic Anarchism," similarly struggles with his inheritance and what it means to belong to an exiled people. Bennahum studies the inextricable link between a body and a language, and specifically the weight of history which Galván's surname carries. The name "Galván" "belongs to exiled peoples with Sephardic origins and conquered peoples of foreign origins," an ambivalent identity of which Galván is acutely aware. The performer is also aware that his namesake holds the two central principles of modern dance: time and space. Using the nuances of his inheritance, Galván choreographs his many identities into his dancing so as to accentuate his destruction, a state that Bennahum describes as akin to Gastón Gordillo's definition of "rubble," or the "layered sediment, the rings of a tree, buried time: constellations of objects understood in relationship to historical processes." Galván uses his struggle, his ambivalence, to create a new form, a new way of reading the body in motion. Through Galván's body, Bennahum's essay extends Hayes' argument surrounding the displaced body of flamenco and the ways in which both our history and our training "shapes systems of organizing the body and its vocabulary" by excavating Galván's layered aesthetic.

"Israel Galván's Aesthetic Anarchism" also lays the foundation for Michael Sakamoto and Christopher-Rasheem McMillan's performative piece

12   Introduction

"Brown and Black," which demonstrates two more hybrid choreographies of displacement. The radical dance artists of the 1960s inspire McMillan's re-envisioning of Trisha Brown's *Locus* (1975). "Do you think blackness matters to white women in spandex?" he asks Sakamoto as part of their performance. In a quest to insert his own homosexual blackness into 1960s postmodern dance, McMillan creates *Black Lōkəs* (2017), substituting Brown's number system with the names of people who have recently been killed by the police as well as biographical information. Re-dubbing Brown's "Locus" as "*Lōkəs*" highlights a sense of cool and allows McMillan to use the text in addition to the body to "evacuate hegemony from the premises and move in…" so that the "master's tools" restructure the house rather than dismantling it (Lorde 2007: 112); this move quite literally turns Brown's relaxed approach on its head. Sakamoto reconciles his desire to be black with his Japanese ancestry, using his upbringing in a predominantly Mexican part of Los Angeles, and his personal affinity to hip hop to develop his own style. His style, a butoh-Zen-b-boying panache, helps him, like McMillan, to reimagine an American dance body and develop strategies for creating effective space for difference in contemporary dance. The work of these bodies—Pappas, Galván, McMillan, and Sakamoto—all ask their audiences to *see* them as products of diaspora, as both processes and conditions.[19]

The authors of this collection's last two essays ask their readers to examine a few "accepted" narratives of the female body. The first challenges an historic (Western) gendering of space and stillness, while the second contests the Orientalized and feminized rhetoric of time and text. Gwyneth Shanks questions how the musical revival of Sondheim's *Sunday in the Park with George* "refracts" narratives of art historical legacies through "a distinctly gendered understanding of visual cultures and modernity." Researching the historic gendering of urban space and an emerging consumerism, Shanks expands upon the ways in which feminist scholars have understood women's roles and labor in the private sphere. Looking closely at the revival's costuming and staging she uncovers the "corporeal conditions of posing." Reading Dot's stillness as a different kind of staged temporality, Shanks offers a viewing of the lead female character as a distinctly feminized dance of labor and agency.

Melissa Melpignano shows how the language of nineteenth-century dance librettos conveys the corporeal presence of dancing bodies whose memory would otherwise be lost. The shifts in Armida's corporeality—the female Arab sorceress who became a hallmark of European theatre, and in particular nineteenth-century ballet—Melpignano argues, correspond to shifts in the conceptualization of the female body. From femininity which coincides with pagan sin and un-granted redemption to patriarchal conceptions of hysteria, the text presents a series of rhythmic and emotional interventions, in a sense choreographing Armida's dance and thus her femininity. Where feminine labor—or the lack thereof—is characterized by Dot's *stillness* in Shanks' essay, hysterical movement and excessive emotion represent Melpignano's European woman, an exotic counter to nineteenth-century standards of female comportment in France. While the repressive codes seem to mirror one another in both Shanks and Melpignano's essays, Dot and Armida's dances are portrayed with a completely different speed and intensity, clearly delineating the difference between Western codes and the perceived exoticism and moral ambiguity of an "Oriental" woman.

This final essay brings us full circle back to the book's first section on "Writing the Body," and in particular Hill's "Flowers of Menace." Though in many ways "radical," Petronio's textual enhancements and staging of sexual ambiguity is not that dissimilar to the nineteenth-century re-writings of Armida. As Hill points out, while Petronio is likely unaware, his projected text in *Full Half Wrong*—one iteration of his *Rite* series—constructs an "Orientalized" subject. That is, in exploring the "fluidies of sexual behavior," Petronio constructs his female subject as primitive, irrational, and fanatic, amongst other things.

Conclusion

Reading this collection in the order it is presented in the table of contents is but one suggestion. Feel free to jump around or skip one and return to another. The order of things is not as important as the meaning you derive from them. "The production of history is a physical endeavor"[20] and so too is the reading of it. Bear in mind that I do not consider the essays in this book to be a comprehensive representation of the bodies which History has displaced. This represents but a sampling of Others and authors of their histories.

One more thing to keep in mind as you are reading the text: It is with the best of intentions that the contributors to this text author these narratives. Still the process of writing about dance and recovering the other is a difficult task. On the one hand, it is the "great difficulty of reducing the realities of motion to verbal formulas"[21] and on the other hand the challenge we face when writing about any individual or group of people. I leave you with a final question which I hope you will consider as you read through the essays in this book and then take with you out into the real world: *In trying to recuperate the body of the Other, in what ways do we create new divides?*

How do we reconcile our own search for a body, a voice, a self and those who are placed in the margins? Michelle Summers touches upon this in her discussion of the USA Freedom Kids' feeling targeted by the liberal media and moreover how whiteness guarantees these girls a certain privilege which is then compromised by both their femaleness and their youth. Likewise, we see this bind in the work of Petronio wherein, according to Valis Hill, he creates an Orientalized primitive while trying to debunk contemporary notions of femaleness and inserting male-on-male sexuality and the AIDS epidemic into the dialogue. If the "abject and abjection are the primers of my culture"[22] as Kristeva posits, *how do we make space for one self without creating a new radical other?*

NOTES

1. Susan Leigh Foster, *Choreographing History* (Bloomington: Indiana University Press, 1995), 5.
2. Thomas DeFrantz, *Dancing Many Drums: Excavations in African American Dance* (Madison: University of Wisconsin Press, 2002).
3. Diana Taylor, *The Archive and the Repertoire: Performing Cultural Memory in the Americas* (Durham: Duke University Press, 2003).
4. Susan Leigh Foster, *Choreographing History* (Bloomington: Indiana University Press, 1995), 3.
5. *Ibid.*, 15.
6. Susan Leigh Foster, *Choreography and Narrative* (Bloomington: Indiana University Press, 1996), xvi.
7. Sara Ahmed, *Queer Phenomenology: Orientations, Objects, Others* (Durham: Duke University Press, 2006), 3–5.
8. André Lepecki, *Of the Presence of the Body: Essays on Dance and Performance Theory* (Middletown, CT: Wesleyan University Press, 2004), 130.
9. See Jacques Derrida, "Cogito and the History of Madness," in *Writing and Difference*, trans. A. Bass (London: Routledge, 1978).
10. *Ibid.*, 133.
11. Homi Bhabha, "The Other Question: Difference, Discrimination, and the Discourse of Colonialism," in *Black British Cultural Studies: A Reader*, ed. Houston Baker, Manthia Diawara, and Ruth Lindeborg (Chicago: University of Chicago Press, 1996), 105.
12. Judith Butler, *Gender Trouble: Feminism and the Subversion of Identity* (New York: Routledge, 1990), 278.
13. José Esteban Muñoz, *Disidentifications: Queers of Color and the Performance of Politics*, Cultural Studies of the Americas, Vol. 2 (Minneapolis: University of Minnesota Press, 1999).
14. Karen Shimakawa, *National Abjection: The Asian American Body Onstage* (Durham: Duke University Press, 2002), 2.
15. This is the philosophical distinction between types, unique things/events/people, and tokens, concrete particulars that may be reproductions of the same type. My decision to capitalize the word "Other" throughout this introduction is based in this distinction where "Other" with a capital "O" represents the token and the non-capitalized form of the word signifies the type, or singular other (e.g., Billy Kersands or José Otero Aranda). While this distinction remains consistent throughout this introduction, each chapter reflects the individual author's personal preference.
16. Karen Shimakawa, *National Abjection: The Asian American Body Onstage* (Durham: Duke University Press, 2002), 2.
17. *Ibid.*, 3.
18. *Ibid.*
19. Tiffany Ruby Patterson and Robin D. G. Kelley, "Unfinished Migrations: Reflections on the African Diaspora and the Making of the Modern World," *African Studies Review* 43, no. 1 (2000): 11–45.
20. Susan Leigh Foster, *Choreographing History* (Bloomington: University of Indiana Press, 1995).
21. Ellen W. Goellner and Jacqueline Shea Murphy, *Bodies of the Text: Dance as Theory, Literature as Dance* (New Brunswick, NJ: Rutgers University Press, 1995), 5.
22. Julia Kristeva and Leon S. Roudiez, *Powers of Horror: An Essay on Abjection* (New York: Columbia University Press, 1982), 2.

BIBLIOGRAPHY

Ahmed, Sara. *Queer Phenomenology: Orientations, Objects, Others*. Durham: Duke University Press, 2006.
Bhabha, Homi. "The Other Question: Difference, Discrimination, and the Discourse of Colonialism," in *Black British Cultural Studies: A Reader*. Ed. Houston Baker, Manthia Diawara, and Ruth Lindeborg. Chicago: University of Chicago Press, 1996.
Butler, Judith. *Gender Trouble: Feminism and the Subversion of Identity*. New York: Routledge, 1990.
DeFrantz, Thomas. *Dancing Many Drums: Excavations in African American Dance*. Madison: University of Wisconsin Press, 2002.
Derrida, Jacques. "Cogito and the History of Madness," in *Writing and Difference*. Trans. A. Bass. London: Routledge, 1978.
Foster, Susan Leigh. *Choreographing History*. Bloomington: Indiana University Press, 1995.
_____. *Choreography and Narrative*. Bloomington: Indiana University Press, 1996.
Goellner, Ellen W., and Jacqueline Shea Murphy. *Bodies of the Text: Dance as Theory, Literature as Dance*. New Brunswick: Rutgers University Press, 1995.
Kristeva, Julia, and Leon S. Roudiez. *Powers of Horror: An Essay on Abjection*. New York: Columbia University Press, 1982.
Lepecki, André. *Of the Presence of the Body: Essays on Dance and Performance Theory*. Middletown, CT: Wesleyan University Press, 2004.
Muñoz, José Esteban. *Disidentifications: Queers of Color and the Performance of Politics*. Cultural Studies of the Americas, Volume 2. Minneapolis: University of Minnesota Press, 1999.
Patterson, Tiffany Ruby, and Robin D. G. Kelley. "Unfinished Migrations: Reflections on the African Diaspora and the Making of the Modern World." *African Studies Review* 43, no. 1 (2000): 11–45.
Shimakawa, Karen. *National Abjection: The Asian American Body Onstage*. Durham: Duke University Press, 2002.
Taylor, Diana. *The Archive and the Repertoire: Performing Cultural Memory in the Americas*. Durham: Duke University Press, 2003.

PART 1
Writing the Body

The production of history is a physical endeavor.
—Susan Leigh Foster

Mouth Over Matter

Writing Early Tap Dance in the Margins of the Black Body

BRYNN WEIN SHIOVITZ

On Wednesday December 4, 1878, the *Atchison Daily Globe* announced, "Billy Kersands isn't as good looking as Sam Lucas, but he has a heap bigger mouth."[1] Contrary to the use of such language, the author of this white newspaper was neither describing the events of last night's pie-eating contest nor was he commenting on Kersands' vocal qualities. Sam Lucas, the "unrivaled negro character artist, unmatched in his songs and dances,"[2] and Billy Kersands, the "unequaled plantation humorist, first rate in his grotesque specialties,"[3] set a new standard on the black minstrel stage. At the same time that they popularized dances such as the "essence of old Virginny,"[4] and helped to define the buck-and-wing, their idiosyncratic performances garnered attention from white journalists eager to brand the black body during a period of Reconstruction.

This essay tracks the performance careers of three lesser-known black minstrels, as archived by popular white newspapers. While Sam Lucas,[5] Billy Kersands,[6] and Tom McIntosh[7] were highly individualized performers, each with a distinct song and dance aesthetic, language and synecdoche collapsed these three men into a set of fixed facial expressions, specifically that of the big mouth. Not only did their supposed grotesque features sell ticket stubs, but their individual pay scales were directly affected by the size of their mouths. Heavily influenced by Bourdieu's theory of capital,[8] this essay analyzes the ramifications of writing about the dancing body in terms of discreet parts. These writers' oral fixations took the place of more descriptive reviews leaving the archive barren of clues as to what these popular dances actually looked like; these writers have "choreographed" the archive to a certain

> **CALLENDER'S
> ORIGINAL**
> # GEORGIA MINSTRELS
> **THE GREAT SOUTHERN SLAVE TROUPE.**
>
> They have given over 3,000 performances in England and America, appearing to hundreds of thousands of delighted auditors. Their recent appearance in New York City and Boston occasioned the greatest success ever known in minstrelsy, calling forth the most lavish praise from the entire press. For four weeks in Boston, Beethoven Hall was filled nightly with an audience of 1,500 persons, throngs being constantly turned away unable to secure seats.
>
> **FOUR END-MEN! TWENTY ARTISTS!**
>
> **SAM LUCAS,** The Unrivaled Character Artist, in his New Acts, Songs and Dances.
>
> **BILLY KERSANDS,** The Unequaled Plantation Humorist, in his Grotesque Specialties.
>
> **PETE DEVONEAR,** The Fine Protean Artist and Eccentric Comedian.
>
> **JAMES GRACE,** The Comic Bone End-Man, and Character Comedian.
>
> | **DICK LITTLE,** | **WILLIE LYLE,** | **AL SMITH,** |
> |---|---|---|
> | The Great Basso and Banjo King. | The Graceful Prima Donna. | The Humorous Plantation Oddity. |
>
> **MASTERS ALBERT and EDWARD, the PLANTATION JUVENILES.**

Callender's Original Georgia Minstrels Advertisement (courtesy New York Library for the Performing Arts, Billy Rose Theatre Division).[9]

degree. It is this mentality, along with the great success—economic, social, and cultural—garnered by Kersands and others' participation in the minstrel enterprise, *and* the way that nineteenth-century white newspapers recorded their features rather than their dancing, which I argue have affected the trajectory of mainstream tap dance.

Bourdieu's understanding of capital as both "linked to the body" and presupposing embodiment offers this analysis something Marx's more contemporaneous theory would have lacked: Marx's theory of capital neglects the symbolic world and thus fails to provide an analysis of modes of symbolic production such as dance and journalism. Bourdieu's analysis makes space for cultural production and allows me to analyze the minstrel stage in terms of representational struggle. Using direct quotes from newspapers spanning the second half of the long nineteenth century, this essay highlights the way white journalists elided tap's aural nature in favor of sensationalist, and certainly racist, language that described specific physical attributes of its performers. By abjecting the black body through text, chroniclers of the form wrote a subtext of tap dance. That is, these journalists materialized a Kristevean understanding of the "jettisoned object,"[10] treating the black body as an exorbitant outside, his mouth "ejected beyond the scope of the possible, the tolerable, the thinkable."[11] In writing about black tap dancers in this way, these authors inscribed a simultaneous *jouissance* and repulsion into the genre's narrative. Thus, most nineteenth-century accounts of tap dance transform the stage into a kind of carnivalesque auction block, urging the public to come and see the black body for its parts. I am arguing that the words these writers used were responsible for these dancers' marketability: Americans bought into the sale of Billy Kersands and others' big mouths. Fast forward 150 years and this "logic of transmission"[12] continues to feed the American public, creating limitations not only for the black body that dances, but also for forms historically referred to as "black."

I have also included an extensive biography of Billy Kersands in the notes to this essay. With the exception of Tom Fletcher's 1954 book, *100 Years of the Negro in Show Business*,[13] and Frank Cullen's[14] 2007 compendium, *Vaudeville Old and New*, as well as short mentions of the performer in Robert Toll's 1976 text,[15] and the *Oxford African American National Biography*,[16] this is the most comprehensive history written of Kersands to date; some of the facts I present here differ from previous accounts. I have synthesized several conflicting newspaper clippings—some from black newspapers and some from white—and conducted extensive research to arrive at this biography. I have also made note of facts that still remain ambiguous. A possibility exists, as it often does, that the individual authors writing these accounts of Kersands had specific motives in mind. In writing this history, I hope to accomplish three things: first, this essay as a whole seeks to recuperate the lives of three talented and prolific black minstrels who seem to have disappeared from history books; second, I want to consolidate the bits and pieces of Kersands' history that do exist into a single text, and moreover, try to make sense of its ambiguities and mutual exclusivities; finally, I hope that my "digging" establishes new pathways for scholars working within minstrel and performance studies, so that they may dig even deeper, and help to re-write this history.

* * *

The highest paid black performers on both vaudeville and minstrel circuits during the long nineteenth century were those men who were willing to subject themselves to crass stereotypes about the black body and the imagined day-to-day way of life for black Americans. This included everything from a long list of visual grotesquerie—e.g., the Venus Hottentot or the deranged Jim Crow, to the "razor-toting,"[17] "watermelon eatin,'" gambling type. While it was initially white actors in blackface who created such caricatures, by the mid–1860s, many black artists were reinforcing such stereotypes through a mode of performance designated black-on-black minstrelsy, a type of entertainment geared towards the white working-class. Lucas, McIntosh, and Kersands were some of the most popular performers on the minstrel circuit, in part because the public believed that "real Ethiopians" were more "authentic." By the mid–1870s, Kersands—who never learned to read or write—was the highest paid entertainer, black or white, earning a salary equivalent to more than $2300 per week by today's standards. McIntosh, by 1890, made roughly the same.

Writers made these salaries public knowledge, advertising not only when one of these performers moved up the pay scale, but also publicizing the cost of admission, likely because a cheaper ticket confirmed the success of a venue.[18] Additionally, these dancers would flaunt such wealth on stage. Kersands, for example, bedazzled himself in diamonds. Such financial independence, and the access it proffered, opened doors that had historically been unfathomable to black performing artists; this Janus-faced freedom, however, reinforced slave-trade logic. He, along with a handful of other "authentic Ethiopians," profited economically off of the very cultural capital[19] that supported an imagined relation between the black body, grotesque human form, and tap dance.

The juxtaposition, albeit on separate stages, of "real Ethiopian specimens"[20] and white performers pretending to be "black" raised the bar for white minstrel performers. The ensuing competition between white performers and black led to a rise in the number of articles proclaiming ties to either "mother Africa" or the Southern plantation, as well as representations of the black body that were often even more unfavorable than they had been in the antebellum South, as one performer was always

Newspaper Clipping from the *Lawrence Daily*, 1890.

trying to out-"coon" another. The better white performers became at pretending, and the better audience members became at deciphering, the more each group—black performers and white—had to up the ante, embellishing the stereotype and exaggerating its fictions. As black minstrel troupes began advertising their realness, either harkening back to their African roots or alluding to their status as "genuine" ex-slaves, they legitimized not only their performing bodies in the moment, but also validated a popular American sentiment of glorifying the Southern plantation and the happy-go-lucky slave life of the past. This excerpt from the Times speaks to the value that was placed on "genuine" blackness: "The troupe which will occupy the Academy this week have an advantage in one respect. They require no make-up. They are the right color to begin with.... They include about all of the end men and specialty actors among the colored people and ought to give a pretty laughable entertainment."[21] Thus, blackface minstrelsy actually peaked in popularity *after* the Civil War, during the so-called Reconstruction Era and through the first few years of the twentieth century.

The ways in which post-emancipation performance spoke to a largely nostalgic white American audience are many. However, I will focus here on the ways that two primary tap dances—performances which, at the time, would have been classified as belonging to a category of "Nigger dancing," "Ethiopian Dancing," or, by the late 1880s, "coon dancing"—commemorated slavery and the old South while pairing tap dance with a low or grotesque aesthetic.[22] These two tap dances are buck dancing (later revised and referred to as the buck-and-wing[23]) and the essence of old Virginia—or Virginny—depending on who performed the dance and where.

The first designates a broad category of tap dancing that is said to have started on slave plantations. While it has morphed over time, it is safe to say that the buck dance of the mid to late 1800s was a percussive dance which included a lot of shuffle steps and was often presented as a competition. By "shuffle" I am referring to a dragging step, like that found in the African American ring shout, where the feet remain planted, rather than the more contemporary notion of shuffling, which requires that one pick up the foot: brush, spank. "Buck dancing" was a generic term—much like the word "hoofing" is today—describing a flat-footed dance that was often performed in 2/4. According to one advertisement, "mobile buck dancing" began in 1820 and required extra heavy shoes that had no heel; this gave the dance its distinctive quality.[24] This grounded technique stands in contrast to a lighter, more upright clogging style that contemporaries such as Marcus Doyle or Barney Fagan[25] performed on white minstrel stages; the press often described Doyle's specialty, the "Coachman's Clog," for example, as "graceful," "refined," "handsome," and "attractive." Besides the racial implications present in these descriptions of the clogging rhythm—typically a 6/8 or 3/4—those writing about these dances might have experienced the rhythm as more Western, and thus familiar, than the 2/4 of buck dancing.[26] Like the cakewalk or walk around—a strut performed in 2/4 to a piece of ragtime—of the same period, buck dancing was often reserved for the end of a program. Two or more buck dancers might be pitted against one another, competing for the audience's approval. The atmosphere was similar to what we might in the twenty-first century call a cutting contest.[27] According to a few sources, the buck-and-wing was frequently performed to a Southern tradition, "patting Juba."[28] While the term "buck dancing" has always designated a category of dancing, rather than one specific step, some variations of the name imply a certain specificity. "Mobile buck dancing," for example, denotes a traveling step and "buck-and-wing" refers to a buck dance which includes a wing step, requiring that the performer lift the feet off of the ground.

The essence of Virginia, on the other hand, was a specific dance, almost always performed solo, and not overtly competitive like other tap dancing of the day. I will return to the details of this dance later on in the essay, but since it was mostly improvised and went through a series of rhythmic shifts, it is hard to know if the dance Kersands performed visually resembled today's version of the dance; I have much more clarity with regards to the rhythmic transformations it made over time, and the approximate dates of those shifts. Because the history of both buck dancing and the Virginia essence has either been transmitted through the body over a period of about 150 years or by means of white newspapers, we have to assume that our very methods of *knowing* this dance have been staged to a certain degree, at least from a visual stand point; a brief elucidation of the way in which writers choreographed specific dances—e.g., the buck-and-wing and essence of ol' Virginny—is in order.

One way that authors choreographed the buck-and-wing was to create a narrative of the black body as strong, resilient, and born with a natural inclination to entertain. An article in the *Estherville Daily News* published in April 1894 entitled "Nimble Negroes: Some of the Queer Dances of the Southern Negro"[29] attempts to give its white readership some understanding of buck dancing, a kind of dance it attributes to the old Southern plantations' terpsichoreans. It reads: "Before the war, a plantation negro who could not dance a few steps of that particularly characteristic Ethiopian Dance known as the 'buck dance' was very hard to find."[30] A similar article published in the *Evening Democrat* around the same time announces that "it was a matter of great surprise to the stranger traveling through the South to see with what precision and versatility young darkies executed the many difficult steps that characterize 'buck' and 'wing' dancing."[31] Together these two reports promote the durable nature of the black body, and when combined with articles like this one from the *Daily Chronicle*, which goes on to say that it is "impossible for anyone *but a negro* to put that distinctive personality into the dance which makes it so *fascinating* to the beholder,"[32] such writing distinguishes the black body from the white; it abjects.

While writers frequently remarked on the potency of black masculinity, it placed the black woman on the opposite extreme: languid and fleshy. An advertisement in the *Fort Wayne Gazette* tries to sell "The Second and Last Week of Big Eliza"[33]—supposedly the largest and most grotesque woman alive—by promising a "Mobile buck dance." The announcement reads: "Imagine a mountain of human flesh doing a dance. This will be the richest treat of the season." Furthermore, the paper offers five dollars in gold to any gentleman willing to dance with Big Eliza in full view of the audience.

Thus, with minimal print, these widely-read newspapers purported a history of the buck dance, which, while conceivably favorable to the *dancing* of black men, also firmly linked buck dancing to the black "plantation negro" or grotesque woman, both in referencing the figure of the slave and by quite literally selling the performance as a "fascinating" spectacle, ready for white viewers' consumption. For all of those nostalgic Americans who missed watching their property dance, going to see genuine ex-slaves or a 900-plus-pound black woman perform *for* them might have been the next best thing, especially if the show these black performers concocted was designed specifically to appease a white audience's desire to have the black body confirmed as "fascinating," "exotic," and/or other-worldly. Such racism was pervasive and often manifested most clearly on the minstrel stage.

These articles' focus on individual qualities—i.e., the man's robust muscles or the woman's fleshiness—sets a precedent for future articles that fetishize the black body by writing about it in terms of its discreet parts. Here capital is both "linked to the body" and presupposes embodiment as Bourdieu would have it, and signals what Kristeva might refer to as creating and occupying a liminal space. Here the newspapers' margins disturb the body, both of the writer (the "I") and of the Other, creating not a dancer but a spectacle: "a conjunction of waste and object of desire, of corpse and life" (Kristeva 185).[34] Thus writers sold not black dancers' dancing, but instead their labor; as in chattel slavery, journalists, like auctioneers, capitalized on their parts.

Writers have left almost no evidence of what steps were entailed in a typical buck dance other than a shuffle "as seen on slave plantations" and a jig-like step that goes unspecified. What most accounts *do* contain are descriptions of the body performing the buck dance as well as some of the stylistic choices made. According to that same *Daily Chronicle* article referenced earlier, the buck dance was a contest for only the fittest of the fit and would usually start with "two robust young darky boys." Such descriptions of the buck dance elide specific movement vocabulary in favor of socially mediated qualifiers of the black body and its behavior. "Robust" boys, after all, make the best manual labor.

Some famous buck-and-wing dancers, like Master Henry Williams, were hired for their grotesque dancing abilities as bait to bring in large audiences to new shows and jubilees. A clipping advertising the New National Theatre reads, "'Who Stole the Chicken?' A Real Negro Jubilee introducing, HENRY WILLIAMS, the greatest of all living buck dancers."[35] Obtaining labels such as "greatest of all living," "champion,"[36] and "unequaled" took time to acquire, but for men possessing very few freedoms, such "accumulated labor" translated directly into capital—a taste of sovereignty—even if it meant reinforcing false beliefs about the institution of slavery.

These shows often staged the black man's dancing as confirmation that slaves were a happy people. One particularly telling review evidencing such logic comes from the *San Antonio Light* in 1882: "Even now the homogenous minstrelsy of that musical people can be no longer found at the fireside in the citizen or in the halls of the planter; but must be sought upon the stage, where, with slight exaggerations, are preserved the true characteristics of the inevitable darkey."[37] The author of this Texas newspaper evangelizes the authenticity of the stage, claiming that the exaggerations are only "slight." Here the allusions to slavery manifest on two levels simultaneously: on the one hand writers emphasize these black bodies' labor and on the other hand, they remark on their happy dispositions, thereby establishing the ways in which these two qualities coexist.

Had journalists focused on the actual steps and rhythmic qualities of buck dancing, readers would have been confronted with the harsh realities of slavery rather than a narrative of the black body as physically resilient, indefinitely happy, and entertaining. As performance studies scholar Diana Taylor has noted, "writing has paradoxically come to stand in for embodiment."[38] That is, the "archive" has superseded what she labels the "repertoire," or the more embodied practices often ignored during acts of transfer. Slave narratives imply that the buck dancing that slaves were doing on Southern plantations was not meant to entertain white audiences so much as a way to physically and spiritually revitalize themselves under such oppressive conditions. Furthermore, buck dancing was itself an act of transmission between diasporic peoples. In addition to rendering a uniquely American, hybrid form, buck dancing preserved a series of movements and rhythmic culture that traveled from across the black Atlantic and also recreated the body's kinesthetic memory of bound movements aboard the slave ships; buck dancing spoke directly to a diasporic process and condition. For the most part, buck dancing is self-contained: an infinite number of sounds produced through the hands and feet, executed primarily on the vertical plane. Thus, the movement itself, the repertoire so to speak, narrates a history of migration in and through the black Atlantic as well as tells a story of the slaves' current conditions. By objectifying the black body and speaking to its physical qualities and so-called *responsibility* to dance, chroniclers of such dancing wrote both the artistic hybridity and a very specific transatlantic history out, while writing a racial grotesquerie in, to buck dancing. As a result of this archive, much of the repertoire—or at least the episteme derived from such an archive—has gone missing.

At the same time that seasoned minstrel performers were furthering their careers with buck-and-wings, ensembles often hired small children versed in this style of dancing to join their companies as pickaninnies, or "picks," as they were called for short. Pickaninnies populated all sorts of stages; theatre companies, minstrel troupes, comedy duos, and soubrettes hired them. While writers often remarked on their dancing, they, too, were qualified as less than human. Many of the reviews I found expressed pity for the cute "people of Ham."[39] Some articles refer to picks as "dusky mites,"[40] and some go so far as to liken their dancing to demons as in this review of Mamie Remington's 1901 performance: "Mamie Remington and her two pickaninnies also went well. Miss Remington is a splendid coon song singer and her assistants dance like demons."[41] Authors continuously re-staged this narrative of black dancing as something parasitic and cursed. These assertions about the black race, combined with advertisements selling a "lively" buck-and-wing, promised entertainment and their ensuing success at bringing people into the theatre; these advertisements exemplify the ways in which cultural capital is almost always linked to the person. Here abjection assists in the transmission of cultural capital. Abjection is rooted in the body (in a Kristevean sense) and ambivalent: the press (through the use of certain textual qualifiers) positions it as outside the nation, but its value as a form of capital prevents its *complete* rejection. Writers perpetuated the sale of tap dance qua a form of cultural capital by defining its materiality in relation to the black body; i.e., the dance as a form of cultural capital, defined only in relation to embodied capital. Over time, the archive—through its focus on the abject—disguised the "repertoire," invisibilizing the integrity of both the dance form and the very bodies tied to its value.

* * *

The text authors used to describe the dancers of the buck-and-wing was not the only way they inscribed otherness into the art. The spacing of text within the paper was another effective way of choreographing the history of tap dance and the black body. Imagine seeing an advertisement for Billy Kersands alongside an invitation for a watermelon-eating contest, managed by "all of our best colored people," and promising a lively buck dance. Another example might be found a few columns over from an advertisement for Kersands' famous essence of old Virginia: "In addition there is to be a greased pig Thursday night, while Saturday six 'coons' with mouths as big as Billy Kersands' will contest in a pie-eating match."[42] In both examples, the newspaper feeds its readers a narrative of watermelons and pies alongside the dancing black man who we will soon come to know as the man with the "capacious" mouth, one so big, that it "resembles the crater of Vesuvius."[43] Many of these ads emerged after the Civil War during a period of alleged emancipation, yet these papers bound blackness to a slave narrative, perpetuating the sale of enslaved bodies.[44] However the buck-and-wing was not the only dance that writers positioned alongside ads of consumption: writers staged the Virginia essence as well and perpetuated its sale through the same abject pairings and material signifiers.

The Virginia essence seems to have served a different purpose than that of buck dancing. Newspapers point to the idea that the latter became a common street performance, whereas the former was something the public most frequently observed in a performance context. A typical Virginny essence would have been danced vertically with both feet on the ground in a wider-than-hip-width stance, knees bent just slightly—*spank touch ball change, spank touch ball change*—and the body hunched over enough to assist in finding the groove of a nice swung rhythm.[45] One of the only descriptions we have of Kersands' *dancing* comes from an obituary written by Sam Lucas for the black newspaper, the *New York Age*. Lucas recalls, "His main specialty was his dance, 'The Essence of Old Virginia.' In that dance, he would lie flat on his stomach and beat first his head and then his toes against the stage to keep time, and then, looking out at the audience, he would say, 'Ain't this nice?'"[46] He would famously roll his eyes back in his head, perhaps not unlike the performances of Josephine Baker that would sweep the nation in just a few decades. While Kersands' horizontal version departs from a typical Virginny essence, his comic variation speaks to both his mastery of the form, and also the way in which the essence lends itself to improvisation. But beyond the spatial configurations, and visual display of plantation caricatures, the archive leaves little in the way of explanation of this popular minstrel dance. The archive thus teaches us about Kersands' unique approach to dancing and the series of "inimitable plantation drolleries,"[47] which patrons would have observed in conjunction with his essence of old Virginia.

While the black press frequently commented on how *successful* was the performance of Kersands, white newspapers made his repressive idiosyncrasies their focus. Furthermore, these writers used language that clearly delineated the stylistic differences between black performance and white. One advertisement reads: "Among them are Billy Kersands, the man with the biggest mouth in the United States, called the black Billy Emerson ... also Dick Little the coal-pit bass-singer whose voice is as big as your leg, and as deep as an oil well, and warranted to penetrate the bowels of the earth one hundred and fifty feet lower than anything yet discovered."[48] Here the author draws attention to Kersands and Little's blackness with words like "coal-pit," and qualifies Little's bass-singing with adjectives that signify more than a tonal quality; these descriptors signal a dark unknown, a subterranean space. And then of course, the *Reading Times*' account of Kersands' performance centers in on his mouth, and only his mouth. The *Indianapolis News* referred to Kersands as "he of the large and loud mouth."[49] *He of the large and loud mouth.* This is the phrase that one finds over and over again. Billy Kersands was "The Great!" "The Only!" Man of the fantastical mouth. His name would usually appear next to or underneath the words "Essence of Ole Virginia," but his mouth always seemed to take center stage. The press went so far as to speak of his mouth as a member of the cast: "Billy Kersands is with the company, and so is his mouth, and the latter organ is a well-spring of delight to most of the onlookers."[50]

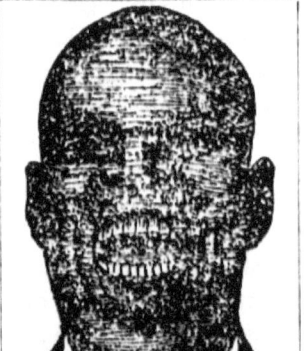

Billy Kersands advertisement from the February 28, 1897, issue of the *Wilkes Barre Times.*[51]

In addition to publishing monstrous caricatures of Kersands, the reviews of his performances often compared his features to those of animals. One reviewer described him as a "species of public entertainment"[52] He was known as the "man with the mouth bigger than a mule"[53] and likened to circus hippos. In a review entitled "Billy Kersands and his Face," the author identifies the dancer as "marked with the largest mouth ever seen on a human being," and then proceeds to describe the attraction (i.e., his mouth) as comparable to the hippos at the circus: "You've seen circus posters, the gaily-colored ones that adorn or at least are posted on the big bill boards, and there you've seen a picture of the hippopotamus that the menagerie does not contain. This hippo always has his mouth open and when you look at the picture and into the mouth which yawns in front of you, you have the sensation of looking into a deep cave. Well, Billy Kersands can discount the hippo on an exhibition of the mouth."[54]

The general American public was so fixated on Kersands' mouth that his name became part of the vernacular, impressed in language for years following his death. If something was Billy Kersands–like, it meant that it was large, frightening, and enticing, such as Miss Johnson's choristers, "flashily dressed slang flingers ... dressed in swallow-tail coats, low-cut waistcoats ... their pretty faces unrecognizably besmeared with burnt cork, and their tempting lips daubed with the reddest paint, making their mouths look like a great gash in the face."[55] It was used in courts and in witness statements to describe real life events. A gambler recalls a recent shooting: "There was a tunnel made by a forty-five-caliber ball as big as Billy Kersands' mouth."[56] Kersands' mouth thus became fact, a unit of measurement by which others could measure divisions of grotesque size. Just as the *Light* reminds us that there is no greater source of truth regarding the characteristics of the plantation slave than the stage, so too did using Kersands' body as a quantifier of size qualify the institution of minstrelsy as a reflection of facts.

As evidenced by these numerous advertisements and reviews, journalists seldom seemed to care about Kersands or any others' dancing abilities. It was the curiosity of the black body that brought people into the theatre. A part of this phenomenon resulted from this logic of transmission I have been discussing; that is, the way in which the archive has been mediated or choreographed by journalists in a position of power. Here writers, through grotesque realism, reduced the black body down to its main orifice. According to Bakhtin, the mouth is a primary site for the making of a grotesque body: "The stress is laid on those parts of the body that are open to the outside world, that is, the parts through which the world enters the body or emerges from it, or through which the body itself goes out to meet the world. This means that the emphasis is on the apertures or convexities, or on various ramifications and offshoots: the open mouth, the genital organs.... The body discloses its essence as a principle of growth which exceeds its own limits only in copulation ... eating, drinking, or defecation."⁵⁷ Emphasizing the size and quantities of objects consumed by these performers, writers stressed their "archaic" nature. Not only did the reviews of black tap dancers stage the carnivalesque *vis-à-vis* the performers' mouths, but their descriptions of voices—"deep as an oil well, and warranted to penetrate the bowels of the earth"—choreographed the dancers' relationship to the outside world, an underworld. Another component of this, however, is the outcome of the ways in which black performers fed into this logic, upholding various stereotypes *vis-à-vis* performance in pursuit of survival and recognition.

Kersands became known for performing the Virginia essence, buck-and-wing, or cakewalk,⁵⁸ while trying to fit larger and larger objects in his mouth: he tried everything from billiard balls to cups and saucers.⁵⁹ Journalists began focusing on these behaviors rather than his actual dancing which not only affirmed his reputation as "he of the large mouth" but also validated the circulation of Southern nostalgia, such as the "Jolly Nigger Bank," a cast iron object that would devour money and roll its eyes back upon being fed, popular from the late nineteenth up through the mid-twentieth centuries. These material signifiers bring Bourdieu's theory of capital full circle, as cultural capital literally becomes a vessel for economic capital and vice versa. Wealthy patrons paid Kersands to emphasize and continually "transgress his own limits."⁶⁰ Kersands thus acquired economic and social capital at the expense of his living body. The emphasis authors placed on his open orifice limited the impact of his dancing and formed the basis of objectified forms of cultural capital like the "Jolly Nigger Bank." Wealthy patrons could purchase these banks and participate in the embodied act of feeding the open mouth, an artifact of Kersand's grotesque materiality, frozen in time. Thus, this logic worked in

Newspaper drawing of Billy Kersands, c. 1880s

two directions: Kersands' consumption of objects supported the writers' focus on the size of his mouth and the way he "ate" his performances. One review reads: "After Billy Kersands eats the song, 'You've Got Your E-String Tuned One Note Too High,' pepsin gum will be on sale to revive those who have been overcome by super-vibrated diaphragms."[61] Reviews such as this reinforced Kersands' appetite for more, as they promised a fuller house, more economic stability, and fame.[62]

"Jolly Nigger Bank," c. 1890 (author's collection).[63]

Bourdieu posits three forms of cultural capital: embodied, objectified, and institutionalized.[64] Viewing the attention of the press given to Kersands as a form of institutionalized cultural capital—i.e., the publicity he received, the various titles he was given, and even the insertion of his name into the vernacular—makes clear how Kersands' body acted as a locus of power for a whole minstrel enterprise. The words these writers used to describe the performer make up a linguistic market; Kersands' performance in itself a form of embodied capital; and his mouth, both in the flesh and fetishized by the press, a model of objectified cultural capital at its finest. Here the black body itself becomes a cultural commodity and creates for the performer of color a habitus steeped in this idea that one's physical features are one's greatest asset. In other words, to have a "feel for the game"[65] in this sense is to identify not one's talent, but instead one's most abject feature.

Following Kersands' rise to fame, many black minstrels capitalized on their mouths. For example, "The Big Mouth Comedian, TOM McINTOSH," stepped into Kersands shoes in the mid–1890s and became "the highest salaried colored artist in the world."[66] By playing this game, performers like Kersands and McIntosh were able to elevate their individual class status at the cost of the tap genre's reputation.

Performers like Kersands, Lucas, and McIntosh played a huge role in shifting both the social and economic playing fields of the performing arts for the black community. While their performances did not directly impact the discrepant pay scales or living conditions of fellow black performing artists, these highly successful individuals showed America a set of possibilities. An article in *The Colored American* praises these men for their labor: "The field then graced by but a few who could boast of the proportions of a 'star' was led by such well-known people as Sam Lucas, Billy Kersands, Tom McIntosh" and a few others, "whose every appearance created a profound sensation."[67] In 1893 the *Colored American* and the *New York Age* in 1897 both wrote articles lauding[68] the ways in which these men helped their fellow black Americans climb the social ladder. All three represented the "rags to riches"[69] narrative, proving that not only could a black man rise up from nothing, but that he could become a household name, defy segregation laws, and make more money than many of his white contemporaries. Bourdieu speaks about this in terms of embodied cultural capital, or "external wealth converted into an integral part of the person,"[70] which one must acquire over time through a process of assimilation. He writes, "The work of acquisition is work on oneself (self-improvement), an effort that presupposes a personal cost … an investment above all of time, but also of that socially constituted form of libido, *libido sciendi*, with all the privation, renunciation, and sacrifice that it may entail."[71] The sacrifice these men made procured an external wealth and a habitus, one that, over time, both extended and limited certain freedoms. A surge in articles placing the work of these performers on a pedestal overlaps with the work of thinkers like W.E.B. Du Bois who, at this time, were circulating ideas of the talented tenth and double consciousness.[72]

Du Bois begins his famous "Talented Tenth" essay, "The Negro race, like all races, is going to be saved by its exceptional men."[73] While Du Bois speaks specifically to the responsibility of "college-bred" men to "save the race," interviews with some of the most talented black artists from around this same period speak to a more generalized "duty" perceived by the race's most successful. In reflecting upon the work of Kersands, the Colored Benevolent

Association speaks of him in the following way: "One great characteristic trait in him was his duty to the public. He was asked at one time, when blessed with abundance of this world's goods, why he did not retire from the stage. His answer was like him, ever mindful of his duty to the world, 'I can't. Not that I need the money, but the world needs me and I must obey.'"[74] This excerpt is interesting in that it gestures to a talented tenth mentality that extends beyond the nation's black scholars. While we have no way of knowing the motivation behind such a mindset, such a statement reflects the presence of a certain double consciousness likely experienced by successful performers such as Kersands.

* * *

The focus on grotesque features paired alongside tap dancing has come to haunt early cinema and today's Hollywood as it did the minstrel stage and early vaudeville. Bill Robinson and his lucrative Hollywood career as a tap dancing butler marks one example. Honi Coles noted once of Robinson, "Bo's face was about forty percent of his appeal" (quoted in Stearns and Stearns 188).[75] Just as Billy Kersands, the leading tap dancer of the day, profited primarily off of his big mouth, so too did the face of the greatest living tap dancer of the 1930s help to draw in the American public. As Margaret Morrison contends, Hollywood used close-up shots of the face alongside tap dance in order to construct narratives of race. She writes, "The editing technique of intercutting close-up shots of smiling teeth with full-body shots of tap dancing creates multilayered narratives of power, intimacy, and submission."[76] Such extreme close-ups severed body part from dancing subject, only to be juxtaposed with full body tap solos. In pairing virtuosic tap dance with minstrel tropes, filmmakers replicated the work of nineteenth-century journalists who saw Kersands' mouth as an entity unto itself. Directors of these Hollywood musicals thus re-cast "tambo and bones" through a different medium, but their principal message was roughly the same. Follow this logic up through the present day and televised dance competitions and the same pairings that existed in nineteenth-century newspapers manifest on Fox Network's all-American dance competition show, *So You Think You Can Dance* (*SYTYCD*).

Tap dance aired on mainstream platforms continues to contain within it strategic modes of display. *SYTYCD* premiered in July of 2005 and was still one of the most well-liked shows on television over a decade later. A brief overview: dancers from a variety of disciplines audition and when chosen by a panel of popular judges, advance to the next round until only 20 top competitors remain. Dancers must prove their competency in a variety of styles by masterfully working with their assigned choreographer and partner. Not until the show's seventh season did a tap dancer make the cut. Melinda Sullivan was the first tap dancer to stay until the top 20 and thus the first to bring tap to the televised competition. Not until this seventh season—which aired in 2010—was tap considered an art form worthy of such a mainstream platform by Fox. Yet despite her overall success on the show—Sullivan finished in ninth place—the judges continued to see what she did as "interesting" or "weird,"[77] not all that different than reviews from the 1880s that sold tap dance as a mix of "eccentricities and oddities."[78]

From newspaper headlines, to extreme close-ups, to the powerful words of Nigel Lithgoe on prime-time television, the legacy of tap dance has survived as a relic of the minstrel stage. With the judges in a position of power, their words have guided audience perception and ultimately shaped Sullivan's fate—and that of tap dance—on the show. During a press conference following Sullivan's dismissal, executive producer and judge Lithgoe, remarks: "Sometimes being different alienates you."[79] Lithgoe, like the white writers who commented on Kersands' performance, instructs viewers to see tap dancers as different.

Tap dancers on *SYTYCD* are not cast to win; tap dance exists outside the rules of the game. The playing field can never be level when tappers are expected to be proficient in all forms, but everyone else gets a pass on the tap dance portion of the competition. That is, someone like Sullivan is expected to contend in all forms, yet other competitors are spared having to tap dance. Furthermore, rarely are the judges trained to critique tap; this genre cannot be judged by the same standards as ballet or ballroom, and yet judges constantly compare the form to the same Western standards. Rather than evaluating the musical expression of the tap dancer's footwork, or their bodies' relationship to gravity, judges assess the tap dancer according to "classical" standards useful for upright forms such as ballet. Since the alleged "experts" inform audience feedback, audiences are trained to see tap dance through this lens, one that contradicts the aesthetic system in place for *SYTYCD*'s other genres. Shows like *So You Think You Can Dance* endorse the work of nineteenth-century journalists and succumb to the same minstrel mindset as those in a position of power did 150 years ago.

While Melinda Sullivan helped to bring mainstream attention to tap dance, one has to wonder what role her whiteness played in this process, and also examine the ways in which her white body was still subjected to many of the same narratives as Kersands' when dancing an Africanist form. *So You Think You Can Dance* is but one example of the ways in which the pairing of tap dance with grotesque imagery has, over time, limited the public's ability to see the art in any other light. What will it take to sidestep this logic of transmission that tap has carried in its wake? Televised competition dance is not the only vestige of nineteenth-century journalism and the minstrel stage, but it is a place to start. I conclude with this idea as an invitation, a suggestion for how we might read the past as a primer of our relevant present, and what we might do to ensure that certain corporealities are not lost in the writing of history. There was a time when Kersands, Lucas, and McIntosh were the most popular performers on the black minstrel circuit. What words caused these dancers' disappearance? There was a time when the United States took pride in a dance form created on American soil. What authors took this away? The way that one writes the body determines its present and its future: we cannot change the past, but we can resist it, and write the future differently.

NOTES

1. Associated Press, *Atchison Daily Globe* (Atchison, KS), Wed., Dec. 4, 1878.
2. Georgia Minstrels Advertisement, c. 1876, MWEZ nc 11,947, New York Library for the Performing Arts, Billy Rose Theatre Division.
3. *Ibid.*
4. It is worthwhile to note that this tap dance shared a name with a well-liked wheat whiskey. A.M. Bininger's "Essence of Old Virginia" was a popular spirit between roughly 1859 and 1880 and was known for depicting happy slaves dancing on its label.
5. Sam Lucas was born on August 7, between 1840 and 1850. He started working as a cab driver before a manager discovered him in New York. He performed with Lew Johnson's Plantation Minstrels from 1871 to 1873, Callendar's Original Georgia Minstrels from 1872 to 1876, and Sprague's Georgia Minstrels from 1878 to 1879 in Havana, Cuba. After a successful career in blackface minstrelsy, he branched out and took on more serious roles, performing in ground-breaking works such as *The Creole Show* and *A Trio to Coontown*. He caught the attention of James Weldon Johnson who labeled him the "Grand Old Man of the Negro Stage" (see James Weldon Johnson, *Black Manhattan* [New York: Da Capo Press, 1968], 90). He died on January 5, 1916, from chronic liver disease and pneumonia. See *The New York Age*, August 5, 1915, and Bernard Peterson, *Profiles of African American Stage Performers and Theatre People, 1816–1960* (Westport, CT: Greenwood Press, 2001), 170–71.
6. Billy Kersands was born a slave in either Baton Rouge, Louisiana, or Louisville, Kentucky, in 1842. Unfortunately, I have been unable to locate facts surrounding his escape from slavery to freedom. He moved to New York at a very young age and quickly became a bootblack on the Bowery. He soon became known for singing and dancing barefoot, especially Hoedowns and Pattin' Juba. A minstrel performer who went by the name of "Craddock" noticed Kersands' dancing and hired him for a part as a slave boy at $12 a week. Still unclear is whether he performed this part as a pick or whether he bypassed this common title. One source cites Kersands' first performance as a slave boy in *Cudjoe's Cave* at the Old Bowery Theatre in 1866. After only two weeks on stage, Craddock raised Kersands' salary to $24 a week, and just two weeks after that, upped this to $50 a week. Different newspapers attribute slightly different pay scales to Kersands, but the fact that every obituary ever published about Kersands includes mention of his salary is significant. I speak to this more in the body of the essay.

In a very short amount of time, Kersands became a featured performer, known for his antics, including that of fitting an entire saucer in his mouth. Different clippings offer a slightly different story, but more sources point to Kersands' stint with Callendar's Minstrels, an all-black troupe, beginning around this time. His first performance was, according to the *National Republican*, a double song and dance with fellow minstrel, Alf Smith. As their star comedian, he brought in $75 a week. According to Sam Lucas in the *New York Age*, Kersands could neither read nor write (i.e., he was completely illiterate), and fellow cast members like Lucas had to teach him his music verbally. Callendar's Minstrels was then sold to J.H. Haverly and re-named Haverly's Minstrels. Many of Haverly's performers were dissatisfied with their pay and left the company. A man by the name of Charles Frohman saw Kersands perform this saucer trick with Haverly's and invited him to take part in a negro minstrel show that

36   Part 1: Writing the Body

would be touring in Europe. Three weeks after this initial meeting, Kersands signed a contract to tour Europe for $250 a week, and Frohman became the manager of Haverly's Minstrels. When Kersands returned to America, he organized his own troupe, the Hicks and Kersands Minstrels, which had a successful five-season run and then dissolved. Kersands then joined the Richard and Pringle Show, and stayed with the troupe for many years as the leading comedian. When Richards and Pringle died, Rosco and Holland took over the company and retained Kersands as a featured comedian. When this partnership dissolved, Kersands once again formed his own company, the Kersands Minstrels, which had a successful four-season run. Kersands extensively toured Europe. This included a special invitation from Queen Victoria (c. 1907–1909) to perform in her home. He also toured all of Asia, East Asia, and North Africa, and performed in Australia and West Fercogo, a region near the South Pole. His last tour abroad, which began in Honolulu, was with the Hugo Brothers' Minstrels. Kersands spent the last years of his life performing with his wife, Madame Louisa Strong (married in 1895), in their own minstrel troupe, the Dixie Minstrels, under the management of Nigro and Stevenson. Kersands died in Artesia, New Mexico, on June 30, 1915 from heart failure. The net worth of Kersands at the time of his death is unknown, but his salary has been recorded as the highest made by any stage performer of that era and we know that in his prime he made enough money to buy out a private railway carriage so that he could travel with the same freedoms as a white man.

See Associated Press, *National Republican* (Washington City [D.C.]), July 25, 1885; Associated Press, *New York Age* (New York, NY), July 15, 1915 and August 5, 1915; Associated Press, *The Donaldsonfield Chief* (Donaldsonfield, Louisiana), July 24, 1915.

7. Tom McIntosh was born in Lexington, Kentucky, in 1840. He was best known as a singer, musician, dancer, and comedian and performed in the leading minstrel troupes of the day, including Callendar's Georgia Minstrels, the Original Georgia Minstrels, Haverly's Genuine Colored Minstrels, and Callendar's Collossal Consolidated Colored Minstrels. His time with many of these troupes overlapped with Kersands. McIntosh died of a stroke in Indianapolis, Indiana while on tour with the Smart Set in 1904. Of the three "lost" minstrels I will be discussing, Tom McIntosh's name has the most extensive literature attached to it. For a more detailed account of his life, see Lynn Abbott and Doug Seroff, *Ragged But Right: Black Traveling Shows, Coon Songs, and the Dark Pathway to Blues and Jazz* (Jackson: University of Mississippi Press, 2007).

8. See Pierre Bourdieu, "The Forms of Capital *in Handbook of Theory of Research for the Sociology of Education*, ed. J. E. Richardson (Westport, CT: Greenwood Press, 1986).

9. Georgia Minstrels Advertisement, c. 1876, MWEZ nc 11,947, New York Library for the Performing Arts, Billy Rose Theatre Division.

10. See Julia Kristeva, *Powers of Horror: An Essay on Abjection* (New York: Columbia University Press, 1982), 2.

11. *Ibid.*, p. 1

12. See Pierre Bourdieu, "The Forms of Capital," in *Handbook of Theory of Research for the Sociology of Education*, ed. J. E. Richardson (Westport, CT: Greenwood Press, 1986).

13. See Tom Fletcher, *100 Years of the Negro in Show Business: The Tom Fletcher Story* (New York: Burdge, 1954).

14. See Frank Cullen, Florence Hackman, and Donald McNeilly, "Billy Kersands," *Vaudeville Old & New: An Encyclopedia of Variety Performers in America* (New York: Routledge, 2007).

15. See Robert C. Toll, *Blacking Up: The Minstrel Show in Nineteenth-Century America* (New York: Oxford University Press, 1976).

16. See Kevin Byrne, "Kersands, Billy," *African American National Biography*, ed. Henry Louis Gates, Jr., and Evelyn Brooks Higginbotham. *Oxford African American Studies Center*, accessed Sun., Nov. 19, 15:17:05 PST 2017: http://www.oxfordaasc.com/article/opr/t0001/e1605.

17. A reference to May Irwin's "The Bully Song," 1896.

18. The more well-off the performer was, the cheaper the ticket was for the audience member. Associated Press, *Donaldsonville Chief* (Donaldson, IN), July 24, 1915.

19. See Pierre Bourdieu, "The Forms of Capital," in *Handbook of Theory of Research for the Sociology of Education*, ed. J. E. Richardson (Westport, CT: Greenwood Press, 1986).

20. Miscellaneous clipping, date unknown, MWEZ nc 11,947, New York Library for the Performing Arts, Billy Rose Theatre Division.
21. Associated Press, *Times* (Philadelphia, PA), Sun., Dec. 24, 1882.
22. Advertisements made the same sorts of connections between slavery and music with statements such as "Music of America's rice, cotton, and sugar lands." *Harrisburg Telegraph* (Harrisburg, PA), Sat., Nov. 25, 1882, 5.
23. I will talk both about buck dancing and the buck-and-wing.
24. Associated Press, "The Leading Buck and Wing Dancers of Dixie," source unknown, MWEZ nc 6,959, New York Library for the Performing Arts, Billy Rose Theatre Division.
25. Barney Fagan was known for dancing to European music: "Barney Fagan's clog dance to the gavotte, 'Secret Love,' was another noticeable feature. Fagan's costume was a pale blue satin and pink silk, and his dancing the perfection of grace." Associated Press, *Daily Arkansas Gazette* (Little Rock, AR), Tue., Jan. 16, 1883, 5.
26. Associated Press, *Salt Lake Herald* (Salt Lake City, Utah), Thu., Mar. 4, 1886, 4.
27. Also known as a tap dance challenge. See Constance Valis Hill, "Stepping, Stealing, Sharing, and Daring: Improvisation and the Tap Dance Challenge," in *Taken by Surprise: A Dance Improvisation Reader*, ed. Ann Cooper Albright and David Gere (Middletown, CT: Wesleyan University Press, 2003).
28. See for example, *San Francisco Call* (San Francisco, CA), Feb. 4, 1900, 7.
29. Associated Press, *Estherville Daily News* (Estherville, Iowa), Thu., Apr. 19, 1894.
30. *Ibid.*
31. Associated Press, *Evening Democrat* (Warren, PA), Apr. 19, 1894.
32. Associated Press, *Dalles Daily Chronicle* (The Dalles, OR), July 14, 1894.
33. Associated Press, *Fort Wayne Gazette* (Fort Wayne, IN), February 23, 1890.
34. Julia Kristeva, *Powers of Horror: An Essay on Abjection* (New York: Columbia University Press, 1982), 185.
35. Associated Press, *New National Theatre Clipping*, MWEZ nc 11,947, New York Library for the Performing Arts, Billy Rose Theatre Division.
36. *The Scranton Tribune* reads: "The wonderful dancing of Master Henry Williams, the champion buck, wing, and grotesque dancer of the country, heading a band of pickaninnies, will be a feature of the performance." November 26, 1894, 3.
37. Associated Press, *San Antonio Light* (San Antonio, TX), Wed., May 17, 1882, 1.
38. Diana Taylor, *The Archive and the Repertoire: Performing Cultural Memory in the Americas*. (Durham: Duke University Press, 2003),16.
39. This is a reference to Noah's curse, a story in Genesis 9, where he punishes the descendants of his son Ham with servitude for the sin of seeing him drunk and naked. Many have used this as justification for slavery and thus members of the African diaspora have historically been referred to as "people of Ham," or a group cursed by Noah and born into slavery. See Felicia Lee, "From Noah's Curse to Slavery's Rationale," *New York Times*, Nov. 1, 2003, and David Goldenberg, *The Curse of Ham: Race and Slavery in Early Judaism, Christianity, and Islam* (Princeton: Princeton University Press, 2005).
40. Associated Press, *Morning Democrat* (Davenport, Iowa), Thu., Feb. 21, 1895, 3.
41. Associated Press, *Boston Post* (Boston, MA), Tue., JunE 11, 1901, 7.
42. Associated Press, *St. Paul Daily Globe*, March 06, 1892.
43. Associated Press, *Opelousas Courier* (Opelousas, LA), January 20, 1894.
44. This stands in contrast to the way in which authors positioned ads for the coachman's clog right above things like the "intelligence column." *Quad-City Times* (Davenport, Iowa), Tue., Oct. 13, 1885, 1.
45. Note that dancers were performing the Virginy essence long before the birth of jazz, but that these early, syncopated tap steps were where the jazz groove, in the most basic sense, began.
46. Sam Lucas, *New York Age*, Thu., Aug. 5, 1915, 6.
47. Miscellaneous clipping, date unknown, MWEZ nc 11,947, New York Library for the Performing Arts, Billy Rose Theatre Division.
48. Associated Press, *Reading Times* (Reading, PA), Thu., Nov. 23, 1882.
49. Associated Press, *Indianapolis News*, Jan. 22, 1880.

38   Part 1: Writing the Body

50. Associated Press, *Opelousas Courier* (Opelousas, LA), October 4, 1890.
51. Associated Press, *Sunday News* (Wilkes-Barre, PA), Sun., Feb. 28, 1897, 5.
52. Associated Press, *Reading Times* (Reading, PA), Thu., Nov. 23, 1882.
53. Associated Press, *Great Bend Tribune* (Great Bend, KS), Wed., Aug. 4, 1915, 2.
54. Associated Press, *Decatur Herald* (Decatur, IL), Sat., Aug. 29, 1903, 5.
55. Associated Press, *St. Louis Post-Dispatch*, Fri., May 1, 1896, 4.
56. Associated Press, *Montana Standard* (Butte, MT), Sat., Jan. 5, 1884, 3.
57. Mikhail Bakhtin, *Rabelais and His World*, trans. Helene Iswolsky (Bloomington: Indiana University Press, 1984), 26.
58. Not much exists on Kersands and the cakewalk, but enough evidence exists in the archive to suggest that his oral antics were present for this dance as well. For example: "The olio introduces ... Billy Kersands and a cake-walk that is simply too funny to describe." *Meridional* (Abbeville, LA), Jan. 5, 1895.
59. Associated Press, *Donaldsonville Chief* (Donaldsonville, LA), Sat., July 24, 1915, 4.
60. Mikhail Bakhtin, *Rabelais and His World*, trans. Helene Iswolsky (Bloomington: Indiana University Press, 1984), 26.
61. Associated Press, *Washington Standard* (Olympia, Washington Territory), Dec. 28, 1900.
62. *The Democrat and Chronicle* writes, "Tom McIntosh and S.H. Dudley were a minstrel troupe in themselves and were just as funny with their mouths closed as when they were 'in action.'" (Rochester, NY), Fri., Sept. 15, 1899, 12.
63. Author's personal collection.
64. See Pierre Bourdieu, "The Forms of Capital," in *Handbook of Theory of Research for the Sociology of Education*, ed. J. E. Richardson (Westport, CT: Greenwood Press, 1986).
65. See Pierre Bourdieu, *The Logic of Practice* (Stanford: Stanford University Press, 1990), 66.
66. Associated Press, *Topeka State Journal*, Thu., July 31, 1890, 8.
67. See *Colored American* (Washington, D.C.), Nov. 1, 1902.
68. Another example reads: "The singing is good and is blending between grand opera and rag-time, and the dancing is very superior, the buck and wing dancing surpassing anything of its kind ever seen in the successful business..." *Colored American*, May 14, 1904, 2.
69. Associated Press, *The Sun* (New York, NY), Sun., June 17, 1888, 9.
70. Pierre Bourdieu, "The Forms of Capital," in *Handbook of Theory of Research for the Sociology of Education*, ed. J. E. Richardson (Westport, CT: Greenwood Press, 1986), 48.
71. Ibid.
72. W.E.B. Du Bois, "The Talented Tenth" and *"The Souls of Black Folk Revisited,"* in *The Souls of Black Folk: Centennial Reflections*, ed. Tamara Brown, Ida Jones, and Yohuru Williams (Trenton, NJ: Africa World Press, 2004).
73. W.E.B. Du Bois, "The Talented Tenth," in *The Souls of Black Folk: Centennial Reflections*, ed. Tamara Brown, Ida Jones, and Yohuru Williams (Trenton, NJ: Africa World Press, 2004).
74. Associated Press, *New York Age*, July 15, 1915.
75. Marshall and Jean Stearns, *Jazz Dance: The Story of American Vernacular Dance* (New York: Macmillan, 1968), 188.
76. Margaret Morrison, "Tap and Teeth: Virtuosity and the Smile in the Films of Bill Robinson and Eleanor Powell," *Dance Research Journal* 46, no. 2 (2014): 21–22.
77. *So You Think You Can Dance*, season seven, episodes 1–11, aired May 27–July 1, 2010, Fox Television.
78. Associated Press, *Reading Times* (Reading, Pennsylvania), Thu., Nov. 23, 1882.
79. *So You Think You Can Dance*, season seven, episodes 1–11, aired May 27–July 1, 2010, Fox Television. https://www.youtube.com/watch?v=F0lpNut36yM.

Bibliography

Abbott, Lynn, and Doug Seroff. *Ragged but Right: Black Traveling Shows, Coon Songs, and the Dark Pathway to Blues and Jazz*. Jackson: University of Mississippi Press, 2007.
Associated Press. *Atchison Daily Globe* (Atchison, KS), Wed., Dec. 4, 1878.

Associated Press. *Boston Post* (Boston, MA), Tue., June 11, 1901, 7.
Associated Press. *Colored American* (Washington, D.C.), Nov. 1, 1902.
Associated Press. *Colored American* (Washington, D.C.), May 14, 1904, 2.
Associated Press. *Daily Arkansas Gazette* (Little Rock, AR), Tue., Jan. 16, 1883, 5.
Associated Press. *Dalles Daily Chronicle* (The Dalles, OR) July, 14, 1894.
Associated Press. *Decatur Herald* (Decatur, IL), Sat., Aug. 29, 1903, 5.
Associated Press. *The Donaldsonfield Chief* (Donaldsonfield, Louisiana) July 24, 1915.
Associated Press. *Donaldsonville Chief* (Donaldson, IN) July 24, 1915.
Associated Press. *Estherville Daily News* (Estherville, Iowa), Thu., Apr. 19, 1894.
Associated Press. *Evening Democrat* (Warren, PA), Apr. 19, 1894.
Associated Press. *Fort Wayne Gazette* (Fort Wayne, IN), Feb. 23, 1890.
Associated Press. *Great Bend Tribune* (Great Bend, KS), Wed., Aug. 4, 1915, 2.
Associated Press. *Harrisburg Telegraph* (Harrisburg, PA) Sat., Nov. 25, 1882, 5.
Associated Press. *Indianapolis News*, Jan. 22, 1880.
Associated Press. "The Leading Buck and Wing Dancers of Dixie," Source Unknown, MWEZ nc 6,959. New York Library for the Performing Arts, Billy Rose Theatre Division.
Associated Press. *Meridional* (Abbeville, LA), Jan. 5, 1895.
Associated Press. *Montana Standard* (Butte, MT), Sat., Jan. 5, 1884, 3.
Associated Press. *Morning Democrat* (Davenport, Iowa), Thu., Feb. 21, 1895, 3.
Associated Press. *National Republican* (Washington City [D.C.]) July 25, 1885.
Associated Press. *New National Theatre Clipping*, MWEZ nc 11,947. New York Library for the Performing Arts, Billy Rose Theatre Division.
Associated Press. *The New York Age* (New York, NY) July 15, 1915.
Associated Press. *The New York Age* (New York, NY) Aug, 5, 1915.
Associated Press. *Opelousas Courier* (Opelousas, LA), Oct. 4, 1890.
Associated Press. *Opelousas Courier* (Opelousas, LA), Jan. 20, 1894.
Associated Press. *Quad-City Times* (Davenport, Iowa), Tue., Oct 13, 1885, 1.
Associated Press. *Reading Times* (Reading, PA), Thu., Nov. 23, 1882.
Associated Press. *St. Louis Post-Dispatch* (St. Louis, MO), Fri., May 1, 1896, 4.
Associated Press. *St. Paul Daily Globe* (St. Paul, MN), Mar. 6, 1892.
Associated Press. *Salt Lake Herald* (Salt Lake City, Utah), Thu., Mar. 4, 1886, 4.
Associated Press. *San Antonio Light* (San Antonio, TX), Wed., May 17, 1882, 1.
Associated Press. *San Francisco Call* (San Francisco, CA), Feb. 4, 1900, 7.
Associated Press. *The Scranton Tribune* (Scranton, Ohio), Nov. 26, 1894, 3.
Associated Press. *The Sun* (New York, NY), Sun., June 17, 1888, 9.
Associated Press. *Sunday News* (Wilkes-Barre, PA), Sun., Feb. 28, 1897, 5.
Associated Press. *Topeka State Journal* (Topeka, KS), Thu., July 31, 1890, 8.
Associated Press. *Times* (Philadelphia, PA), Sun., Dec. 24, 1882.
Bakhtin, Mikhail. *Rabelais and His World*. Trans. Helene Iswolsky. Bloomington: Indiana University Press, 1984.
Bourdieu, Pierre. "The Forms of Capital," in *Handbook of Theory of Research for the Sociology of Education*, ed. J. E. Richardson. Westport, CT: Greenwood Press, 1986.
Bourdieu, Pierre. *The Logic of Practice*. Stanford: Stanford University Press, 1990.
Byrne, Kevin. "Kersands, Billy." *African American National Biography*, ed. Henry Louis Gates, Jr., and Evelyn Brooks Higginbotham. *Oxford African American Studies Center*, accessed Sun., Nov. 19, 2017, 15:17:05 PST, http://www.oxfordaasc.com/article/opr/t0001/e1605.
Cullen, Frank, Florence Hackman, and Donald McNeilly. "Billy Kersands." *Vaudeville Old & New: An Encyclopedia of Variety Performers in America*. New York: Routledge, 2007.
Du Bois, W.E.B. "The Talented Tenth" and "*The Souls of Black Folk Revisited*," in *The Souls of Black Folk: Centennial Reflections*, ed. Tamara Brown, Ida Jones, and Yohuru Williams. Trenton, NJ: Africa World Press, 2004.
Fletcher, Tom. *100 Years of the Negro in Show Business: The Tom Fletcher Story*. New York: Burdge, 1954.
Goldenberg, David. *The Curse of Ham: Race and Slavery in Early Judaism, Christianity, and Islam*. Princeton: Princeton University Press, 2005.
Hill, Constance Valis. "Stepping, Stealing, Sharing, and Daring: Improvisation and the Tap

Dance Challenge," in *Taken by Surprise: A Dance Improvisation Reader*, ed. Ann Cooper Albright and David Gere. Middletown, CT: Wesleyan University Press, 2003.
Johnson, James Weldon. *Black Manhattan*. New York: Da Capo Press, 1968, 90.
Kristeva, Julia. *Powers of Horror: An Essay on Abjection*. New York: Columbia University Press, 1982.
Lee, Felicia. "From Noah's Curse to Slavery's Rationale." *New York Times*, Nov. 1, 2003.
Morrison, Margaret. "Tap and Teeth: Virtuosity and the Smile in the Films of Bill Robinson and Eleanor Powell." *Dance Research Journal* 46, no. 2 (2014): 21–22.
Peterson, Bernard. *Profiles of African American Stage Performers and Theatre People, 1816–1960*. Westport, CT: Greenwood, 2001.
So You Think You Can Dance. Season seven, episodes 1–11. Aired May 27–July 1, 2010. Fox Television.
Stearns, Marshall, and Jean. *Jazz Dance: The Story of American Vernacular Dance*. New York: Macmillan, 1971.
Taylor, Diana. *The Archive and the Repertoire: Performing Cultural Memory in the Americas*. Durham: Duke University Press, 2003.
Toll, Robert C. *Blacking Up: The Minstrel Show in Nineteenth-Century America*. New York: Oxford University Press, 1976.

Spain in the Basement
Dancing Race and Nation at the Paris Exposition, 1900

KIKO MORA *and* K. MEIRA GOLDBERG

At the Exposition Universelle of 1900 in Paris, Auguste and Louis Lumière shot two short films of Spanish dance.[1] The film industry had already produced several motion pictures: Thomas Edison had shot a short reel titled "La Carmencita" (1894) for kinetoscope in his studio in West Orange, New Jersey, protagonized by Carmen Dauset Moreno, the most famous Spanish dancer in the United States at the end of the nineteenth century. British director Robert W. Paul had filmed sisters Amparo and Margarita Aguilera dancing an "Andalusian Dance" (1896) in Lisbon, and in 1898, the Lumière company traveled to Seville to shoot twelve short films in the *Járdines del Alcázar*.[2] All of the films previous to the 1900 films shot by Lumière in Paris record the *escuela bolera*, the bolero school, or the Spanish school of classical dance.

The first of the two films (1123) shot by Lumière during the Paris Universal Exposition is the first cinematic document in which a *cuadro*, an ensemble of flamenco singers, dancers, guitarists, and *palmas*, or clapping percussion, appears. We the authors—Kiko Mora, a semiotician and researcher in the transatlantic migrations of late nineteenth- and early twentieth-century Spanish music and dance, and K. Meira Goldberg, a flamenco dancer and historian who has worked on the 1917 films which Léonide Massine made of the legendary dancer Juana Vargas, "La Macarrona"—initially began studying this footage simply because of its significance in this regard.[3] Our initial assumption as we began our research was that, as several prominent scholars making brief comments about this movie had written, and as had been echoed in several specialized flamenco blogs and in press notices, the male dancer in the film was José Otero Aranda, a famous bolero school maestro from Seville.[4]

Otero had already appeared in the Lumière films shot in Seville in 1898, but Mora had serious doubts regarding his presence in the 1900 Lumière film, for the simple reason that the dancer in that film is black.[5] Further, none of the research prior to that presented here provided any sources documenting Otero's presence in Paris in 1900. It may be that scholars took the film itself as evidence, but a film can be useless when the eye is treacherous. Mora has convincingly documented that the dancer performing in the Lumière film is not Maestro José Otero, but rather Jacinto Padilla, "a celebrated mulatto" known as "El Negro Meri."[6] Padilla, of Afro-Cuban origin, was a multifaceted artist who hardly appears in the annals of flamenco historiography.[7] Since he began his performing career in Spain in the 1870s, El Negro Meri made a living as a circus equestrian, bullfighter, and flamenco singer, dancer and guitarist. In 1900, Meri was performing at the Paris Exposition as a member of the cast enlivening the evenings at La Feria, a restaurant located in the basement of the Spanish Pavilion. His appearance in the first of the two films shot by Lumière (1123 and 1124), both titled *Danse espagnole de la Feria*, is due to this fact.[8]

But there is another fact which contextualizes El Meri's appearance in the film, along with his subsequent erasure from the historical record. On June 27, 1900, just a few days before the Lumière films were shot, the governments of France and Spain signed the Paris Treaty, severely reducing Spanish sovereignty in the African colonies of Sahara and Western Guinea. For Spain, the Treaty confirmed once again the death throes being suffered by the once-great Spanish empire after the Spanish-American War, known in Spain as the "Disaster of '98," in which Spain lost the last of its colonial possessions in America and Asia: Philippines, Puerto Rico, and the so-called "jewel of the Spanish empire," Cuba. The French, victors in the 1900 Treaty, perceived this armistice differently: Paris, its status as a global capital ratified, began to enthusiastically welcome the Afro-American sounds and rhythms of cakewalk and ragtime.[9] In this cultural and geopolitical context, the Lumière Company filmed a reel in which El Negro Meri represented the epitome of flamenco—and, ultimately, of Spain—in the eyes of the world. The first man ever filmed dancing flamenco was not a Spanish Roma, or "Gypsy," as might be presumed according to the stereotype, but a black man of Cuban ancestry. How can this double irony of history be explained? The flamenco world's collective inability to recognize El Negro Meri's presence resembles an anecdote told by Umberto Eco about Albrecht Dürer's woodcut "Rhinoceros" (1515): the scrap books of explorers and zoologists who travelled in Africa during the following decades included sketches and illustrations of rhinos whose design was more faithful to the German artist's model than to the real specimens that they encountered.[10] Like the rhinos drawn by sixteenth-century travelers, Maestro Otero's appearance in Lumière's 1900 film is a mirage, a phantom who, sneaking into the take, turns it into a palimpsest.

Universal Exhibitions: Sites for Nation-State Making

In 1900 Paris, Julius Lessing, art historian and first director of the Museum of Decorative Arts in Berlin, lamented the loss of much of the world's expositions' original character, the "enthusiasm" of their free-trade capitalism having been dampened by the "cool calculation" and "protectionism" of nation-state policies. "Whereas in 1850," writes Lessing, "the ruling tenet was that the government need not concern itself" with the affairs of international commerce, "the situation today is so far advanced that the government of each country can be considered a veritable entrepreneur."[11] During the earliest phase of these world's fairs organized throughout Europe (1791–1850), private exhibitors sought to display raw materials, manufactured articles, mass-produced goods, technological innovations, and inventions in a very limited space inside the city, with the aim of invigorating the flow of commodities in the international market. During that era, art and leisure were offered sporadically, mainly as adjunct events staged at theaters and concert-halls in host cities. But, as urban historian H. Hazel Hahn has noted, "the collusion of the interests of the state, Parisian Government, commercial establishments, and the publishing industry" at work in the Paris expositions of 1834, 1837, 1839, and 1844 already evidenced a broadening in the fairs' offerings.[12] State governments, utterly aware that being agents in the organization of the exhibitions implied, according to Lessing, "a sort of representation," began to promote the massive display of cultural artifacts, ranging from buildings, works of art, literature, engravings, photography, and cinema, to performances of music, dance, and theater.

As a report on the 1862 London Exposition in Kensington Palace stated proudly, "even our national collection [the National Gallery] ... is small and imperfect compared to the great gathering which presented itself in the Exhibition."[13] The scale and scope of world exhibitions grew steadily, becoming micro-cities within cities, mosaics of nations in which governments advertised an "official" image of their countries, aiming to aggrandize colonial imperialism and thus stimulate the circulation of symbolic exchange on a large scale.

Nation-theorist Benedict Anderson has observed that, during the eighteenth century, the novel and the newspaper "provided the technical means for 're-presenting' the *kind* [his italics] of imagined community that is the nation."[14] A century later, the census, the map, and the museum "illuminate the late colonial state's style of thinking about its domain": the core of this thinking, according to Anderson, being "a totalizing classificatory grid, which could be applied with endless flexibility to anything under the state's real or contemplated control: peoples, regions, religions, languages, products, monuments, and so forth."[15] In this sense, world fairs as well as early cinema were the visual epitomes of these systems of classification—and were, as such, manifestations of new tactics of surveillance, control and domination. But, as we will presently examine in the two 1900 Lumière films of Spanish dance, they also became discursive products of the states' new nation-building policies, symbolic containers expressing tensions between the representational strategies of bourgeois "official nationalisms" and those of imperial, mainly dynastic and aristocratic, forms of colonialism.[16]

The 1900 Spanish Pavilion: A Semiotic Ambivalence

In the 1900 Paris Exposition Universelle, as anthropologist José Antonio González Alcantud has explained, the Spanish government built a Royal Pavilion in the Neo-Renaissance style. This design choice aimed to counteract the reigning Romantic image of Spain as an underdeveloped country of bullfighters, smugglers, *Gitanos* (Spanish Roma, so-called "Gypsies") and Moors, as well as to reinforce the idea of a nation whose modernization was linked to its glorious imperial past.[17] The four facades of architect José Urioste Velada's building, inspired by the University of Salamanca, the Palace of Monterrey (also in Salamanca), the University of Alcalá (near Madrid), and the Alcázar (fortress) of Toledo, symbolized sixteenth-century Spain's achievements in scientific knowledge, its industrial advancements, and the splendor of its art and military power (see figure 1). Evoking four monuments of Castilla, the seat of the renaissance kingdom which had "reconquered" Spain from the Moors in 1492 and ruled an empire upon which the "sun never set," the Spanish Pavilion paid homage to the period of Habsburg rule (1516–1700) following the 1479 marriage between Isabela I of Castilla and Fernando II of Aragón, which unified the Christian kingdoms and set the stage for the emergence of a Catholic empire.[18]

In addition to the Spanish Pavilion's program of countering Romantic stereotype and re-asserting Spanish cultural status, there were imperial political considerations at play in the building's design. Following a brief and convulsive period of liberal democratic rule, the monarchy had been restored to the throne in 1875.[19] In symbolizing Castilla, the building not only evoked the power of a strongly centralized state, but also the alliance of the Habsburg monarchy, the ruling dynasty of the Spanish renaissance now restored to the throne under the regency of Alfonso XII's queen, Maria Cristina of Habsburg, with its Habsburg cousins, the Austro-Hungarian Empire.[20]

Spain, unlike neighboring France, was still ruled by a monarchy, and Spaniards suffered the resulting strictures and privations imposed by an ineffective and corrupt government, a dysfunctional system of education which left three quarters of the population illiterate, and a bankrupt military, which had lost the War of 1898.[21] The often-violent political and economic upheavals leading up to this painful juncture are reflected, we will argue, in the Spanish Pavilion's architecture—and also in its use.[22]

Like the restored monarchy, the Spanish intelligentsia, the intellectuals and writers of the so-called "Generation of '98," also sought to resuscitate the nation's glorious past. However, and in opposition to the Spanish Crown, the Generation of '98 saw in the Habsburg renaissance an era, writes Hispanist Edward Inman Fox, of "absolutism, the abolition of representative institutions, intolerance, and economic decline."[23] The late-nineteenth-century elaboration of Spanish nationalism, Fox argues, must be understood as a critique of "the decadence of Spain in the sixteenth and seventeenth centuries." Aspiring to the tenets of liberal capitalist democracy, the Generation of '98 looked back beyond the Habsburg renaissance, to medieval Iberia, when epic heroes like El Cid led the charge of a Christian federation against Muslim invaders for control of the peninsula. The medieval era, Fox continues, was "unified by a religious spirit, but with a legislated common law"—and "characterized by a sense of individual liberty as a sign of social progress."[24]

The racial politics at play here are subtle yet important for understanding the countervailing tensions working through the Spanish Pavilion as well as through the two Lumière films recording the performances staged there. As Fox observes, the luminaries of the Generation of '98, such as Miguel de Unamuno, who published *En torno al casticismo* (*On Castilian National Essence*) in 1902, chafed at Romanticism's casting of Spain as a land of Andalusian *Gitanos*—closer to the "primitivity" of Africa than to the hegemonic Whiteness of European nations—just as the monarchist government did.[25] Nonetheless, there is a crucial distinction between the pro-monarchy position of the Spanish government and the thinking of the Spanish intelligentsia of this period. The Generation of '98 sought to reclaim Spain's national essence from the period *before* the establishment of Habsburg absolutism: these intellectuals critiqued what is often called the "*Siglo de Oro*," or Golden Age, by attempting to recover and discover "*what could have been*" (Fox's italics): "regenerating Spain by reviving a more promising past."[26]

Still, we should bear in mind that both the state and the intellectual elite looked to Castilla as a symbol of Spain's "pure-blooded" Whiteness, postulating a lineage and a history which was only legitimate to the degree that it

46 Part 1: Writing the Body

was seen as untainted by Moorish and Jewish ancestry.²⁷ As Goldberg argues in *Sonidos Negros: On the Blackness of Flamenco*, Andalucía, in contrast, loomed large in the Spanish imagination as the locus of both the Muslim past and, as the decaying center of an expired empire, of Spain's lost colonial future. For the Spanish state and intelligentsia alike, then, these were the

Figure 1. The Spanish Pavilion on La Rue des Nacions, 1900. Located on the left bank of the Seine, between the Eiffel Tower and the Esplanade des Invalides, the Spanish Pavilion was near the British, Belgian, Norwegian, and German pavilions (courtesy Brown University Library).

racialized politics regarding flamenco—which comes from Andalucía—as emblematic of national identity.

Despite the Castilian symbolic agendas advanced by both the monarchy and the Generation of '98, the General Committee of the Exposition licensed the establishment of attractions pandering to the marketability of the Romantic image of Spain. *L'Andalousie au temps des Maures* (Andalucía in the Days of the Moors) played at the Palais du Trocadéro, a Moorish-styled concert hall which had been erected for the 1878 exposition. And in the basement of the Spanish Pavilion, French impresarios A. Henri and H. Cauderon set up La Feria restaurant, where "suggestive and intoxicating female shopkeepers ... carefully selected from the most authentic Andalucía," "brunette Spanish women," their "buns studded with red flowers," decorated a Sevillian patio setting, selling "liquors and *bibelots*" (trinkets) beneath the horseshoe of a faux Cordovan arch.[28]

Describing the show at La Feria, the French press thoroughly catalogues the components of this imaginary: dancing in "a circle of mandolin players," a "Spanish girl" is "a poem of love made of curving torso and suggestive hips," her gestures evoking "red-hot passions that flower in the burning midday sun."[29] Her initial graceful modesty "tempers the ardor she incubates in the flames of her glance," the "luminous pools" of whose pupils are "illuminated like savage torches." The "shudders" and "explosions" of her movement, spreading and mounting like "an invisible aphrodisiac fluid," are sonorous as well: "a beating of the hands, regular, muted, but like a secret call" drums out a "lascivious march," as "the orchestra wreaks havoc, the cries explode and their stridence is both exciting and punishing: the mandolins grate, the guitars resound, the castanets beat and, in this crazy whirlwind loosed by hoarse interjections, the body of the dancer is no more than a supreme vibration; her feet strike the floor with fury, her hips move, her arms entwine and, upon these convulsive shocks rides an attractive, dominating calm, the enigmatic smile of a woman."[30] At the opening of the Royal Pavilion on May 9, the French press noticed the building's semiotic ambivalence, representing Spain as Castilian empire on top and subordinate Andalucía below: "The Royal Pavilion is a world unto itself. Downstairs, the Spain of *seguidillas* [an ancient Spanish dance-song] and warriors; upstairs, the genius of Spain, solemn and proud of its glorious past."[31]

Aside from a few artistic and literary bohemians, the small number of Spanish visitors to La Feria, mainly members of the industrial, political and cultural elite, saw things differently. Accounts of the show performed in the *cafe-chantant* La Feria in the Spanish press voiced the official criteria promulgated by the Spanish Commission's Pavilion. Reviews published in Spain revolved around two entwining arguments: first, the Spanish elite's critique of the commodification of Spanish folklore as distorting the national repertoire into a kind of cheapened product, a tourist souvenir for export to the urbanized mass-market, and second, a critique of the image of Spain, anchored to the notions of backwardness, moral inferiority, and intellectual poverty promoted by these souvenirs.[32]

Thus, the Spanish Pavilion and La Feria restaurant manifested tensions between the image of Spain which the government sought to project and the circulation of opposing signs offered by private business. After the loss of its last colonies in 1898, and against the government's efforts to promote Spain as a modernizing state transforming itself with dignity from a vast empire to a single nation, La Feria restaurant capitalized on the exotic aural and visual imaginary of timeless Andalucía. In this sense, the architectural configuration of the Spanish Pavilion resembled the structure of a Freudian dream: the main building represented the manifest discourse of Castilla as the oneiric portrait of Spanish nation-making, while the basement, where La Feria was located, represented Andalucía as the hidden transcript of its latent content. Andalucía, consciously excluded from Spanish nationalism, emerged as a sort of Freudian "return of the repressed."

On the Dialectics of Flamenco and Bolero Schools

In his memoirs, flamenco singer Rafael Pareja (1877–1965) recounts how in late nineteenth-century Sevilla, in the illustrious *Café del Burrero*, one of the principal stages upon which flamenco was germinating, a "famous *bailaor*" (flamenco dancer) and a renowned maestro of the bolero school, Spain's classical or academic tradition, made a bet.[33] In a contest to decide which genre was the most difficult, they would dance in each other's styles. Maestro Moreno, the bolero school maestro, would dance six flamenco "*escobillas*": "little brooms," or brushed steps, *escobillas* are footwork combinations of percussive sound with dexterous and intricate step patterns. Pepe Ronda, the flamenco dancer, would perform the "*baile inglés*," the English dance, which Pareja deems "the most difficult of the bolero dances."[34] Pareja does not give the first name of Maestro Moreno, whom he describes as directing a small company of bolero dancers at the Burrero.[35] Neither does he give a date for the anecdote, although it may have taken place after 1888, when the Burrero, having moved from its original location to the Calle Sierpes, entered its halcyon days.[36] The protagonist of Pareja's story might be Vicente Moreno "El Valenciano," a bolero school maestro whom Otero calls the best "Spanish classical dancer who I have ever met"—none of the maestros of Sevilla could compete with his virtuosic footwork, Otero writes.[37] Despite Moreno's trained virtuosity, however, Pareja recounts how Ronda—to be a flamenco dancer in this era was by definition to be "unschooled"—learned *El baile inglés* in just three days, and when he performed it at the Burrero "his jumps and beats were not only better than those of the bolero group but beat even the maestro himself," who publicly admitted defeat after Ronda's demonstration.[38]

It is fair to say that flamenco itself emerged victorious from this contest, since it flourishes today, whereas the bolero school inhabits a far more circumscribed platform. Nonetheless, the competition between flamenco and bolero dancers like Moreno and Ronda, in which they learned each other's steps, is illustrative of flamenco's cross-fertilization with the bolero school on stages such as the Burrero—*cafés cantantes* regularly alternated bolero and flamenco *cuadros* in their lineups.[39]

One salient comparison between these two forms, as the Ronda/Moreno contest demonstrates, is their shared degree of technical difficulty. In fact, as we will describe presently, flamenco and the bolero school overlap a great deal: not only in steps, but also in rhythmic and musical forms, instrumentation, and so on.[40] And yet, like the program of the Spanish Pavilion at the exposition, the model of contest between the two forms reflects class rivalries between dance academies, whose beginnings lie at court, and popular forms.

How would "unschooled" flamenco *escobilla* and the "schooled" *baile inglés* have differed? In her book on *The Bolero School*, Marina Grut explains that *El baile inglés* was created by Ángel Pericet Carmona, a virtuoso dancer in the bolero style and founder of the Pericet dynasty, for a theater engagement in 1897 Sevilla.[41] In his 1912 treatise Otero discusses foreign dances from the polka and the waltz to the cakewalk and maxixa, but does not mention *El baile inglés*, nor is this piece among the stage dances such as the *Vito*, *Petenera*, and *La malagueña y el torero* filmed by the Lumière brothers in 1898 Sevilla.[42] *El baile inglés* may have referenced the English hornpipe, a dance of sonorous footwork in seventeenth- and eighteenth-century England, which by the nineteenth century had been embellished using steps and techniques from ballet, and was thus an audience-pleasing athletic and virtuosic form that still bore the imprimatur of Europeanness and classicism.[43] It also may have referenced the sonorous footwork of a jig, as implied by an 1863 announcement of a performance in the Circo Barcelonès in which a generic "*baile inglés*" was titled "*Jig*."[44] Under the influence of blackface minstrelsy imported to Europe from the United States, Spain set English dance in racialized contrast to non-white dance—just as flamenco *escobilla* and the bolero *baile inglés* were juxtaposed in the Pareja anecdote, the 1863 *Jig* was paired with a *tango de negros*—a black tango, which for this era was an Afro-Cuban dance. We will see this racialized polarity active in the 1900 Lumière films as well.

Lumière's *Danses Espagnoles* at the 1900 Paris Exposition

Ever since the Lumière Brothers had begun producing commercial motion pictures in 1895, dance had been a recurring subject, screened in café-concerts, music-halls and all kinds of theaters. Of the 1422 items that make up Lumière's complete filmography, 103 of them can be classified as "dancefilms" recording social, popular, ritual, and theater dances, from ballet to vaudeville and cabaret performances from around the world; of these, fourteen record Spanish dance.[45] The majority of them were produced before 1900; the number of dancefilms declined significantly from that year until Lumière closed its doors in 1905.

The two silent motion pictures shot at the Paris Exposition in 1900, both titled *Danse espagnole de la Feria*, total just over a minute in length. Both clips, whose titles would seem to indicate that the dancers are performers

Figure 2. Meri jumping. "Danse Espagnole de la Feria Sevillanos," *Vue* no. 1123, screen shot 1900. Left to right: José Fernández, director of the string ensemble in *Vue* no. 1124, Jacinto Padilla, "El Negro Meri," mid-jump, an unidentified woman seated, guitarist Eduardo Salmerón Clemente (obscured behind Reguera), Anita Reguera, "Anita de la Feria," dancing, another unidentified woman and man, seated. Filmed at the Exposition Universelle in Paris by the Frères Lumière between July 1 and July 8, 1900 (© Institut Lumière).

from the fashionable La Feria restaurant in the basement of the Spanish Pavilion on the Rue des Nations, were filmed on the same smooth outdoor platform on a terrace of the Palace of Horticulture, overlooking the Seine.[46] According to Michelle Aubert and Jean-Claude Seguin, the films were shot between April 15th and July 8th, although Mora has determined, based on his identification of the female dancer in both films as Anita Reguera, "Anita de la Feria," that they must have been filmed during the first week of July.[47] The fact that their subtitles are reversed in the Lumière catalog—1123 should be *Quadro Flamenco*, while 1124 should be *Sevillanos* (sic)—is perhaps indicative of French audiences' lack of discernment regarding Spanish dance.[48] Was the French company aware of the heated debates within Spain regarding modernization, and dance as a tool of nation-making? Cristina Cruces Roldán, who has written a rich survey of Spanish dance films from 1894 to 1910, argues that early non–Spanish filmography ignores the nation's diversity of identity, propagating instead a monolithic and folkloric image in which "Spanish" and "Andalusian" appear as undifferentiated labels—exactly the demeaning stereotypes of Romanticism which both the Spanish intelligentsia and the government, in placing Castilla on top in the Spanish Pavilion, sought to counter.[49] We think it more likely that the decision to draw a contrast between flamenco and bolero dancing in the two Lumière films came from the artists themselves.

As a pair, these two films not only highlight the continuities and gradations between bolero and flamenco styles—a single woman dances in both—but, as a different man dances in each clip, they also reproduce the dance battle between Pepe Ronda and Maestro Moreno at the Café del Burrero. The flamenco clip is significant as the first film record of this form, but the classical clip, including for the first time an ensemble of mandolins and guitars in the shot, yields important insights as to how the dancers adapted their choreography for the camera.

In the classical clip, as indicated in the title, a couple dances "sevillanas," a dance which exists today in both flamenco and bolero school repertories. Mora has identified the male dancer in this clip as Virgilio Arriaza, son of bolero maestro Domingo Arriaza and brother of New York émigré Aurora Arriaza, with whom he performed in Belasco's *Rose of the Rancho* in 1906 New York.[50] Arriaza is tall, slim, elegantly-coiffed and mustachioed (see figure 3). His dancing references the *sevillanas boleras*: the *sevillanas* of the bolero school. He plays castanets and performs intricate beats and foot patterns while keeping his body controlled and upright, often inclining in a gallant diagonal toward his partner in the manner of ballet.

Like all flamenco dances, *sevillanas* are structured by the *copla*, or verse. *Sevillanas* verses have three parts; each section opens with a *"sevillanas* step," in which partners curtsey to each other, the first two sections end with a

Figure 3. Arriaza. "Quadro Flamenco," *Vue* no. 1124, screen shot 1900. Background: a string ensemble (led by José Fernández, hidden), wearing *trajes de la estudiantina* (student's costumes). Dancers: Anita Reguera, "Anita de la Feria," and Virgilio Arriaza. Filmed at the Exposition Universelle in Paris by the Frères Lumière between July 1 and July 8, 1900 (© Institut Lumière).

pasada, the step with which the partners change places, and the verse closes with a *bien parado*: the couple freezing in a pose on the final beat.[51] In the Lumière film, we enter the scene mid-verse and leave it likewise. Counting the phrases before the *bien parado*, we know that musically we are in the second and third phrases of the verse, but neither dancer conforms to the choreography taught in dance schools today. As the film opens, rather than dancing the second and third phrases, Arriaza dances the first and second phrases of the first verse of *sevillanas boleras*: *assemblé battu, sissone, pas de bourée*, followed by *attitude sauté* (hops in low attitude crossing his partner), or perhaps these are *careos* (something like a traveling *pas de basque*).[52] Reguera, meanwhile, dances two second phrases: *matalaraña*, the second step of the first *sevillanas copla*, and *pas de basque* or *esplante*, the second step of the second *sevillanas* verse today.

The couple finishes the *copla* facing the camera. Reguera stops in a typical Spanish dance pose, head turned profile, arching backward, with one arm overhead and one arm in front of her chest, framing her face. But Arriaza finishes behind her, holding her waist and peeking at her over her right shoulder. Neither his touch to her waist nor his coquettish glance is typical of Spanish dance; rather, both are reminiscent of the cakewalk's play.

The cakewalk did not arrive in Spain until 1902, but Spain was certainly aware of international fashion: in 1889 New York, Sam T. Jack had sparked a trend with *The Creole Show*, featuring a beautiful African American artist, Dora Dean, dancing a cakewalk; by 1897 Ada Overton Walker had danced the cakewalk in England, in Isham's revue *Oriental America*.[53] In between verses, Arriaza politely takes Reguera's hand to change sides. Like his ending pose, this detail is unusual. Perhaps it is adopted from the social dances taught alongside Spanish dances in academies such as those directed by Otero and Arriaza's father Domingo in this era.

In fact, much of what we see in this short film is choreographic pastiche—decisions that reveal the performers' subtle responses to current trends in dance, and that seem advantageous for the camera. At the beginning of the second *copla*, Reguera performs *seasé con tres pasos por detrás*, which is (correctly in today's terms) the first phrase of the second copla of sevillanas.[54] Arriaza finishes his first phrase with a pencil turn, an elegant drawing of the body together while spinning, often used in mid–twentieth-century masculine Spanish dance. Then, almost as if they had measured and rehearsed how long the camera take could last, the couple truncates their verse, skipping forward to the last phrase of the third *copla* of the *sevillanas boleras* (the last phrase of the fourth *copla* in flamenco *sevillanas*): *careos*.[55] This is a showy step to end the clip, and they perform it with deep bends in the waist, almost touching lips as they pass by one another, in what might be read as a gesture to the nineteenth-century Romantic imaginary of Spain.

If the performative elements deemed to be most appealing to the camera—deep bends at the waist, near kisses, and virtuoso jumps and beats—are evident in the classical clip, perhaps the same may be said for the flamenco clip as well. Lumière 1123, which should be titled "Quadro Flamenco," opens with performers sitting in a semi-circle, just as flamenco cuadros do today. At left is José Fernández, who also leads the musical ensemble in Lumière 1124, wearing a *traje de estudiantina*, the costume of fourteenth-century university students who raised money to pay their tuition as street musicians; by the turn of the twentieth century these musicians were professional groups touring internationally.[56] Seated to Fernández's left is Reguera, and next to her is Jacinto Padilla, "El Negro Meri," who, as flamencos often do, both sings and dances in the film. Mora has identified the guitarist as Eduardo Salmerón Clemente, and two more as-yet unidentified women and another man also appear, all dressed in theatrical costumes.[57] All of the seated performers act as percussionists, playing *palmas*, traditional flamenco handclaps; Padilla, in the manner of illustrious singers of past generations, such as Silverio Franconetti, instead marks rhythm on the edge of his chair with a small bamboo stick.[58]

As Goldberg describes in *Sonidos Negros*, the rhythm they are dancing in the flamenco clip is what we now call *bulerías*—the triple-meter rhythm of much Andalusian music. Both Reguera and Padilla dance in a style that is easily recognizable to flamenco dancers of today. Reguera's dancing in both clips highlights both the continuities and the subtle stylistic differences between flamenco and bolero school styles: for example, she lifts slightly more through her back and through the back of her neck in the flamenco clip, whereas she bends more deeply at the waist more often in the classical clip.[59]

Padilla's dance, however, clearly has a different technical basis from that of his counterpart in the classical clip, Virgilio Arriaza: rather than playing castanets, which, tied to the thumbs, preclude a dancer from playing *palmas*, Padilla marks the rhythm with *pitos* (finger snaps). Likewise, in contrast to Arriaza's vertically-oriented jumps, Meri's spectacular jumps are propelled not only from deep bends in the knees, but also at the hips. As Goldberg discusses in *Sonidos Negros*, if his jumps recall the virtuosity of a circus equestrian, and the valor of a bullfighter, they also predict the daring jumps of Antonio Montoya Flores "El Farruco" (1935–1997), a Roma artist, and one of the greatest male flamenco dancers of the twentieth century. In *Sonidos Negros*, Goldberg also discusses clues that, in the course of several tours to France, Meri may have established a close relationship with another black Cuban clown, the famous "Chocolat" (see figure 4). Rafael "Chocolat" was born enslaved in Cuba around 1865. He was brought to Spain as child, where he escaped bondage and, through a series of chance encounters, landed in Paris, forming a duo with British clown George Foottit, which would gain him great celebrity in France.

On Becoming: The Dialectics of Race in Flamenco

We should pause here to wonder over the complex politics of the fact that in the 1900 Lumière film, a Spaniard of Afro-Cuban descent figures as the epitome of male flamenco dance, a doubled irony holding, in the words of W. E. B. Du Bois, "two unreconciled strivings"—French perceptions of Spain and Spanish views of themselves—within "one dark body."[60] France and Spain, waning empires both, were equally haunted, terrified, and fascinated by the specter of the rising Americas. And yet Chocolat—albeit in racist terms—is remembered in France, while El Negro Meri is forgotten in Spain. Why?

Spain in the Basement (Mora and Goldberg) 55

Figure 4. Chocolat. George Footit and Rafael Padilla "Chocolat," photograph before 1917 (courtesy Bibliothèque nationale de France).

Along with the new cinematographic technologies deployed at the 1900 World's Fair, the forerunners of jazz had arrived in Paris with John Phillip Sousa's ragtime performances at the exposition. Jazz strongly influenced the international Parisian art scene of the first decades of the twentieth century, providing the soundscape as well as the visual language for Modernism. As art historian Jody Blake observes, primitivism and modernism were mirror images; the silvery, stripped-down modernity of jazz could not be separated from the sun-blasted and seemingly boundless possibilities of black Africa. Likewise, the figure of the racialized primitive, determined at this fluid moment by its exchange value in a capitalist society, was conflated with an image of the mechanized body. As theorist Louis Onuorah Chude-Sokei argues, the mechanized or "cyborg" body obviates the need for the uncompensated labor of the enslaved—and yet the terror of the modern robot (presaged in Mary Shelley's *Frankenstein*, 1818), is directly related to white terror that those enslaved will rise up against their torturers.[61]

Thus, "the discovery of things '*negre*' by the European avant-garde," Blake writes, "was mediated by an imaginary America, a land of noble savages simultaneously standing for the past and future of humanity."[62] In Paris, by 1905, painters such as Spaniard Pablo Picasso (1881–1973), and Frenchmen Henri Matisse (1869–1954) and Andre Derain (1880–1954) began to study and acquire African masks and figures, seeking to clear away superficial mannerisms of decoration and sentiment and intuit the essential and un-mediated tenets of aesthetic form.[63]

In contrast to Paris, then the center for avant-garde Modernism, where imitating and appropriating *negritude* was in vogue, during the first decade of the twentieth century in Spain, flamenco, as a nativized response to newer cultural invasions from the now politically dominant Americas, substituted for *negritude*. As Goldberg has argued, having lost the last vestiges of empire in 1898 and stymied in its efforts to control its symbolic image internationally, Spain absorbed the cakewalk into the flamenco repertoire, and the modernist imagery of jazz into the long-extant figure of a nativized and minstrelized imaginary *Gitano*.[64] Nonetheless, the ambivalence with which Spain viewed flamenco—and its popularity on foreign stages—had everything to do with the racialized hierarchy declaimed by the Spanish Pavilion itself: with Castilla on top, and dark, Africanized Andalucía in the basement.

Ever since its occupation by Muslim forces in 711, Spain's uniquely ambiguous situation *vis-à-vis* the rest of Europe had been defined by its troubled relationship to Africa. Flamenco, springing up in the same moment as the racist theories of writers such as Arthur de Gobineau (whose notorious *Essai sur l'inégalité des races humaines* was published in 1853–55), embodied a new iteration of Spain's long-standing racist discourse. For the Spanish elite, the bolero school's balleticized stagings of Spanish folk dances (such as *La Jota*, filmed by

Lumière in 1898 Sevilla), preserving the ethos of aristocracy in a nation which had for centuries certified social hierarchy according to degrees of whiteness, embodied the true "soul of the nation."[65] In contrast, flamenco, professionalized by the "Andalusian *wage earner*," was seen as pseudo-artistic "tuneless howls and *cancan*-esque contortions"—a salaried and "studied voluptuousness" spuriously derived from Spanish tradition, debasing the aesthetic purity of Spain's native folklore while turning a profit for French theatrical managers.[66]

In 1900 Paris, flamenco, like Afro-American performance (such as the cakewalk), ironically reminded Spaniards of their subordinate, exoticized status in the eyes of Europe: Spaniards *shared* with Americans of African descent the "aspiration and inability to be European and modern."[67] For Spain to participate in what Spanish novelist and essayist Emilia Pardo Bazán in 1889 called "the enormity of beginning," it had first to shuck off the stereotypes which placed it in the basement of European hierarchies. The exclusion of flamenco, a synecdoche of Andalucía just as Andalucía was a synecdoche of Spain, from the nationalist image constructed in the Royal Pavilion operated through this reasoning. From the perspective of the Spanish elites, flamenco, "an uncultured spectacle," promulgated "a poor idea of the morality and the capacity" of the Spanish people—it could neither represent the longed-for winds of progress nor the "essence" of the "national soul."[68] Writer Luis Bonafoux thus recorded the disgusted reaction of renowned Spanish violinist Pablo de Sarasate, who at the 1900 exposition happened into La Feria just as the flamenco show was reaching its climax. "But," Sarasate exclaimed, "why such bawling? It seems they are not human beings!"[69]

And yet the Spanish elite needed to somehow reckon with American cultural imports, whose appropriation of blackness threatened to swallow Spain, identified with those swarthy so-called Gypsies, in its gigantic maw. This is the agenda, shared by both monarchists and the Generation of '98, at play in the Spanish Pavilion of 1900, and likewise evidenced in Spain's nativizing response to the cakewalk in terms of flamenco. The descriptions of the 1900 Paris exhibition written by Spanish journalists and artists such as Luis Bonafoux and Pablo de Sarasate delineate a paradigmatic perception of a fragmented and ruptured flamenco body—noisy, chaotic, dislocated, lustful, primitive, and animalized. This corporeal frame was porous and receptive, permeable—the audience came inside—and this is the key to Spain's ambivalent denigrations and appropriations of flamenco. Flamenco dancing was an "exercise of dislocation"—a breaking of the gravitational center, a magnetic field now directed away from the elevated stages upstairs and toward the tavern of the basement netherworld.[70]

Dance theorist Melanie Kloetzel argues that one of the markers of dance-film consists in "the use of alternative contexts," "moving the body outside the studio or theater space for the film shoot."[71] While La Feria restaurant was located beneath the Spanish Pavilion, the Lumière company filmed the artists on a terrace of the Palace of Horticulture, on the other side of the Seine. As a French newspaper stated, the restaurant was decorated as "a set representing a courtyard in a [Sevillian] inn, with green-painted balconies and a red-tile ceiling."[72] But the cinematograph located the Spanish dances in outdoor Paris, against the background of one of its most emblematic symbols: an ephemeral reproduction of the Louvre Tower.

With this gesture, Lumière pulls flamenco out of the basement and sets it at a nodal point of cosmopolitanism. Seen through Lumière's lens, the bitter contests within Spain for control of the narrative of its own national identity, transcribed in the semiotic ambiguity between the Royal Pavilion's Castilian upper structure and its Andalusian and flamenco basement, are rendered blurry and insignificant. In the acquisitive view of the Parisian consumer, Spain's Africanness—seen as both timeless primitivism and up-to-the-minute fashionability—is its most marketable characteristic. Thus, in Lumière's lack of distinction between the flamenco and bolero schools, as in the hegemonic view of Spain implicit in the French filmmakers' choice not to distinguish between black and white flamenco performers, we read the script of Meri's erasure from flamenco history. As cultural theorist Stuart Hall admonishes in his seminal analysis of Frantz Fanon's *Black Skins, White Mask*, scholars must grapple with: "the as-yet deeply unresolved question ... as to how to reconcile—or at least hold in a proper balance—...*both* Fanon's spectacular demonstration of the power of the racial binary to *fix*, and [Homi] Bhabha's equally important and theoretically productive argument that all binary systems of power are nevertheless, *at the same time*, often if not always, troubled and subverted by ambivalence and disavowal. Our dilemma is how to *think together* the overwhelming power of the binary ... *and simultaneously* the ambivalences, the openings, the slippages which the suturing of racial discourse can never totally close up."[73] Although Jacinto Padilla, "El Negro Meri," was erased from flamenco history for over a century, his dance vocabulary—those circus-derived jumps—but perhaps more importantly his ironic yet audacious sensibility, define the heart center of flamenco aesthetics today. Exemplified by Roma artists such Farruco, flamenco's play at the edges of control razed the rigid standards that would petrify the bolero school and condemn it to obscurity, while catapulting flamenco onto the Modernist stage. Like jazz, flamenco has been shaped by race mimicry, and the stereotypical flamenco dancer is still, to this day, white. But an appreciation of Meri's power as an artist reveals how much of flamenco's soul emanates out of the Africanist ethos.

Notes

1. For a close reading of the first of these films (1123), see "Tightropes and Wild Horses: The Dance of the Blackface Clown," in K. Meira Goldberg, *Sonidos Negros: On the Blackness of Flamenco* (New York: Oxford University Press, 2018). K. Meira Goldberg,

Kiko Mora, and Cristina Cruces Roldán are among the scholars exploring the complex and fascinating intersections between classical Spanish dance, early cinema, developing modern dance, flamenco, and vaudeville during this period. See Mora, "La representación contra-hegemónica de la negritud: La Perla Negra, entre la rumba y la danza moderna (1913-1928)," *Sinfonía Virtual* 32 (invierno 2017), 1-36; on Carmencita, Mora, "Carmencita on the road: Baile español y vaudeville en Los Estados Unidos de América (1889-1895)," *Asociación Lumière* (2011), http://www.elumiere.net/exclusivo_web/carmencita/carmencita_on_the_road.php (accessed August 25, 2017); and Mora, "Carmen Dauset Moreno: primera musa del cine estadounidense," *Zer. Revista de estudios de comunicación* 19 (2014): 13-35; on Carmencita, see also Ninotchka D. Bennahum, "Early Spanish Dancers on The New York Stage," in Ninotchka D. Bennahum and K. Meira Goldberg, eds., *100 Years of Flamenco in New York* (New York: New York Public Library for the Performing Arts, 2013), 26-57; and Bennahum, *Antonia Mercé La Argentina. Flamenco and the Spanish Avant Garde* (Middletown, CT: Wesleyan University Press, 2000), 51-53. For an overview of early films of Spanish dance, see Cruces Roldán, "Bailarinas fascinantes; géneros y estereotipos de 'lo español' en el cine primitivo (1894-1910)," *Bulletin of Spanish Visual Studies* 1, no. 2 (2017): 161-192.

2. "Carmencita," Library of Congress, Motion Picture, Broadcasting, and Recorded Sound Division, http://www.loc.gov/item/00694116/ (accessed August 27, 2014); Robert W. Paul, "Andalucían Dance," *The Collected Films 1895-1908* (London: British Film Institute, 2007); Antonio J. Ferreira, *A fotografia animada em Portugal: 1894-1897* (Lisbon: Cinemateca Portuguesa, 1986); Francisco Griñán, *Las estaciones perdidas del cine mudo en Málaga* (Málaga: Diputación de Málaga. 2009), 25-35.

3. In addition to Mora on Carmen Dauset Moreno, "Carmencita," and on "La Perla Negra," see José Luis Ortiz Nuevo, Ángeles Cruzado, and Kiko Mora, *La valiente: Trinidad Huertas "La Cuenca"* (Sevilla: Libros con Duende, 2016); and Mora, "Sounds of Spain in the Nineteenth Century USA: An Introduction," in Goldberg and Antoni Pizà, eds., *The Global Reach of the Fandango in Music, Song and Dance: Spaniards, Indians, Africans and Gypsies* (Newcastle upon Tyne: Cambridge Scholars, 2016), 270-307. On Macarrona, see Goldberg, "The Latin Craze and the Gypsy Mask: Carmen Amaya and the Flamenco Gypsy Aesthetic, 1913-1963," in Bennahum and Goldberg, eds., *100 Years of Flamenco in New York*, 68-99; "Juana Vargas, 'La Macarrona': A Flamenco Treasure," New York Public Library Blogs, January 21, 2015, http://www.nypl.org/blog/2015/01/21/juana-vargas-la-macarrona-flamenco (accessed January 21, 2015); Goldberg, "Jaleo de Jerez and Tumulte Noir: Primitivist Modernism and Cakewalk in Flamenco, 1902-1917," in Goldberg, Bennahum, and Michelle Heffner Hayes, *Flamenco on the Global Stage* (Jefferson, NC: McFarland, 2015), 124-42; and "Jaleo de Jerez and Tumulte Noir: Juana Vargas 'La Macarrona' at the Exposition Universelle, Paris, 1889," in Goldberg, *Sonidos Negros*.

4. José Luis Ortiz Nuevo, *Coraje. Del maestro Otero y su paso por el baile* (Sevilla: Libros con duende, 2013), 9; José Luis Navarro García, *La danza y el cine*, vol. 1 (Sevilla: Libros con Duende, 2014), 28; José Antonio González Alcantud, "Andalucía 'en el tiempo de los moros,'" *Andalucía en la Historia. Dossier resistencias cotidianas* 14, no. 52 (2016): 50-2.

5. Kiko Mora has also identified another bolero maestro, José Segura, in the 1898 films. Kiko Mora, "'El otro' que baila en las películas de Lumière (Sevilla, 1898)," *Cadáver Paraíso* (blog), July 6, 2017, https://cadaverparaiso.wordpress.com/2017/07/06/el-otro-que-baila-en-las-peliculas-de-lumiere-sevilla-1898/ (accessed November 27, 2017).

6. Jacinto Padilla was called "El Negro Meric," "El Negro Meri," "El Mulato Meric," "El Mulato Meri," "El Americano Merit," etc., in the press. For consistency, we refer to him by his given name, as "El Negro Meri," or "Meri." Kiko Mora, "¡Y dale con Otero!... Flamencos en la Exposición Universal de París de 1900," *Cadáver Paraíso* (blog), June 11, 2016, https://goo.gl/YCTtSJ (accessed June 18, 2016); Kiko Mora, "Who Is Who in the Lumière Films of Spanish Song and Dance at the Paris Exposition, 1900," Le Grimh (Groupe de reflèxion sur l'image dans le monde hispanique), https://www.grimh.org/index.php?option=com_content&view=article&layout=edit&id=2745&lang=fr#4 (accessed September 21, 2017).

7. Mora, in "Who Is Who," provides documentation indicating that Padilla was born in Algeciras: "Plaza de Toros de Murcia," *Las provincias de Levante* (June 10, 1887), 4. But Mora also notes a brief chronicle of the life of bullfighter Francisco Arjona, "Curro "Cúchares,"

60   Part 1: Writing the Body

stating that he and El Mulato Meri fought bulls together in Havana before 1868 (Don Crispín, April 11, 1932), 5.

8. "L'Exposition," *Le Matin* (May 27, 1900), 1; Enrique Gómez Carrillo, "La Exposición de París al día," *Nuevo Mundo* (May 30, 1900), 12; Santiago Romo-Jara, "La Exposición de París," *La opinión* (May 6, 1900), 2; Santiago Romo-Jara, "La Exposición de París," *El álbum iberoamericano* (July 22, 1900), 314.

9. John Phillip Souza brought ragtime to Paris in the 1900 Exposition Universelle, and the cakewalk arrived with the review *Joyeux Nègres* in 1902. See Goldberg, "Primitivist Modernism and Cakewalk."

10. Umberto Eco, *La struttura assente. La ricerca semiotica e il metodo strutturale* (Milan: T. Bompiani, 1980 [1968]), 119–120.

11. Cited in Walter Benjamin, *The Arcades Project* (Cambridge: Belknap Press of Harvard University Press, 2002), 183.

12. H. Hazel Hahn, *Scenes of Parisian Modernity. Culture and Consumption in the Nineteenth Century* (New York: Palgrave Macmillan, 2009), 34.

13. Tal. P. Shaffner and W. Owens, *The Illustrated Record of the International Exhibition of the Industrial Arts and Manufacturers, and the Fine Arts* (London: The London Printing and Publishing Company Limited, 1862), 289.

14. Benedict Anderson, *Imagined Communities: Reflections on the Origin and Spread of Nationalism*, rev. ed. (London: Verso, 1991), 25.

15. Anderson, *Imagined Communities*, 184.

16. In *Imagined Communities*, Anderson (159) defines "official nationalism" as a style of nationalist discourse "emanating from the state, and serving the interests of the state first and foremost." On the conflicts between official nationalism and imperialism, see 83–111.

17. José Antonio González Alcantud, "Andalucía 'en el tiempo de los moros,'" *Andalucía en la Historia. Dossier resistencias cotidianas* 14, no. 52 (2016): 50–52.

18. For more on this topic, see for instance E. Inman Fox, "Spain as Castile: Nationalism and National Identity," in David T. Gies, ed., *The Cambridge Companion to Modern Spanish Culture* (Cambridge: Cambridge University Press, 1999), 21–36.

19. In political terms, the period from 1875 to 1902, during which King Alfonso XII was enthroned and monarchy restored following the turbulent and revolutionary years from 1868 to 1874—the first, short-lived Spanish Republic governed 1873 to 1874—is known in Spanish historiography as the first stage of *"La Restauración"* (The Restoration).

20. After the early death of Alfonso XII (reigned 1874–85) in 1885, his wife Maria Cristina was proclaimed regent of Spain until 1902, when Alfonso XIII assumed the throne, ruling until the establishment of the Second Spanish Republic (1931–39).

21. Fox, *Spain as Castile*, 21.

22. See, for example, Javier Varela, "Crisis de la conciencia nacional en torno al 98," in Antonio Morales Moya, Juan Pablo Fusi Aizpurúa, and Andrés Blas Guerrero, eds., *Historia de la nación y del nacionalismo español* (Barcelona: Galaxia Gütemberg/Círculo de lectores, 2013), 543–62.

23. Fox, *Spain as Castile*, 27–28.

24. Fox, *Spain as Castile*, 27–28. For a fascinating analysis of the integration of the tenets of liberal capitalist democracy, national identity, and Christian-derived notions of "race" and "progress," see Max Hering Torres, "Order and Difference in Mid-Nineteenth Century Colombia," in Goldberg, Walter Clark, and Antoni Pizà, eds., *Spaniards, Natives, Africans, and Roma: Transatlantic Malagueñas and Zapateados in Music, Song and Dance* (Newcastle upon Tyne: Cambridge Scholars, forthcoming).

25. Fox, *Spain as Castile*, 2–8. Miguel de Unamuno, *En torno al casticismo* (Madrid: F. Fé, 1902); translation of the title is from Javier Irigoyen-García, *The Spanish Arcadia: Sheep Herding, Pastoral Discourse, and Ethnicity in Early Modern Spain* (Toronto: University of Toronto Press, 2014), 3. For a discussion of the racial politics of Gitano representation at the 1889 Paris Exposition, see "Jaleo de Jerez and Tumulte Noir: Juana Vargas 'La Macarrona' at the Exposition Universelle, Paris, 1889," in Goldberg, *Sonidos Negros*.

26. Fox, "Spain as Castile," 30–31.

27. On *"limpieza de sangre"* or purity of blood, the ethno-religious categorizations

which governed social hierarchy in both Old and New Worlds, see María Elena Martínez, David Nirenberg, and Max-Sebastián Hering Torres, *Race and Blood in the Iberian World* (Zürich: Lit, 2012).
   28. We have not found any image of the restaurant interior. "Au Pavillon d'Espagne," *Le Matin* (May 9, 1900), 1; "L'Exposition," *Le Matin* (May 27, 1900), 1.
   29. "Au fil de la Semaine," *La justice* (July 12, 1900), 1. All translations are by the authors, unless otherwise noted.
   30. "Au fil de la Semaine," *La justice* (July 12, 1900), 1. All translations are by the authors, unless otherwise noted.
   31. "Au Pavillon d'Espagne," *Le Matin* (May 9, 1900), 1.
   32. The same argument was also raging inside the flamenco genre itself, in attempts to differentiate between "cante jitano" (Gitano song), considered to be of Roma origin, and "cante flamenco" (flamenco song), its commercialized version, as it is explained by nineteenth century folklorist and foundational flamenco scholar Antonio Machado y Álvarez in his prologue to *Colección de cantes flamencos recogidos y anotados por Demófilo* (Sevilla: El Porvenir, 1881).
   33. Juan Rondón Rodríguez, *Recuerdos y confesiones del cantaor Rafael Pareja, de Triana* (Córdoba: Tipográfica Católica, 2001), 71.
   34. On Pepe Ronda, José Blas Vega and José y Manuel Ríos Ruiz, *Diccionario enciclopédico ilustrado del flamenco y maestros del flamenco* (Madrid: Cinterco, 1988), 774.
   35. Rondón Rodríguez, *Recuerdos y confesiones*, 71.
   36. Daniel Pineda Novo, *Juana, "la Macarrona" y el baile en los cafés cantantes* (Cornellà de Llobregat [Barcelona]: Aquí + Más Multimedia, 1996), 21; José Blas Vega, *Los cafés cantantes de Sevilla* (Madrid: Cinterco, 1987), 47; Daniel Pineda Novo, *Silverio Franconetti: noticias inéditas* (Sevilla: Giralda, 2000), 71–5.
   37. Otero, *Tratado*, 211.
   38. Juan Rondón Rodríguez, *Recuerdos y confesiones*, 71–72. On flamenco dance in the academy, see Clara Mora Chinoy, "The First Academy of Flamenco Dance: Frasquillo and the 'Broken Dance' of the Gitanos," in Goldberg, Bennahum, and Hayes, eds., *Flamenco on the Global Stage* (Jefferson, NC: McFarland, 2015), 143–56.
   39. K. Meira Goldberg, *Border Trespasses: The Gypsy Mask and Carmen Amaya's Flamenco Dance* (doctoral dissertation, Temple University, 1995), 152–53. On the 1889 Exposition in Paris, see Goldberg, *Sonidos Negros*.
   40. Faustino Núñez, *Guía comentada de música y baile preflamencos (1750-1808)* (Barcelona: Ediciones Carena, 2008); Peter Manuel and María Luisa Martínez Martínez, "El Murciano's 'Rondeña' and Early Flamenco Guitar Music: New Findings and Perspectives," in K. Meira Goldberg and Antoni Pizà, eds., *The Global Reach of the Fandango in Music, Song and Dance: Spaniards, Indians, Africans and Gypsies*. *Música Oral Del Sur* 12 (2015): 249–72 and (Newcastle upon Tyne: Cambridge Scholars, 2016), 153–81; José Miguel Hernández Jaramillo, *La música preflamenca. Aproximación a la formación y evolución musical de los diferentes estilos del flamenco a través de la documentación musical escrita* (Sevilla, Consejería de Relaciones Institucionales, Junta de Andalucía, 2002).
   41. Marina Grut, Alberto Lorca, Ángel Pericet Carmona, Eloy Pericet, and Ivor Forbes Guest, *The Bolero School: An Illustrated History of the Bolero, the Seguidillas and the Escuela Bolera: syllabus and dances* (Alton: Dance Books, 2002), 162.
   42. José Otero Aranda, *Tratado de Bailes de Sociedad, regionales españoles, especialmente andaluces, con su historia y modo de ejecutarlos* (Seville: Tip. de la Guía Oficial, Lista núm. 1, 1912). The 1898 Lumière films shot in Sevilla are *El Vito* (843), *Estrella de Andalucía* (844), *La Jota* (845), *Boleras Robadas* (846), *Bolero de Medio Paso* (847), *Las Peteneras* (848), *Las Manchegas* (849), *Boleras Robadas* (850), *La Malagueña y El Torero* (851), *Bolero de Medio Paso* (852), *La Sal de Andalucía* (853), and *El Ole de la Curra* (854); more information, and images from each clip, can be found in the Lumière catalog, https://catalogue-lumiere.com/?s=1er+mai+1898+seville (accessed July 21, 2016).
   43. For a history of the hornpipe in seventeenth-century English music, beginning with John Playford's *The English Dancing Master* (1621), a collection of English country dances, see Elizabeth Aldrich, Sandra Noll Hammond, and Armand Russell, *The Extraordinary Dance Book T B. 1826: An Anonymous Manuscript in Facsimile* (Stuyvesant, NY: Pendragon Press,

2000), 5–6, 9–13. For more on the political significations of sonorous footwork versus balletic beats and interlacings of the feet, see Goldberg, *Sonidos Negros*.

44. "Diversiones públicas," *El Lloyd español* (March 8, 1863), 4. On the relationships between the Irish jig and developing tap dance in the United States, see Constance Valis Hill, *Tap-Dancing America: A Cultural History* (New York: Oxford University Press, 2009).

45. For the Lumière's Brothers complete filmography, see Michelle Aubert and Jean-Claude Seguin, *La production cinématographique des Frères Lumière* (Paris: Centre national de la cinématographie/BIFI, 1995). Dancefilm has been defined "as a modality that appears across various types of films including the musical and experimental shorts and is characterized by filmic performance dominated by choreographic strategies or effects [and which have] particular common approaches or themes: an interest in gesture or close-up, in the corporeal facility of the dancing 'Star' or the transference of movement across peoples and things." Erin Brannigan, *Dancefilm. Choreography and the Moving Image* (Oxford: Oxford University Press, 2011), vii.

46. On the "fashionable" La Feria: "A L'Exposition," *Le Journal* (July 27, 1900), 2; "A L'Exposition," *Le Journal* (September 11, 1900), 3. For a detailed study of the location of these films, see Kiko Mora, "Flamencos en la Exposición de París 1900 (II): El lugar de filmación de las películas de Lumière," *Cadáver Paraíso* (blog), June 3, 2016, https://goo.gl/kcSAOf (accessed July 6, 2016).

47. Aubert and Seguin, *La production cinématographique*, 185–6; Mora, "Who Is Who."

48. The Lumière clip featuring El Negro Meri can be seen most clearly at the Forum des Images, http://collections.forumdesimages.fr/CogniTellUI/faces/details.xhtml?id=VDP 13353 (accessed November 8, 2016). For an in depth examination of the cast of performers, see Kiko Mora, "Who Is Who."

49. Cristina Cruces Roldán, "Bailes boleros y flamencos en los primeros cortometrajes mudos. Narrativas y arquetipos sobre 'lo español' en los albores del siglo XX," *Revista de dialectología y tradiciones populares* LXXI, no. 2 (2016): 462.

50. "Christmas at the Belasco," *New York Times* (December 25, 1906), 7; Mora, "Who Is Who."

51. For more on the structure of sevillanas, see Thomas Baird, K. Meira Goldberg, and Paul Jared Newman, "Changing Places: Toward the Reconstruction of an Eighteenth Century Danced Fandango," in K. Meira Goldberg and Antoni Pizà, eds., *Spaniards, Indians, Africans, and Gypsies: The Global Reach of the Fandango in Music, Song, and Dance, Música Oral del Sur* 12 (2015): 628–65 and (Newcastle upon Tyne: Cambridge Scholars, 2016), 579–621; and Ana María Durand Viel, *La sevillana: datos sobre el folklore de la baja Andalucía* (Seville: Biblioteca de Temas Sevillanos, 1983).

52. Anna de la Paz, to whom we are grateful for making these identifications, notes that Arriaza's sevillanas differ somewhat from the sevillanas boleras of the Pericet school, notated in Grut, et al., *The Bolero School*, 377–85. Marcellus Vittucci, "Matteo," with Carola Goya, *The Language of Spanish Dance* (Norman: University of Oklahoma Press, 1990), *matalaraña*: 133, *esplante*: 80–81, *careo*: 51.

53. Marshall W. and Jean Stearns, *Jazz Dance: The Story of American Vernacular Dance* (New York: Macmillan, 1968), 78, 86, 117–18; Jayna Brown, *Babylon Girls: Black Women Performers and the Shaping of the Modern* (Durham: Duke University Press, 2008), 137, 133; Thomas L. Riis, "The Experience and Impact of Black Entertainers in England, 1895–1920," *American Music* 4, no. 1 (1986): 52–53. For more on the cakewalk in Spain, see K. Meira Goldberg, "Jaleo de Jerez and Tumulte Noir: Primitivist Modernism and Cakewalk in Flamenco, 1902–1917," in K. Meira Goldberg, Ninotchka Bennahum, and Michelle Heffner Hayes, eds., *Flamenco on the Global Stage* (Jefferson, NC: McFarland, 2015), 124–42.

54. Marcellus Vittucci, "Matteo," with Carola Goya, *The Language of Spanish Dance* (Norman and London: University of Oklahoma Press, 1990), 223–4.

55. Matteo, *The Language of Spanish Dance*, 51–52; Grut et al., *The Bolero School*, 38–40.

56. For more on José Fernández, see Mora, "Who Is Who." For more on the Spanish students' tour in the United States, see Mora, "Sounds of Spain." For more on the *estudiantinas*, see Félix Martín Sárraga, *Mitos y evidencia histórica sobre las tunas y estudiantinas* (Lima: Cauce, 2016). For more on the birth and tours of the Spanish students, see Michael Christoforidis, "Serenading Spanish Students on the Streets of Paris: The International Projection

of the *Estudiantinas* in the 1870s," *Nineteenth Century Music Review* 14, no. 2 (2017): 1–14; Félix Martin Sárraga, "Apuntes sobre las giras europeas de la estudiantina Fígaro," *Tunae Mundi* (blog), July 24, 2015 (updated June 6, 2017), http://www.tunaemundi.com/index.php/component/content/article/7-tunaemundi-cat/630-apuntes-sobre-las-giras-europeas-de-la-estudiantina-figaro (accessed December 14, 2017); and Félix Martín Sárraga, "Análisis comparado de los integrantes de la estudiantina española Fígaro (1878–1892)," *Tunae Mundi* (blog), July 29, 2015 (updated June 11, 2017) http://www.tunaemundi.com/index.php/publicaciones/sabias/7-tunaemundi-cat/632-analisis-comparado-de-los-integrantes-de-la-estudiantina-espanola-figaro-1878-1892 (accessed December 14, 2017).

57. Mora thinks that the seated women may be sisters, either Juana and Felisa Peña, or Margarita and Amparo Aguilera. Mora, "Who Is Who."

58. This is in the manner of Silverio Franconetti, who in turn followed the great flamenco singers "El Loco Mateo" and "El Nitri." Pineda Novo, *Silverio Franconetti: noticias inéditas* (Sevilla: Giralda, 2000), 71.

59. Goldberg explored stylistic differences in posture as signifying race and authenticity in flamenco in a presentation given in June 1998 at the Second Biennial Flamenco History Conference, University of New Mexico: "From Gautier to Hurok: The International Public and Notions of Authenticity in Flamenco."

60. W. E. B. Du Bois, *The Souls of Black Folk* (1903; rpt. New York: Vintage/Library of America, 1990), 3.

61. Louis Onuorah Chude-Sokei, *The Sound of Culture: Diaspora and Black Technopoetics* (Middletown, CT: Wesleyan University Press, 2015).

62. Jody Blake, *Le Tumulte Noir: Modernist Art and Popular Entertainment in Jazz-Age Paris, 1900–1930* (University Park: Pennsylvania State University Press, 1999), 40.

63. Jody Blake, *Le Tumulte Noir*, 40; Léonide Massine, *My Life in Ballet* (London: Macmillan, 1968), 106. For more on the European migrations of the Ballets Russes "neo-primitivist" choreographic techniques, combining "authenticity and stylized gesture," and "reworking ... a familiar nineteenth-century story as modernist narrative," see Lynn Garafola, "The Choreography of *Le Tricorne*," in Vicente García-Márquez, Yvan Nommick, and Antonio Alvarez Cañibano, eds., *Los Ballets Russes de Diaghilev y España* (Granada: Archivo Manuel de Falla, 2012), 89–95.

64. Goldberg, "*Jaleo de Jerez* and *Tumulte Noir*," 124–42.

65. On the convergence of the bolero school dances and the regional characters of Spain, see for instance "Bailes españoles," *Diario de Córdoba*, January 21, 1898, 1.

66. Castro y Serrano, "París en 89. IV. La España flamenca," *La ilustración española y americana*, October 30, 1889, 246–7.

67. Woods Peiró, *White Gypsies*, 108.

68. Moreno de la Tejera, "Cante y baile flamenco," *La ilustración ibérica* (February 15, 1890), 10.

69. "Sarasate en la Exposición," *El Heraldo de Madrid*, June 2, 1900, 1.

70. Vicente Moreno de la Tejera, "Cante y baile flamenco."

71. Melanie Kloetzel, "Bodies in Place: Location as Collaborator in Dance Film," *International Journal of Performance Arts and Digital Media*, 2014, 2–3.

72. "Notes de la semaine," *Les annales politiques et littèraires*, July 1, 1900, 2.

73. Stuart Hall, "The After-Life of Frantz Fanon: Why Fanon? Why Now? Why *Black Skin, White Masks*?" in Alan Read, ed., *The Fact of Blackness: Frantz Fanon and Visual Representation* (London: Institute of Contemporary Arts, 1996), 27–28.

BIBLIOGRAPHY

Aldrich, E. *1826: An Anonymous Manuscript in Facsimile*. Stuyvesant, NY: Pendragon Press, 2000.
Anderson, Benedict. *Imagined Communities. Reflections on the Origin and Spread of Nationalism*, rev. ed. London: Verso, 1991.
Aubert, Michelle, and Jean-Claude Seguin. *La production cinématographique des Frères Lumière*. Paris: Centre national de la Cinématographie/BIFI, 1995.

Baird, Thomas, K. Meira Goldberg, and Paul Jared Newman. "Changing Places: Toward the Reconstruction of an Eighteenth Century Danced Fandango." In K. Meira Goldberg and Antoni Pizà, eds., *Spaniards, Indians, Africans, and Gypsies: The Global Reach of the Fandango in Music, Song, and Dance. Música Oral del Sur* 12 (2015): 628–65, and (Newcastle upon Tyne: Cambridge Scholars, 2016), 579–621.

Bejarano Robles, Francisco. *Cafés de Málaga (...y otros establecimientos)*. Málaga: Bobastro, 1989.

Bennahum, Ninotchka D. "Early Spanish Dancers on the New York Stage." In Ninotchka D. Bennahum and K. Meira Goldberg, eds., *100 Years of Flamenco in New York*. New York: New York Public Library for the Performing Arts, 2013, 26–57.

Bennahum, Ninotchka D. *Antonia Mercé La Argentina. Flamenco and the Spanish Avant Garde*. Hanover, NH: Wesleyan University Press, 2000.

Benjamin, Walter. *The Arcades Project*. Translated by Howard Eiland and Kevin McLaughlin. Cambridge: Harvard University Press, 2002.

Blake, Jody. *Le Tumulte Noir: Modernist Art and Popular Entertainment in Jazz-Age Paris, 1900–1930*. University Park: Pennsylvania State University Press, 1999.

Blas Vega, José. *Los cafés cantantes de Sevilla*. Madrid: Cinterco, 1987.

Blas Vega, José, and Manuel Río Ruiz. *Diccionario enciclopédico ilustrado del flamenco y maestros del flamenco*. Madrid: Cinterco, 1988.

Brannigan, Erin. *Dancefilm. Choreography and the Moving Image*. Oxford: Oxford University Press, 2011.

Brown, Jayna. *Babylon Girls: Black Women Performers and the Shaping of the Modern*. Durham: Duke University Press, 2008.

Caddy, Davinia. "Parisian cake walks." *19th Century Music* 30, no. 3 (Spring 2007): 288–317.

Christoforidis, Michael. "Serenading Spanish Students on the Streets of Paris: The International Projection of the *Estudiantinas* in The 1870s." *Nineteenth Century Music Review* 14, no. 2 (2017): 1–14.

Chude-Sokei, Louis Onuorah. *The Sound of Culture: Diaspora and Black Technopoetics*. Middletown, CT: Wesleyan University Press, 2015.

Cruces Roldán, Cristina. "Bailarinas fascinantes; géneros y estereotipos de 'lo español' en el cine primitivo (1894–1910)." *Bulletin of Spanish Visual Studies* 1, no. 2 (2017): 161–92.

Cruces Roldán, Cristina. "Bailes boleros y flamencos en los primeros cortometrajes mudos. Narrativas y arquetipos sobre 'lo español' en los albores del siglo XX." *Revista de dialectología y tradiciones populares* LXXI, no. 2 (2016): 441–65.

Cruz Gutiérrez, José. *La Córdoba flamenca (1866–1900)*. Córdoba: El Páramo, 2010.

De Unamuno, Miguel. *En torno al casticismo*. Madrid: F. Fé, 1902.

Demófilo, [Antonio Machado y Álvarez]. *Colección de cantes flamencos escogidos y anotados por Demófilo*. Sevilla: El Porvenir, 1881.

Durand Viel, Ana María. *La Sevillana: datos sobre el folklore de la baja Andalucia*. Sevilla: Biblioteca de Temas Sevillanos, 1983.

Eco, Umberto. *La struttura assente. La ricerca semiotica e il metodo strutturale*. Milan: T. Bompiani, 1980 (first ed. 1968).

Fauser, Annegret. *Musical Encounters at the 1889 Paris World's Fair*. Rochester: University of Rochester Press, 2005.

Ferreira, Antonio J. *A fotografia animada em Portugal: 1894–1897*. Lisbon: Cinemateca Portuguesa, 1986.

Fox, E. Inman. "Spain as Castile: Nationalism and National Identity." In David T. Gies, ed., *The Cambridge Companion to Modern Spanish Culture*. Cambridge: Cambridge University Press, 1998, 21–36.

García-Márquez, Vicente, Yvan Nommick, and Antonio Álvarez Cañibano, eds. *Los Ballets Russes de Diaghilev y España*. Granada: Archivo Manuel de Falla, 2012.

Goldberg, K. Meira. *Border Trespasses: The Gypsy Mask and Carmen Amaya's Flamenco Dance*. Doctoral dissertation, Temple University, 1995.

Goldberg, K. Meira. "Jaleo de Jerez and Tumulte Noir: Primitivist Modernism and Cakewalk in Flamenco, 1902–1917." In K. Meira Goldberg, Ninotchka D. Bennahum, and Michelle

Heffner Hayes, eds., *Flamenco on the Global Stage: Historical, Critical and Theoretical Perspectives*. Jefferson, NC: McFarland, 2015, 124–42.
Goldberg, K. Meira. "Juana Vargas, 'La Macarrona': A Flamenco Treasure." New York Public Library blogs, January 21, 2015. http://www.nypl.org/blog/2015/01/21/juana-vargas-la-macarrona-flamenco.
Goldberg, K. Meira. "The Latin Craze and the Gypsy Mask: Carmen Amaya and the Flamenco Aesthetic, 1913–1963." In Ninotchka D. Bennahum and K. Meira Goldberg, eds., *100 Years of Flamenco in New York*. New York: New York Public Library for the Performing Arts, 2013, 68–99.
Goldberg, K. Meira. *Sonidos Negros: On the Blackness of Flamenco*. New York: Oxford University Press, 2018.
González Alcantud, José Antonio. "Andalucía 'en el tiempo de los moros.'" *Andalucía en la Historia. Dossier resistencias cotidianas* 14, no. 52 (2016): 50–55.
Griñán, Francisco. *Las estaciones perdidas del cine mudo en Málaga*. Málaga: Diputación de Málaga.
Grut, Marina, Alberto Lorca, Ángel Pericet Carmona, Eloy Pericet, and Ivor Forbes Guest. *The Bolero School: An Illustrated History of the Bolero, the Seguidillas and the Escuela Bolera: Syllabus and Dances*. Alton: Dance Books, 2002.
Hahn, H. Hazel. *Scenes of Parisian Modernity. Culture and Consumption in the Nineteenth Century*. New York: Palgrave Macmillan, 2009.
Hall, Stuart. "The After-life of Frantz Fanon: Why Fanon? Why Now? Why *Black Skin, White Masks*?" In Alan Read, ed., *The Fact of Blackness: Frantz Fanon and Visual Representation*. London: Institute of Contemporary Arts, 1996, 12–37.
Hernández Jaramillo, José Miguel. *La música preflamenca. Aproximación a la formación y evolución musical de los diferentes estilos del flamenco a través de la documentación musical escrita*. Sevilla: Junta de Andalucía: 2002.
Hill, Constance V. *Tap-Dancing America: A Cultural History*. New York: Oxford University Press, 2009.
Irigoyen-García, Javier. *The Spanish Arcadia: Sheep Herding, Pastoral Discourse, and Ethnicity in Early Modern Spain*. Toronto: University of Toronto Press, 2014.
Kloetzel, Melanie. "Bodies in Place: Location as Collaborator in Dance Film." *International Journal of Performance Arts and Digital Media* 11, no. 1 (2015): 18–41.
Manuel, Peter, and María Luisa Martínez Martínez. "El Murciano's 'Rondeña' and Early Flamenco Guitar Music: New Findings and Perspectives." In K. Meira Goldberg and Antoni Pizà, eds., *The Global Reach of the Fandango in Music, Song and Dance: Spaniards, Indians, Africans and Gypsies. Música Oral Del Sur* 12 (2015): 249–72, and Newcastle upon Tyne: Cambridge Scholars, 2016, 153–81.
Martín Sárraga, Félix. *Mitos y evidencia histórica sobre las tunas y estudiantinas*. Lima: Cauce, 2016.
Martínez, María Elena, David Nirenberg, and Max-Sebastián Hering Torres. *Race and Blood in the Iberian World*. Zürich: Lit, 2012.
Massine, Léonide. *My Life in Ballet*. London: Macmillan, 1968.
Mora, Kiko. *Cadáver paraíso. Del espectáculo popular en la modernidad* (blog). https://cadaverparaiso.wordpress.com/.
Mora, Kiko. "Carmen Dauset Moreno: primera musa del cine estadounidense." *Zer. Revista de estudios de comunicación*, no. 19 (2014): 13–35.
Mora, Kiko. "*Carmencita on the Road*: baile español y vaudeville en los Estados Unidos de América (1889–1895)." *Lumière*, October 28, 2011. http://www.elumiere.net/exclusivo_web/carmencita/carmencita_on_the_road.php (accessed December 16, 2017).
Mora, Kiko. "La representación contra-hegemónica de la negritud: La Perla Negra, entre la rumba y la danza moderna (1913–1928)." *Sinfonía Virtual*, no. 32 (2017): 1–36.
Mora, Kiko. "Some Notes Toward a Historiography of the Mid-Nineteenth Century *Bailable Español*." In K. Meira Goldberg, Ninotchka D. Bennahum, and Michelle Heffner Hayes, eds., *Flamenco on The Global Stage: Historical, Critical and Theoretical Perspectives*. Jefferson, NC: McFarland, 2015, 103–123.
Mora, Kiko. "Sounds of Spain in the Nineteenth Century USA. An Introduction." In K. Meira

Goldberg and Antoni Pizà, eds., *Spaniards, Indians, Africans, and Gypsies: The Global Reach of the Fandango in Music, Song, and Dance. Música Oral del Sur*, vol. 12 (2015): 333–62, and Newcastle upon Tyne: Cambridge Scholars, 2016, 270–309.

Mora, Kiko. "Who Is Who in the Lumière Films of Spanish Song and Dance at the Paris Exposition, 1900." Le Grimh (Groupe sur l'Etude de l'Image in le Monde Hispanique), 2017. http://www.grimh.org/index.php?option=com_content&view=article&layout=edit&id=2745&lang=fr#4.

Mora Chinoy, Clara. "The First Academy of Flamenco Dance: Frasquillo and the 'Broken Dance' of the Gitanos." In K. Meira Goldberg, Ninotchka D. Bennahum, and Michelle Heffner Hayes, eds., *Flamenco on The Global Stage: Historical, Critical and Theoretical Perspectives*. Jefferson, NC: McFarland, 2015, 143–56.

Navarro García, José Luis. *La danza y el cine*, vol. 1. Sevilla: Libros con Duende, 2014.

Navarro García, José Luis. *Semillas de ébano. El elemento negro y afroamericano en el baile flamenco*. Sevilla: Portada, 1998.

Núñez, Faustino. *Guía comentada de música y baile preflamencos (1750–1808)*. Barcelona: Carena, 2008.

Ortiz Nuevo, José Luis. *Coraje. Del maestro Otero y su paso por el baile*. Sevilla: Libros con duende, 2013.

Ortiz Nuevo, José Luis, Ángeles Cruzado, and Kiko Mora. *La Valiente. Trinidad Huertas "La Cuenca."* Sevilla: Libros con duende, 2016.

Otero Aranda, José. *Tratado de Bailes de Sociedad, regionales españoles, especialmente andaluces, con su historia y modo de ejecutarlos*. Sevilla: Tip. de la Guía Oficial, Lista núm. 1, 1912.

Pardo Bazán, Emilia. "Carta XXIII Diversiones—Gente Rara, Paris, septiembre 28." In *A los pies de la torre Eiffel, Obras completas*, vol. 19 (1891): 276–91.

Pineda Novo, Daniel. *Silverio Franconetti: noticias inéditas*. Sevilla: Giralda, 2000.

Pineda Novo, Daniel. *Juana, "la Macarrona" y el baile en los cafés cantantes*. Cornellà de Llobregat [Barcelona]: Aquí + Más Multimedia, 1996.

Riis, Thomas L. "The Experience and Impact of Black Entertainers in England, 1895–1920." *American Music* 4, no. 1 (1986): 50–58.

Rioja, Eusebio. "Un pinturero personaje del Flamenco decimonónico: EL NEGRO MERI." April 2004. Published on documents.mx July 15, 2015. http://documents.mx/documents/el-negro-meri.html.

Rioja, Eusebio. *El arte flamenco de Málaga—Los cafés cantantes (V): Una aproximación a sus historias y a sus ambientes*, partes 1 y 2 (2014). http://www.jondoweb.com/archivospdf/loscafescantantesdemalaga5_1.pdf, http://www.jondoweb.com/archivospdf/loscafescantantesdemalaga5_2.pdf (accessed July 22, 2016).

Rondón Rodríguez, Juan. *Recuerdos y confesiones del cantaor Rafael Pareja, de Triana*. Córdoba: Tipográfica Católica, 2001.

Sazatornil Ruiz, Luis. "Andalucismo y arquitectura en las exposiciones universales, 1867–1900." In *Andalucía: una imagen en Europa (1830–1929)*. Sevilla: Fundación Centro de Estudios Andaluces, 2008, 126–42.

Shaffner T. P., and W. Owens. *The Illustrated Record of the International Exhibition of the Industrial Arts and Manufacturers, and the Fine Arts*. London: The London Printing and Publishing Company Limited, 1862.

Stearns, Marshall W., and Jean Stearns. *Jazz Dance: The Story of American Vernacular Dance*. New York: Macmillan, 1968.

Steingress, Gerhard. "El cante flamenco como manifestación artística, instrumento ideológico y elemento de la identidad cultural andaluza. Perspectivas teóricas." In Gerhard Steingress and Enrique Baltanás, eds., *Flamenco y nacionalismo. Aportaciones para una sociología política del flamenco*. Sevilla: Fundación Machado/Universidad de Sevilla/Fundación El Monte (1998), 21–39.

Steingress, Gerhard. *Sociología del cante flamenco*. Sevilla: Centro andaluz de flamenco, 1993.

Varela, Javier. "Crisis de la conciencia nacional en torno al 98." In Antonio Morales Moya, Juan Pablo Fusi Aizpurúa, and Andrés Blas Guerrero, eds., *Historia de la nación y*

del nacionalismo español. Barcelona: Galaxia Gütemberg/Círculo de lectores, 2013, 543–62.
Torres, Max Hering. "Order and Difference in Mid-Nineteenth Century Colombia." In K. Meira Goldberg, Walter Clark, and Antoni Pizà, eds., *Spaniards, Natives, Africans, and Roma: Transatlantic Malagueñas and Zapateados in Music, Song and Dance*. Newcastle Upon Tyne: Cambridge Scholars, forthcoming.
Vittuci, Marcellus, "Mateo," with Carola Goya. *The Language of Spanish Dance*. Norman: University of Oklahoma Press, 1990.
Woods Peiró, Eva. *White Gypsies: Race and Stardom in Spanish Musicals*. Minneapolis: University of Minnesota Press, 2012.

White Dreadlocks
Black Aesthetics in the Work of Louise Lecavalier and La La La Human Steps

MJ THOMPSON

> It is ... necessary to imagine a world of composite elements, without the notion of purity.
> —Jennifer González

How did black aesthetics figure in the production of La La La Human Steps' most groundbreaking work of the 1980s and 90s? I offer a preliminary speculation, focusing on principal dancer Louise Lecavalier's appropriation of dreadlocks, the iconic Africanist hairstyle and a potent symbol of Caribbean culture and power. Though critics, audiences and academics alike referred to Lecavalier's locks repeatedly—wherein the dreadlocks are named and used as a descriptor in the popular press and imaginary—no critical attention was paid to them at the time. Taken as style, it was as if their meaning was self-evident: often code for wildness, otherness, or, to cite Kevin Frank's scholarship, *dread*.[1] Today I read the dreadlocks as a marker of black aesthetics[2] in the work and interpret Lecavalier's appropriation, not as theft, but as part of a challenge to notions of authenticity and purity (*"pure laine"*)[3] infecting nationalist discourse in 1980s Quebec.

Whereas much has been made of the play of gender in the work—particularly, how Lecavalier's performance intersected with feminism; and more recently with queering the gender binary[4]—little has been said about its ethnicity: that is, its whiteness, its blackness and a spectrum of identity positions suggested in between. Indeed, the taking up of the work as a product of Quebec society has become over time its *de facto* identity: evidence of nationhood, a sign of its status as a "distinct society" rooted in French language, culture and always with an ambiguous relation to France as colonizer and colonized within the historical context of "New France."[5] As a producer of culture in the form of choreography and dance, La La La "matches" a national self-image of Quebec as culturally unique, boldly innovative and internationally recognized.[6] While these kinds of national signifiers reside deeply in the work, I wonder what is at stake in a failure to see its diasporic registers? How, for instance, might La La La choreographer Édouard Lock's identity as a Moroccan Jew and an immigrant have shaped the movement aesthetic or otherwise left tangible cues that could disrupt understandings of Quebec as exclusively Euro-centric? Or how might company encounters with break-dancing in 1980s New York have influenced the approach? As principal dancer with La La La for nearly two decades, Lecavalier was central to the company identity and to the crafting of the aesthetic. A speculative analysis of her dreadlocks may help challenge nationalist readings that have prioritized whiteness in the work; I argue that, in addition to its punk, queer aesthetic, that the work is indebted to black aesthetics in ways that are central its power.

Racial Cross-Over, Appropriation, Cultural Relay[7]

I start as an ardent fan of Lecavalier's work and a profound skeptic about what it might mean for a white woman to wear dreadlocks. Around the same time period, particularly in the United States, the term "wigger" slips in and out of popular use as a term for "white youth said to be or claiming to be imitating African-Americans today."[8] Whereas Roediger historicizes the phenomenon as a form of racial-crossover, with the practice shifting uneasily between mimicry and critique, writers like Brent Stapleton and Greg Tate dismiss such attempts as naïve at best. In the words of the comic Paul Mooney, playing the character Junebug in the film *Bamboozled*, "White folks want to be black folks, everyone wants to be black.... I hope they start hanging n---- again, we're going see who's black."[9] With an eye to the lived experience of blackness in a racist context and resonating with the scholarly traditions of feminist, queer and black studies, Mooney's joke gets at the dumb ease of white racial crossings, where the option to return to the dominant group always remains.

A parallel discourse here, unfolding with heightened urgency during the period, focuses on the strategic use of appropriation since colonial times as a way to take and benefit from the lands, resources, and cultural artifacts of colonized peoples. In this tradition, appropriation flows in one direction— the powerful taking things from the less powerful. Problematized in contexts with an asymmetry of power, appropriation is only another form of cultural theft. Writing in 1993, in response to the proverbial flattening of discourse around an ethics of borrowing that tends to favor those in power, the artist/filmmaker Richard Fung explains, "The critique of cultural appropriation is … first and foremost a strategy to redress historically established inequities by raising questions about who controls and benefits from cultural resources."[10] Fung's thinking challenges us towards nuance and the need to privilege context in debates around the politics of appropriation.

Returning to Lecavalier's dreadlocks, I argue that rather than constituting an attempt at racial cross-over or a malignant act of theft, the style constituted a marker of less legible ethnic histories and aesthetics embedded in the work. Here the process might recall what Louis Chude-Sokei describes as the "syntheses, transformation, destruction and affirmation at the heart of creolization."[11] Historian Sean Mills has shown how the French-Canadian sense of itself as a colonized people fueled a strong identification with post-colonial peoples and discourses that resulted in productive acts of solidarity with black communities in Montreal, particularly around the 1974 Haitian deportation crisis.[12] His research has complicated the French-English narrative of culture that has tended to dominate the discourse in Quebec. Whereas Lecavalier and Lock were not overtly political artists, they were nonetheless politicized: Lecavalier, as a working-class French Canadian coming up during the late 1960s and 70s when English interests dominated provincial economics and cultural policy; Lock, as a Moroccan Jew, arriving in the city when Jews could not attend the free public school system. In this context, and alongside their work in challenging gender and disciplinary boundaries in dance, the dreadlocks may read as a productive incursion against *pure laine* Quebec.

Enter the Dread: Material, History, Meaning

Press accounts establish the iconicity of Lecavalier's hair as early as 1984, describing it as a "mop of white-blond hair."[13] Through the 1980s, journalists repeatedly identify the performer by her "punk blonde hair,"[14] "the trademark mane of unruly blond hair,"[15] its "white" or "platinum" color. References to her dreadlocks come later, mentioned in passing, again without remark and in identificatory ways, as in "her famous dreadlocks"[16]; or else assigned over-the-top descriptions, as in "her hair bleached brutally into a white, desiccated tangle."[17] The references to "ses tresses rasta"[18] are ubiquitous, foregrounding attention to her personal style, as well as the hairstyle and its attendant exoticization. Published photographs equally rehearse the association between the dancer and the dreadlocks—for example, a 1994 image of Lecavalier airbound wherein her hair constitutes a halo of light at the center of the image is captioned "Wiltern wildness,"[19] pairing the historic Los Angeles theatre with the company's well-established reputation for seemingly high-risk, wild movement. Dreadlocks, it seems, convey the performer and the company's difference, associated with a pleasure in extreme gesture: "terrifying flights," "high risk" barrel turns, or otherwise "violent" moves.[20]

Kevin Frank has identified how the Medusa story, associated with betrayal, rape and death, has structured reception in popular Hollywood films like *Predator* and *Pirates of the Caribbean*, conveying "dread and Otherness through the hairstyle associated with Rastafarians,"[21] rather than the empowerment and anti-racist messages.[22] Here, dreadlocks constitute "a deadly threat to supposedly utopian America or out-of-the-way societies." In the case of Lecavalier, response hovers between fear and longing, fascination and forgetting, with the style rapidly noted then cast from view. What patterns of recognition and erasure, desire and hate might be operative here? On the one hand, hair may seem reductive as a point of focus—certainly, my own gaze here risks repeating modes of fetishization. On the other, Lecavalier's wearing of the style may constitute a visible marker of histories and aesthetics that animate the work, yet have been erased by the work of whiteness.

Louise recalls first wearing the style with *Infante, C'est Destroy* (1992), a break-out work for the company in terms of international touring and success; and arguably one of the repertoire most closely associated with the dancer. *Infante* casts Lecavalier as a warrior, a powerful character and strength she intentionally sought to bring to the work.[23] With its transcendent, religious themes and imagery, *Infante* imagines Lecavalier as a Joan of Arc figure, alternately in control, and destroyed, yet always at the center—in serial pairings, trios and solos and giant film projections, wherein she remains prime agent and actor.[24] An identifying "signature" work for the company, the wearing of dreadlocks become identifying for Lecavalier as a performer around this time. Though wildness and a speculation about zeitgeist-level violence in the work seem the primary pivot of reception, spirituality and empowerment are certainly another option, begging the question of how to reconcile high energy and a professional technique that renders the movement "safe"; and its devotional, ecstatic aspects, that gesture to Catholic spirituality.

Lecavalier remembers the style developed initially as a function of bleaching her hair, and with an aim to render her work as a performer more fully visible: "I thought, 'I'll make white, white hair and all you will see is the face, what I'm saying with my face. After that, they [media] talked about the dreads—but mainly that started because of the bleached air, it was so dry and I would be dancing these long hours, spinning and spinning—it started to mat, and I didn't undo it." What is striking to me in her account is at once the sense of difficulty she experienced in terms of being seen and understood as an artist; and the consistency of a materialist understanding of her aesthetic, wherein the look came from making movement. Time and again in interviews, her profound discomfort with interest in appearances is palpable; she consistently has articulated frustration with the visual limits of identity, especially for dancers who incarnate a collapse between object and subject, dance and dancer. Coming up in the ballet and modern dance worlds of 1970s Montreal, she felt that teachers, choreographers and audiences could get stuck on the look of a dancer; and how compliance to expectations and aesthetics were expected and enforced. Over time, she developed multiple looks as a strategy to counter limits and create opportunities—that is, transforming her appearance in order to survive and to be seen on stage, in a play with identity at once highly intentional, enabled by white privilege, yet deeply critical of consolidations of authority.[25]

Whereas Kobena Mercer and other scholars have noted the meticulous cultivation that is required to produce a black hairstyle such as this,[26] Lecavalier resists ownership here. Rather than seeing this as a pivot away from agency, her words mark a profound ambivalence around matters of style and the anxiety around the gaze as an ever gendered, raced act. To linger over questions of style, from her perspective, could only shift the attention away from the central project of making movement. It is worth noting here the phenomenology of dreads in performance: for while the Medusa character threatens to steal movement, turning those who gaze to stone, the dreads on stage are all about movement. On stage, her hair assumed a larger, animate quality: airborne, reaching out, radiating into space and full of life. Set in motion through an intentional body, activated particularly via repertoire of neck and head, the dreads claim space and expand the body, following its action, yet with a slight delay and a distinctive weighted quality. Here, dreadlocks extend her reach and challenge the limits of the individual body onstage.

At the same time, the bleached-white locks tend to do funny things under theatrical lighting in terms of how her face appears on stage. For instance, in one filmed version,[27] her face recedes from view entirely in darkness, and her dreads in-and-as light assume the expressivity typically associated with the face. Elsewhere, full lighting makes her hair recede from view, all we see are the details of her facial expression. What remains is a sense of tension between revelation and masking; between sight and coverage, pointing to perception itself as partial and limited.

Context Matters

Kobena Mercer has described how cultural acts such as the cultivation of particular hairstyles were "stylistically cultivated and politically constructed in a particular historical moment as part of a strategic contestation of white dominance and the cultural power of whiteness."[28] Mercer identifies the emergence of dreads in England and elsewhere during the 1970s as symbol of black pride and authority, following the radical discourse of Rastafari culture and the cultural and political insurgence of reggae into the mainstream. More, in tracking the style's ricochet from situated meaning to its dissemination and commodification, he writes that the "back and forth indicates an underlying dynamic of struggle as different discourses compete for the same signs."[29]

What discourses compete for meaning in Lecavalier's wearing of dreads? More recently, as time passes and dancer and company receive multiple honors in the name of nation and state,[30] La La La is folded squarely into a narrative of *Québécité* and cultural distinctiveness. Following Stuart Hall's work on "narratives of nation," in which he notes how national cultures take up forms of representation to gather and affirm unified images of themselves, these awards attest to the place of these artists within the national cultures of Québec, smoothing the work into coherent themes of innovation, achievement and, international recognition that resemble the province's aspirations for statehood while all the while undermining the other productive ambiguities.

It is worth remembering that as beloved as their early work was, its meaning and their stature occupied a much more equivocal position in the early days. Was it dance, or something else? Was it theatre, punk rock, or some kind of "street fight"? Was Louise a muse, or an instigator? A "man" or a "woman"? Or something else entirely: "cyborg," "demon," "animal," "angel." Early press accounts record a preoccupation with Lecavalier, configured repeatedly as a non-human, in ways that were gendered and racialized. Equally, the company aesthetic was understood as threatening, with both Lock and Lecavalier on record defending the work against claims of risk and violence. What audiences saw, based on press accounts, oral history documentation and extended by my own reading of the work, were high-speed moves; high-energy leaps and lifts inspired by contact improv, acrobatics, and breakdancing; communicative registers, including hand gestures recalling sign language, mime and voguing; and a remarkable interdisciplinarity, blending dance, theatre, film, and popular music. At once, it looked like nothing else—and it looked like everything.

If the aesthetic was ambiguous, so was the context. If the province had remained a closed society until 1959, with the passing of right-wing nationalist premier Maurice Duplessis, subsequent decades saw rapid modernization and a "cultural awakening" that lead to the production and recognition of a rich and internationally culturally specific literature and arts. More recently, dance historians have overturned dominant narratives of "a language that took the world by storm" and tended to elide black history in the province. Simultaneous to the Quiet Revolution, unprecedented levels of immigration were reshaping the city of Montreal. Immigrants came especially from French-speaking nations, most notably for our discussion Morocco in the 1950s and 60s, with choreographer Lock's family immigrating in 1957. As David Austin and others have noted, during this period, the political struggles of Africa and its diaspora "served as a metaphor for Quebec identity,"[31] foregrounding white Quebecers' and eliding a history of slavery and racism in the province.

Here, I turn momentarily to scholarship on whiteness, not to consolidate pernicious racial binaries, but to mark the cultural pattern of practices and dispositions that attempt to exploit, control and exclude black life.[32] Historian David Roediger, following Du Bois, sees whiteness as a time-based performative practice of differentiation and othering.[33] In his exploration of how, in the early half of the nineteenth century, "whites reach the conclusion that whiteness is meaningful," Roediger locates whiteness as constructed through process of differentiation between wage and slave labor. He writes, "whiteness was a way in which white workers responded to a fear of dependency on wage labor and to the necessities of capitalist working discipline." Which is to say, whitening takes work. It took effort to fold the contributions of La La La into a story of Quebec as singular. First-draft reception in the popular press, and subsequent historiography, accelerated that effort.[34]

But black aesthetics reside in the work too, and how might we revise our viewing of the dance to acknowledge this presence? By the 1970s, the audience for experimental dance would have been predominantly white, French and English elites.[35] Yet the wider culture in Montreal was changing. Meanwhile, when Lecavalier and Lock began working together in 1982, they were not interested in "dance" audiences per se, but in the popular—they rehearsed to Prince, dreamed of punk ballets, performed with rock bands, presented in clubs rather than concert halls, and lived in the old manufacturing spaces along the Main, Saint-Laurent. A city, and a culture in rapid transition, a revised structure of feeling may have pressured awareness of black aesthetics and signifiers in the work. This feeling, bound by a nationalism that sought singularity, would later take the form of race-baiting, as in the Premier Jacques Parizeau attack on "the ethnic vote" to account for the failure of the 1995 referendum on Quebec independence. Alternately, and particularly for the city's youth culture, a new plurality of citizenship meant new ideas, energy and suggested a kind of openness to culture. Together, these suggest a range ways in which her dreadlocks may have been read.

Dick Hebdige, in his classic text on subcultures, reads post-war youth culture in Britain as a series of responses to black immigration. He writes: "The success of white subcultural forms can be read as a series of deep structure adaptations which symbolically accommodate or expunge the black presence from the host community. It is on the plane of aesthetics: in dress, dance, music; in the whole rhetoric of style, that we find the dialogue between white and black most subtly and comprehensively recorded, albeit in code."[36] Hebdige's work reminds us that style does serious work, performing subtle dialogic traversals between difference in race and class; more, it resonates with Mercer's take on style as a site of discourse resistant to dominant ideologies.

Black Aesthetics, Resistant Discourse

To take seriously Lecavalier's wearing of dreadlocks is to read for diasporic traces in La La La Human Steps and, indeed, Montreal. It is, in the words of Brenda Dixon Gottschild, to begin "digging the Africanist presence" embedded in the work. Gottschild places Africanist aesthetics at the very center of much American modern dance innovation and points to an "invisibilizing" process in hegemonic cultural and historical accounts of that history.[37] She writes: "Although we do not and cannot reduce the intertextuality of the African-American/European equation to a laundry list of sources and influences, we desperately need to cut through the convoluted web of racism that denies acknowledgement of the Africanist part of the whole."[38] In the decades that followed its publication, Gottschild's work set the bar for the discussion and acknowledgment of appropriation and theft in dance. Yet equally significant is the way her framework challenges historians to think in more nuanced ways about productive points of contact and influences that confound ongoing and troubling ideas of purity in culture.

In the case of La La La, where to start? Following the work of historian Iro Tembeck, a pattern of distinctive dance did emerge in Quebec in the 1980s, associated with the collective Groupe Nouvelle Aire who sought to create a dance reflective of Québécois identity and experience. Certainly, the La La La's aesthetic is part of that tradition. But equally it lends itself to a European dance history, likely in the tradition of *ausdruckstanz*,[39] a modernist form of expressive dance. And yet among the company's defining contributions, consider the following: the minute gestural detail, especially emphasizing hands, fingers, facial expression, which index modes of communication and can be read as a refusal of modernist cool. The emphasis on hands to trace, convey and sign meaning, referencing sign language, mime and voguing, fall neatly in line with diasporic refutations of art as mystery.[40]

Or else think of how body orientations redraw space in resistant, unexpected ways as in the off-axis work which countered ballet's axis. Diasporic knowledge may be found explicitly here in La La La references to breakdancing. Founding company member Louis Guillemette, for instance, has described the importance of seeing breakdancing in New York in 1981. He,

Myriam Moutillet and Lecavalier had travelled to the city on a state grant for ongoing training in dance, which involved taking classes in a wide range of styles, at Broadway Dance Centre and elsewhere; but equally visits to the Nuyorican Poets Café. He remembered the importance of breakdance as inspiration for the language they were developing with Lock: "We saw all different stuff ... but what we brought back to the company was break-dance. The street dance, for sure, all of that floor work and way of interacting. Pushing, pulling, slapping each other ... more of that attitude."[41] Guillemettre took care in the interview to mark the importance of breakdancing as its own form, something they learned from but respected as the work of other artists, not for anyone's taking. He recalled company members' determination to find their own voice—to "unform form, to transform form." At the same time, he confirmed the importance of b-boys for reimagining energy, attitude and performer roles.

In early works like *Businessman in the Process of Becoming an Angel* or *Human Sex*, the dancers pace, confront, stare down and in other ways seem "battle" each other or the viewer. And while the use of floorwork as surface and springboard recalls contact improv as well, moments of full-body contact seem to mark particular breakdance moves like the worm. Lock would later describe his technique as "off-axis," a critique of ballet, a refutation of straightness—and a kind of resistant horizontality contrasting ballet's verticality. Equally of significance might be the form's cultural situatedness—the importance of neighborhood and street, Montreal as the larger stage—and the value of non-matrixed performance, that is, letting go of character and playing oneself.

Lecavalier, too, has consistently referenced forms of black dance as a source of inspiration. She remembers:

> I saw breakdancing in the 80s, it was just starting. It wasn't an influence exactly—I knew that it was not the same as what we were doing, and that I couldn't do what they were doing. But they felt close in the way that they were in the street, and I was not far from the street. They were raw, they didn't claim to go—like with a ballet dancer, there's always this escalation about where you go. In the street, there is nowhere to go. It's just here. So there was a link in that way. There was something visceral in what they were making: their life was not easy, it was their only place to move, it was in the street and they used what they had to express who they were. In that way, I felt close to them ... what we were making, it was essential, it was urgent.[42]

Lecavalier's memory resonates with many Quebec artists through the second half of the twentieth century, who sought to distance themselves from elite colonial forms. Elsewhere in the interview, she notes the significance of the form's high energy, use of ground and tough attitude. Whereas her formative training lay in ballet and modern, she has supplemented that over the years with steady physical training including yoga, boxing, swimming, and different styles of African dance, which consolidated ideas about ground, weight and energy.

Such energy dovetails with the heightened speed that has remained striking throughout Lock's choreography to challenge viewer perception and the solidity of form. He has been remarkably consistent in theorizing his movement aesthetic as a way to complicate seeing—what I take to be a challenge to the violence of sight, the visual limits of identity. He recalls, "the only thing I can do is sufficiently disorient the perception of the audience so that, for a while, they're intensely awake and stop thinking and start seeing the detail. We're in essence creating an uncertainty about what the body actually is."[43] Stephen Low has argued for this use of speed as a queering of gender binary, and indeed bodily identity itself.[44] But might this be read as part of a Moroccan Jewish aesthetic?

Aomar Boum has written about the phenomenon of the plastic eye, a Moroccan concept that brings together seeing *and* the intentional ignoring of what is being shown for reasons of political expediency or survival. Noting that "Moroccans are socialized to express their grievances through 'ayn mika, 'the plastic eye,' one of Boum's informants says that "Ayn/mika started probably as a phrase used for things that, though glittery, are unimportant and so should be ignored."[45] Analyzing the representation of Jewish culture within national museums of Morocco, Boum writes: "I argue that the representational complexity of Jews and the Moroccan state's attitude toward these museums are best understood through the concept of the plastic eye, which combines not only the faculty of vision (the 'eye'/'ayn), but also the intentional act of ignoring what is exhibited (the 'plastic'/mika), thus allowing Jewish history to be simultaneously foregrounded and back-grounded when it is politically expedient."[46] Here, for instance, in the context of the Israeli-Palestine conflict, Jewish artifacts are shown but not labeled. While I do not want to discount that need or desire to blend may have been operative in 1980s Montreal, I simply want to point to the parallel between Lock's movement blur and blurring as both critique and survival, what Fred Moten has described as "blackness as the enactment of a blur."[47]

Offering only a preliminary read on these presences, I aim to emphasize the significance of black aesthetics and theory to the development of the La La La *oeuvre*. To be clear: this work is *not* part of what d. Sabela Grimes refers to as the "Black movement continuum,"[48] a phrase I understand to hold all predominantly African Diasporic forms rooted through practice and history. Rather, black movement values live among the sources distilled and elaborated in the work and help account for its cultural impact. Within a context of changing demographics in Montreal, as notions of nationhood shifted post–1976 and the 1980–1995 No referendums, Lecavalier's wearing of dreadlocks served as an important pivot for those interpreting company identity and repertoire.

Coda: Enter the Cyborg

How are we to think, in the end, about the politics that permeate this work? Lecavalier's commitment offers one story here. In rehearsal and performance with La La La over two decades, she remained focused on her craft, known for her work ethic, generosity and the seriousness of her approach to research. Onstage, her partner-work, costuming and astonishing musculature joined forces with repertoire thematics to rock received notions about what constituted the feminine and masculine in dance. In the confusion felt between the performed and the real, viewers speculated openly on her gender status, as racialized readings slipped into more coded terrain. Increasingly, the press associated her with the figure of the cyborg.

Something about this performer seemed machine-like, what was it? Her muscularity, her virtuosity, the play with gender, the play with form and vision, the sheer advancedness of whatever it was she was making and meaning…? The sense of her, circa the 1980s, as *futurity* itself. Here, I offer another story, perhaps oblique, to help think further about the political life in her dance. In 1995, Lecavalier appeared in Katherine Bigelow's *Strange Days*—box office and critical disaster, her only big screen appearance—in the role of bodyguard, cyborg, bit character Cindy Minh. By that year, she had achieved international stardom with La La La and was working arguably at the top of her game, dancing with David Bowie and appearing in advertising campaigns for Absolut Vodka and LA Frames. Her career choices appeared limitless. Here, she takes on a small part in a science-fiction parable about power, mediation and the failure to see.

Strange Days is set a dystopic future—that is, clearly not the present in terms of date and available technology and yet very much like the present, wherein whiteness takes the form of corruption and police brutality. Unfolding in Los Angeles over two days around New Year's 1999, the film follows former-police man, Lenny Nero, who has fallen from grace and become involved in the trafficking of high-intensity, highly addictive virtual-reality technology known as SQUID. In anguish over the loss of a girlfriend, ensnared by the VR-media himself, Lenny uncovers two hideous crimes caught on tape: the first, a rape of a young woman working as a prostitute; the second, the police shooting of Jericho One, prophetic rap musician and cultural hero. The plot unfolds as Lenny discovers the two crimes are linked and must choose between personal or political commitments: to save his ex, or to turn in the tape to authorities.

Or else, this: the plot follows Mace, played by Angela Bassett, as powerhouse limousine driver and bodyguard, single mother and friend to troubled former-cop Lenny. Set in a post-apocalypse, Afro-future marred by addiction and corruption, Mace drives Lenny around town on the last night of the year, the city a kaleidoscope of screens and moving images, in a bid to keep him safe and convince him to do the right thing. All action pivots around Mace, she holds moral center—but will the film be about Mace, or Lenny, in the end? Does she exist to perform narrative labor, or will agency, pleasure and the real be really possible?

I pause over Bassett's character, since Lecavalier's functions narratively as opponent and conceptually as echo to Mace. Both are bodyguards, both are cyborg-like. Lecavalier, onscreen briefly, looks familiar, her look is an assemblage of styles familiarized by punk music, cyberpunk genres of fiction and film, and her own work in dance. As Cindy, her form claims space through multiple body extensions—dreads, tutus, leather, fists—and her weaponry primarily includes movement, musculature and style. Ambiguously human, her gestures are at times awkward, robotic even. Taken together, the pair remap the cinematic cliché of the black male bodyguard, whose typically silent presence exists only to convey danger and consolidate the power of the guarded one. Whereas Basset stays at the heart of the film, Lecavalier's brief appearance seems designed to underscore; together they model a powerful physicality for an emergent feminist heroine.

In his work on Afrofuturism, technology and race, Louis Chude-Sokei has shown how historical forms of technology—automatons, robots and other anthropomorphic forms—repeatedly referenced black life as the threshold between human and non-human. Chude-Sokei points to a colonial imaginary that conflated machines and the future with blackness, long before Caribbean-infused sounds and styles, dub and dreadlocks permeated the aesthetics of cyberpunk film and literature.[49] More, and working with key theoretic texts by Sylvia Wyntner and others, he identifies a tradition of creolization that, in addition to acknowledging the racism and trauma of history, "distinguishes itself by an equal commitment to the possible."[50] Turning back to Lecavalier, Chude-Sokei's work may help account for how hybrid forms and cross-cultural referents come to be read as cyborg-like. If it did nothing else, *Strange Days*, shows Lecavalier fully taking up the figure of the cyborg in performance, in what I take to be a response to the critical paucity of understanding around bodies, identity and vision.

Yet the film does considerably more. Made shortly after the LAPD officers caught beating Rodney King on film received a "Not Guilty" verdict—events that surely mark deteriorating faith in photographic evidence—the film is of particular interest for its development of point-of-view shots. These scenes emulate SQUID technology and, critically, place viewers in the position of the perpetrators in the plot and consumers of the cinematic narrative. Sharply criticized at the time for its gritty depictions of suffering and implied realism, the film is at once an experiment in visual form, an argument for vision as truth and a critique of visuality as oppressive. At its best, the film uses its platform to return the public to the King verdict and speak out against acts of white supremacy.

Strange Days ends at a massive outdoor New Year's party, as Lecavalier backs away from Basset, refusing to do battle, and uttering her only line: "Enjoy the Party." Lecavalier gets out of the way, so that Bassett can kick back at the state and corporate forms of control and entertainment that threaten to overtake the real. If the cross-racial kiss between Mace and Lenny that ends the film prompts a groan as fantasy or (white) millennial hope, there is equally relief for Mace's survival, success and traversal from good mother to lover whose agency and sexual desire can be realized in popular cinema and perhaps in life. For Lecavalier, mistaken as cyborg in and through her danced labor, playing Cindy in *Strange Days* allowed her to grab hold of the cyborgian identity and shape it within the critical project of Bigelow's film.

Whereas much of the literature on the cyborg rests on the gendered nature of the machine, Lecavalier situates herself within a critical discussion of race and an expanded set of agencies. If the cyborg, as Jennifer González has written, appears at moments of anxiety and change "to contain fears and desires of a culture caught in the process of transformation," its hybrid status acknowledges only what has always been true, purity as an impossible ideal. *Strange Days* evidences Lecavalier's ongoing staging of self, as advocate for expanded bodily contours, wherein knowledge categories like black/white, man/woman are surely befuddled. What, then, did critics and viewers mean when they referred in passing to her dreads, as wild or mechanical? Perhaps asking what they felt is more apt. Anxiety, tension, excitement and pleasure, and more—a complex and confusing set of responses, "shared emotional repertoires"[51] as rehearsal towards *not knowing* and an expanded dexterity for participation in a heterogeneous world.

Notes

1. Kevin Frank, "Whether Beast or Human: The Cultural Legacies of Dread, Locks and Dystopia," *Small Axe* 23, 11.2 (June 2007): 46–62.

2. The phrase "black aesthetics" makes it sound easy. My aim, instead, is to use the phrase gesturally towards recognition rather than to suggest fixed or proprietary knowledge about what these words can hold—that is, I use the phrase as a shorthand in this essay to refer to the expansive range of ideas, values and kinaesthetics emerging in and through black social histories. Thomas F. DeFrantz, in his discussion of black dance, notes that "a generation of scholars and artists have pinpointed features of African-derived performance to encompass, at least, a percussive attack; an exploration of concurrent, highly complex rhythmic meters; an engagement of call-and-response between dancers and audiences; sophisticated structures of derision that are simultaneously personal and political; and above all, an overarching cool, palpably spiritual dimension to the performance. These hallmarks—Africanisms—provide

a theoretic framework for the identification and interpretation of diasporic traditions of art-making." Later in this essay, I will try to be more specific with regard to particular elements of La La La's work that are suggestive of particular Africanist sources. Meanwhile, I am indebted to DeFrantz's scholarship: see "The Complex Path to 21st Century Black Live Art," *Parallels: Danspace Project Platform 2012*, 62. Additionally, see DeFrantz and Philipa Rothfield, eds., *Choreography and Corporeality: Relay in Motion* (London: Macmillan, 2016), DeFrantz and Anita Gonzalez, eds., *Black Performance Theory* (Durham: Duke University Press, 2014); and DeFrantz, *Dancing Many Drums: Excavations in African-American Dance* (Madison: University of Wisconsin Press, 2002); Brenda Dixon-Gottschild, *The Black Dancing Body: A Geography from Coon to Cool* (New York: Palgrave Macmillan, 2003) and *Digging the Africanist Presence in American Performance and Other Contexts* (Westport, CT: Greenwood Press, 1996); and Fred Moten, *In the Break: The Aesthetics of the Black Radical Tradition* (Minneapolis: University of Minnesota Press, 2003).

3. *Pure laine*, or dyed-in-the-wool, is a controversial term referring to the earliest French settlers to Quebec in the seventeenth and eighteenth century as "original" or "pure."

4. See Ann Cooper Albright, "Techno Bodies: Muscling with Gender in Contemporary Dance," *Choreographing Difference: The Body and Identity in Contemporary Dance* (Middletown, CT: Wesleyan University Press, 1997), 28–55; Stephen Low, "The Speed of Queer," *Theatre Research in Canada* 31, no. 1 (2016); 62–78; and MJ Thompson, "Two-Way Street: The Icon and the City," in *Performance Studies Canada*, ed. Laura Levinand Marlis Schweitzer (Montreal: McGill-Queens Press, 2017), 287–315.

5. Distinct society.

6. Erin Hurley, *National Performance: Representing Quebec from Expo 67 to Celine Dion* (Toronto: University of Toronto Press, 2010), 23.

7. Thomas F. DeFrantz and Philipa Rothfield make elegant use of the notion of relay to signify encounters between theory and practice, without hierarchy, across "multiple locations, cultures and kinesthetic contexts." "Relay: Choreography and Corporeality," in *Choreography and Corporeality: Relay in Motion*, ed. DeFrantz and Rothfield (London: Palgrave Macmillan, 2016), 6.

8. David R. Roediger, "In Conclusion: Elvis, Wiggers, and Crossing Over To Non-whiteness," *Colored White: Transcending the Racial Past* (Berkeley: University of California Press, 2002).

9. Paul Mooney in *Bamboozled* (2000), dir. Spike Lee, 2h 15min.

10. Richard Fung, "Working Through Appropriation," *FUSE* XVI, nos. 5 and 6 (Summer 1993): 16–24.

11. Louis Chude-Sokei, *The Sound of Culture: Diaspora and Black Technopoetics* (Middletown, CT: Wesleyan University Press, 2015), 190.

12. Sean Mills, "Quebec, Haiti and the Deportation Crisis of 1974," *Canadian Historical Review* 94, no 3, 405–35.

13. Deborah Jowitt, "Honk if you love dogs," *Village Voice*, October 9, 1964.

14. Amanda Smith, "La La La Human Steps," *Dance Magazine*, March 1986, 97.

15. Katherine Greenaway, "La La La dancer undaunted by date with pop idol," *Montreal Gazette*, March 7, 1990.

16. *Montreal Gazette*, October 2, 1995.

17. Nadine Meisner, "Superwoman," *Sunday Times*, October 20, 1996, 2.

18. Radio-Canada, March 29, 2016.

19. *Los Angeles Times*, March 26, 1994.

20. Reception of this order is equally seen in response to the queerness of the work, wherein queerness lives in the work in a variety of ways: for example, in the redistribution of gender-based labor (for example, Louise holds up Marc, rather than the other way around) or in the transformation of her body through effort, labor and intention, of which I've written about elsewhere. See also Low, "The Speed of Queer."

21. Frank, "Whether Beast or Human," 46.

22. Frank, "Whether Beast or Human," 47.

23. Interview with Lecavalier, October 12, 2016, Montreal.

24. Ann Cooper Albright notes the centrality of Lecavalier to the piece and identifies

an ambivalence around the meaning of her physical strength and musculature. She writes, "While her built-up body radically challenges a conventionally feminine body or movement style, Lecavalier's disconnect intentionality reinforces her traditionally gendered role with the spectacle." See Cooper Albright "Techno Bodies," 50.

25. Interviews with Lecavalier: October-November, 2009, 2012, 2016.
26. Kobena Mercer, "Black Hair/Style Politics," *New Formations* 3 (Winter 1987): 33–54.
27. *Infante c'est destroy*, DVD, Société Radio-Canada, 1994. Bibliotéque de la danse Vincent-Warren, Montreal.
28. Mercer, "Black Hair/Style Politics," 40.
29. Mercer, "Black Hair/Style Politics," 52.
30. Lecavalier was named Companion of *l'Ordre des arts et lettres de Québec*, and Edouard Lock has received *Prix de Québec* (2002) and Knight in *l'Ordre des arts et lettres de Québec* (2001). Both are officers in the Order of Canada. *Quebecité* may be understood as a mode of feeling Québecois; it is "a project/process of (re)construction of Quebec that develops in reaction (read: in opposition) to the project/process of the country's Candianization." Jocelyn Létourneau in Erin Hurley, the latter cited here especially for the formative work she does in folding a range of labor forms by women into the history of Quebecois performance. See *National Performance: Representing Québec from Expo 67 to Céline Dion* (Montreal: McGill-Queen's Press, 2011).
31. David Austin, "Narratives of Power: Historical Mythologies in Contemporary Quebec and Canada," *Race and Class* 52, no. 1 (2010): 19–32.
32. David R. Roediger, *The Wages of Whiteness: Race and the Making of the American Working Class* (London: Verso, 2007), 13. See also Cheryl Harris, "Whiteness as Property," *Harvard Law Review* 106, no. 8 (June 1993): 1714. She documents the movement of "whiteness from color to race to status to property as a progression historically rooted in white supremacy and economic hegemony over Black and Native American peoples."
33. See Roediger, *The Wages of Whiteness*, 13, 6.
34. Whereas as Roediger grounds his research in the racial economy of America, to borrow Eric Lott's phrasing, the Quebec context is quite distinct, particularly for how the roles of colonizer and colonized are blurred in the figure of the Quebecois, vis-à-vis their relation to Indigenous peoples and to England. Roediger's work nonetheless remains relevant within a province where slavery remained operational until 1834, supporting the colonial economy. For further reading, see Charmaine Nelson, *Slavery, Geography and Empire in Nineteenth-Century Marine Landscapes of Montreal and Jamaica* (London: Routledge, 2017).
35. Whiteness in Quebec aligns with the English language. French Montrealers, then, descendants of the French explorers and colonizers on the landscape since Samuel Champlain in 1648, were subsequently defeated by the British in 1764. Which is to say, French Canadians experienced marginalization under English rule, a legacy that manifested itself perhaps most famously in the vernacular to "speak white" as per the iconic poem by Michele Lalonde which situates Quebecers alongside, for example, Black southerners as colonized, subjugated people. Though both French and English colonial powers imposed terror on the Indigenous tribes of the area, making them at once colonized and colonizers, French Quebecers have identified more strongly with international Independence movements and with the politics of the Black Atlantic, in ways that are important *and* deeply problematic.
36. Dick Hebdige, *Subculture: The Meaning of Style* (London: Routledge, 1979), 44–45.
37. Brenda Dixon-Gottschild, *Digging the Africanist Presence in American Performance and Other Contexts* (Westport, CT: Greenwood Press, 1996), 78.
38. Dixon-Gottschild, *Digging the Africanist Presence*, 3.
39. *Ausdruckstanz*, or expressive dance, developed in Germany in the early twentieth century through figures like Rudolph Laban and Mary Wigman, whose work privileged a wider range of bodily movements and emotional registers. Use of performer emotions and use of the ground served as key sources for movement creation.
40. See Lillian Allen in Clive Robertson, "Lillian Allan: Holding the Past, Touching the Present, Shining Out to the Future," *Caught in the Act: An Anthology of Performance Art by Canadian Women*, ed. Tanya Mars and Johanna Householder (Toronto: YYZ books, 2004), 103–104.

41. Interview with Guillemettre, November 2014.
42. Interview with Lecavalier, November 2016.
43. Édouard Lock in Daryl Jung, "La La La Human Steps," *Now Magazine*, October 29, 1992; 26.
44. Low, "The Speed of Queer."
45. Aomar Boum, "The Plastic Eye: The Politics of Representation in Moroccan Museums," *Ethnos* 75, no. 1 (March 2010): 53.
46. Boum, "The Plastic Eye."
47. Fred Moten, Lahey Lecture, Writers Read/Department of English, Concordia University, September 29, 2017.
48. Sabela D. Grimes, Dance Class and Lecture, Concordia University, November 10, 2017.
49. Chude-Sokei, *The Sound of Culture*, 129.
50. Chude-Sokei, *The Sound of Culture*, 205.
51. Hurley, *National Performance*, 6.

BIBLIOGRAPHY

Austin, David. "Narratives of Power: Historical Mythologies in Contemporary Quebec and Canada." *Race and Class* 52, no. 1 (2010): 19–32.
Balsamo, Anne. "Reading Cyborgs Writing Feminism." In *The Gendered Cyborg: A Reader*, ed. Gill Kirkup, Linda Janes, Kath Woodward, and Fiona Hovenden. London: Routledge, 2000, 148–158.
Boon, Marcus. "On Appropriation." *The New Centennial Review* 7:1 (2007); 1–14.
Chude-Sokei, Louis. *The Sound of Culture: Diaspora and Black Technopoetics*. Middletown, CT: Wesleyan University Press, 2015.
Cooper Albright, Ann. "Techno Bodies: Muscling with Gender in Contemporary Dance." *Choreographing Difference: The Body and Identity in Contemporary Dance*. Middletown, CT: Wesleyan University Press, 1997; 28–55.
DeFrantz, Thomas F. "The Complex Path to 21st Century Black Live Art." *Parallels: Danspace Project Platform 2012*.
DeFrantz, Thomas F. *Dancing Many Drums: Excavations in African-American Dance*. Madison: University of Wisconsin Press, 2002.
DeFrantz, Thomas F. "I Am Black: (You Have to Be Willing to Not Know)." *Theatre* 47:2 (2017): 9–21.
DeFrantz, Thomas F., and Anita Gonzalez, eds. *Black Performance Theory*. Durham: Duke University Press, 2014.
DeFrantz, Thomas F., and Philipa Rothfield, eds. *Choreography and Corporeality: Relay in Motion*. London: Palgrave MacMillan, 2016.
Dixon-Gottschild, Brenda. *The Black Dancing Body: A Geography from Coon to Cool*. New York: Palgrave Macmillan, 2003.
Dixon-Gottschild, Brenda. *Digging the Africanist Presence in American Performance and Other Contexts*. Westport, CT: Greenwood Press, 1996.
Frank, Kevin. "Whether Beast or Human: The Cultural Legacies of Dread, Locks and Dystopia." *Small Axe* 11, 23.2 (June 2007): 42–62.
Fung, Richard. "Working Through Appropriation." *FUSE* V.XVI, no. 5+6 (Summer 1993): 16–24.
González, Jennifer. "Envisioning Cyborg Bodies: Notes from Current Research." In *The Gendered Cyborg: A Reader*, ed. Gill Kirkup, Linda Janes, Kath Woodward and Fiona Hovenden. London: Routledge, 2000, 58–73.
Hebdige, Dick. *Subculture: The Meaning of Style*. London: Routledge, 1979.
Hurley, Erin. *National Performance: Representing Québec from Expo 67 to Céline Dion*. Montreal: McGill Queen's Press, 2011.
Jermyn, Deborah, and Sean Redmond. *The Cinema of Kathryn Bigelow: Hollywood Transgressor*. London: Wallflower Press, 2003.
Low, Stephen. "The Speed of Queer: La La La Human Steps and Queer Perceptions of the Body." *Theatre Research in Canada* 31, no. 1 (2016): 62–78.

Mercer, Kobena. "Black Hair/Style Politics." *New Formations* 3 (Winter 1987): 33–54.
Pégram, Scooter. *Choosing Their Own Style: Identity Emergence in Haitian Youth in Quebec.* New York: Peter Lang, 2005, 45–46.
Roediger, David R. *Coloured White: Transcending the Racial Past.* University of California Press, 2003.
Roediger, David R. "Guineas, Wiggers, and the Dramas of Racialized Cuture." *American Literary History* 7, no. 4 (1995): 654–668.
Roediger, David R. *The Wages of Whiteness: Race and the Making of the American Working Class.* London: Verso, 2007.
Thompson, MJ. "Two-Way Street: The Icon and the City." In *Performance Studies Canada*, ed. Laura Levin and Marlis Schweitzer. Montreal: McGill-Queens Press, 2017, 287–315.

Flowers of Menace
Stephen Petronio's Rites of Spring
CONSTANCE VALIS HILL

Rarely have I seen an American choreographer meet Stravinsky with such force and deep beauty.
—Wendy Perron

Fucking is an act of affirmation. It's one of the most instinctive animal and creative things humans do on a regular basis. It keeps one fresh and tuned; joins the base physical and metaphysical realms.
—Stephen Petronio

Second-generation postmodern choreographer Stephen Petronio is known for making dances defined by speed, impersonality, and violence. "I am trying to make movement that speaks of the time I am living in," he states. While Petronio's creations reflect the milieu in which he lives, the source materials that inspire him reference early and mid–twentieth-century choreographies that are reinterpreted to reflect modern sensibilities and track the evolution of his shifting worldviews. "I am the bastard child of Steve Paxton and Trisha Brown," he reflected. "My surrogate parents in dance, they instill in me a deep love of pure physics in motion. Their renegade minds are linked with the minimalism of the New York visual art world of the 1970's delights of abstraction and purity of form over the histrionics of theatrical dance."[1]

Petronio's most recent project, BLOODLINES, has been an ongoing initiative to honor an incomparable lineage of pioneering American postmodernists, presenting reconstructions of such seminal works as Merce Cunningham's *Rain Forest* (1968) and *Signals* (1970); Yvonne Rainer's *Diagonal* (1963), *Trio A with Flags* (1970), and *Chair Pillow* (1969); Trisha Brown's *Gla-*

cial Decoy (1979); an excerpt from Steve Paxton's *Goldberg Variations* (1986–1992); and Anna Halprin's *The Courtesan and the Crone* (2000).

The earliest manifestation of Petronio's modernist transformations is a series of dance works that directly draw from and reference Sergei Diaghilev's *Ballet Russes* production of *Le Sacre du Printemps* (*The Rite of Spring*). The ballet premiered on May 29, 1913, at the Théâtre des Champs-Élysées, with choreography by Vaslav Nijinsky, stage designs and costumes by Nicholas Roerich, and a modernist score by Igor Stravinsky. The avant-garde nature of the music and jarring choreography caused a near-riot in the audience. From the first notes of the overture, sounded by a bassoon playing well outside its normal register, Stravinsky's haunting music set the audience of sophisticated Parisians on edge. "The curtain rose on a group of knock-kneed and long-braided Lolitas jumping up and down," Stravinsky remarked of the brutal opening scene, which depicted a virgin sacrifice in an ancient pagan Russia.[2]

The concept of the ballet, developed by Roerich from Stravinsky's outline idea and suggested by its subtitle, "Pictures of Pagan Russia in Two Parts," depicted various primitive rituals celebrating the advent of spring, after which a young girl is chosen as a sacrificial victim and dances herself to death. Catcalls began to issue from the audience as they took in the bizarre scene of sacrifice playing out before them; and outraged members of the audience stomped their feet and beat each other over the head, their strokes synchronized with the beat of the music. The noise became great enough that the orchestra could not be heard from the stage, causing Nijinsky to climb atop a chair in the wings, shouting out instructions to his dancers onstage. Stravinsky sat, fuming, as his music was drowned out by jeers, whistles and members of the audience barking like dogs, while Diaghilev frantically switched the house lights on and off in a futile effort to restore order.

Stravinsky's musical score paved the way for dozens of twentieth-century choreographies; the ballet has been the subject of over 150 productions. But no production of *The Rite of Spring* has been as jarring as a series of "Rite" works made by Petronio from 1991 to 2006. Critics were disturbed by the works' brutal sexuality, the fracturing of Stravinsky's score, the sexual ambiguities in the *mise en scène*, and the textual enhancements that seemed intrusive next to Stravinsky's colossal music. The disjunction of sound, image, and movement, and the depiction of the female "Chosen One," not as a sacrificial virgin but instead a sexually-voracious woman, ruptured the audience's viewing and created a surreal schizophrenia, offering a fractured, dangerous, and sexually-wrought universe.

This essay, constructed from several in-depth interviews with Stephen Petronio,[3] traces each iteration of his *"Rites"* of Spring. Petronio, as a radical postmodern disciple of Sergei Diaghilev, re-envisioned Stravinsky's score to accommodate contemporary ideas about brutality, violence, and sexuality, particularly his representation of Woman simultaneously as earth mother and nymphomaniac. Petronio captured the ritual, almost primordial element of Stravinsky's score, while transforming it into a brutally raw and contemporary vision of the nature cycle, an eternal drama of humankind endlessly replayed.

The first iteration of Petronio's *Rite* choreographies, *Wrong Wrong*, was a collaboration with British ballet choreographer Michael Clark that premiered on September 26, 1991, in Angers, France. The naissance of *Wrong Wrong* was rooted in Petronio's involvement in the AIDS crisis in the early 1980s, his meeting and creative union with Michael Clark in the late 1980s, and Petronio's *MiddleSexGorge*.

MiddleSexGorge

Though he already knew of him, Petronio first met Michael Clark on the streets of Glasgow, Scotland in May of 1989, while he was touring with his dance company in Europe. Dubbed as the dance world's "Prince of Punk," Clark was a ballet prodigy, "a cultural force in motion, riding on the wave of fame alongside collaborators The Fall, British speed-punk rockers, as well as the designs of performance artist and cultural provocateur Leigh Bowery."[4] Fueled by talk that they were American and British counterparts, there was an immediate attraction—Petronio to Clark's flamboyant, unmatched style, and heaving fame; Clark to Petronio's quicksilver intellect and bent on pure speed and motion. They became lovers and dance partners. Their first collaboration, *Bed Piece*, a reimagining of John Lennon and Yoko Ono's peace-and-love pronouncements delivered from their double bed, was an improvised, sexually raw encounter that took place on a bed in British art collector Anthony D'Offay's London gallery, in an installation work titled *Heterospective*.

Their next major appearance together was *MiddleSexGorge*, which Petronio choreographed for his company and in which Clark was prominent. First presented in 1990 at the Lyon Festival in France, its New York premiere at the Joyce Theater on January 16, 1991, had music by Gareth Jones and the British rock group Wire, costumes by H. Petal, and lighting by Ken Tabachnick. The program notes state that the work "explored formal concerns of manipulation, propulsion, locomotion and support," though critical attention was to the promiscuous corporeality of the duo. "All the freakishness of post-punk esthetics comes into play. Two men wear corsets (Madonna should sue) above their bare buttocks. The women look less sexy in tank suits cut off at the thigh that shorten their figures. It was enough to make everyone with green and purple hair in the house feel at home," wrote Anna Kisslegoff. "In *MiddleSexGorge*, they appear and disappear mysteriously, materializing amid the corsets and tank suits with rings of flower petals on their legs, while their white pates and all-white leotards elongate their torsos."[5] While critics praised the distinctive differences between Clark's "classical precision that infuses his spaghetti-style fluency" and Petronio's "precise but more hard-edged quality of movement," the dancers making "a fantastic pair, not to be missed,"[6] Petronio was intent on provoking and enlightening New York's gay and straight communities. "The men wear corsets and no bottoms—they just have dance belts," Petronio stated in an interview, "so their asses are out. Here's my quote for *The Advocate*: 'If we're going to get our asses out, I want the gay community to get their asses out and come to the show.'"[7]

Petronio confessed that there was a time in the early 1990s when he did not talk about the actual dancing: "I felt it was my duty to insert my sexuality, and whomever I was sleeping with at the moment, into the dance interviews I was doing. Then I got very sick of that, after three or four years, when I felt I was becoming more of a poster child for a cause than an artist. I let that part of myself become more of a subtext to my work as opposed to the thing I was screaming."[8] *MiddleSexGorge* premiered in the same period of time as the United States' declaration of the Persian Gulf War, waged by coalition forces from thirty-five nations led by the United States against Iraq in response to its invasion and annexation of Kuwait. "The premier of *MSG* is so personally linked in my mind with the force of physical power, freedom and control about to be born into a world where people continue to die for the ownership of oil," Petronio later reflected: "It came out of such an amazingly alive period of understanding—that thrust of the pelvis and that slash of the arms and that anger, that sexual anger about control and loss of control and all the issues we were fighting for in the late '80s, in terms of recognition of the gay community."[9]

Wrong Wrong

The positive attention and reception of *MiddleSexGorge* forged an "obsessive megalomaniacal dream" to tour a show representing Clark's work and Petronio's work, in collaboration with individual dancers. They began to plan a version of *The Rite of Spring* using the Stravinsky score—"the music of Dionysus, of drunkenness, and ecstasy, abandon, and submission to inevitable change," said Clark—wittily naming it *Wrong Wrong*, "as in two wrongs don't make a Rite. Haha."[10] Leigh Bowery was commissioned to design the costumes and Charles Atlas the lighting. The initial idea of a collaboration between Clark, Petronio, and their respective dancers, however, was "spectacularly flawed," as there was considerable dissent between the dancers and with whom they were aligned. They settled on simply alternating scenes, Petronio doing Section One, Clark Section Two, and so on, and shopped the project to the Centre National de Danse Conteporaine (CNDC) in Angers, where it was premiered on September 26, 1991.

In the opening "L'Adoration de la Terre" (Adoration of the Earth), which Petronio choreographed, we are introduced, not to a 300-year old woman foretelling the future, as presented in the Joffrey Ballet's 1987 reconstruction of Nijinsky's 1913 choreography[11] but to a fully-endowed earth-woman who dances naked, except for a long-braided wig that runs down the spine, along the crease of the buttocks, continues between the legs and up along the perineum to attach at the front of the pubic bone. That this wise earth goddess is fecund and ripe for mating is made apparent, not only in grounding movements of her natural surrounds—attitudinal lunges that flex the feet, spiraling movements originating in the pelvis and traveling up the spine to soak up the sun—but in her attracting, at the end of the scene, a hooded male-bodied dancer who literally "partners" her, leaving their liaison, in a fade to blackout, to the imagination of the audience.

While it is clear that the "Chosen One" is a most powerful tantric goddess representing the divine feminine, *Shakti*, the succeeding scenes alternate between heterosexual couplings and ensembles with same-sex partnering. AIDS was in full crisis when *Wrong Wrong* was conceived. Petronio had worked with ACT-UP for a number of years, from the late 1980s, demonstrating and trying to bring awareness of an epidemic going around. "I got arrested repeatedly in civil demonstrations against NYC's non-existent AIDS policies," Petronio recalled. "Being lifted into the police van by a group of NYC cops within the context of a massive demonstration leaves a searing visceral impression": "It hit me that my work might be more engaged if I try to build the most sensual and abandoned part of my nature into the movement language. So I began to push my pelvis forward, off my leg and into the face of the audience, hips and head rolling, carving arcs, arms and legs slashing out through space."[12]

By the late 1980s, Petronio's activism shifted: "We were traveling and touring a lot. So a lot of my activism was through my relationship with Michael, discussing my relationship as openly and frequently and blatantly as I could. There was a lot of press attention about my work, and I used that vehicle, that public format, to discuss my sexuality and relationship with Michael. So I took my politics into a very personal realm."

Dance historian David Gere, in his seminal book, *How to Make Dances in an Epidemic: Tracking Choreography in the Age of AIDS*, asks the question, "How Can a Dance Say 'AIDS'?" Gere offers a three-part theorization of the elements that a so-called AIDS choreography holds in common. First, a choreography must contain homosexual desire, the depiction of male-male eros, which manifests as eros fulfilled or thwarted. The second necessary condition is the abjection factor, in which the gay man is abject, marginalized, outside the mainstream, insofar as he cannot be a subject (the heterosexual man) or object (the heterosexual woman), and that to American society at large, he is none of the above, simply because he is, or appears to be, gay. The third condition is the element of mourning, ranging from the anticipation of loss to unabashed grieving.[13]

While *Wrong Wrong*, or any of the succeeding *Rite* works, does not quite qualify as an AIDS dance, staging homosexual desire is prevalent in Petronio's repertoire, as is the element of abjection—in *Wrong Wrong*, in the marginalization and fetishizing of the Chosen One—constructing a mysteriously powerful earth goddess who is both feared and desired. For Petronio, this was a momentous elucidation of his sexual identity and aesthetic dharma. As he mused: "Why do I think a moving model of a realistically-blurred gender is something that would be of interest? What can I be thinking when I decided to bring issues centered around my cultural passions and needs into a cool and abstract new world? That the impact of men behaving as virtuosic athletes in adrenalized, tender, or erotic ways, as objects even; that women in functional and subjective action are important to weave into the fabric of my dances as a given? Why are these models important to pursue?"[14] It is not the substantiation of Petronio's *Rite* works as "AIDS dances" that ties them to Gere's paradigm, but instead Petronio's suffering experience of living through the AIDS crisis and his subsequent activism; the choreographies that inscribed and radiated homosexual desire; the construction of personae who were abject; and the paranoid experience of loss and mourning. As he recorded: "It's 1981 and in a year or so people start to disappear from the neighborhood with odd abruptness.... In an unthinkable rush, the news of AIDS is everywhere and no one knows what's going on."[15] Petronio acknowledges that his foray into *MiddleSexGorge* and the succeeding *Rite* works was conducted under the shadow of AIDS, and that he and Clark, as gay men, had already activated a set of aesthetic issues relating to the political concerns of homosexuals and people with AIDS.

Half Wrong

With *Wrong Wrong*, Petronio and Clark traded off choreographing sections of the score. When they split up, Petronio claimed his sections and set out to create his own *Rite* work, calling it *Half Wrong*. It premiered at London's Dance Umbrella on October 27, 1992, to the Stravinsky score. "If I were to do a *Rite of Spring*, I needed to make it relevant to my life," he explained: "So for me, the idea of a virgin sacrifice seemed kind of idiotic. My version has a highly empowered sexual woman dancing to death—or maybe she's dancing to life at the very end. My Virgin was no virgin. It takes a super-empowered woman to move the agricultural cycle forward."

"Why a woman? And why this woman," I asked Petronio. "I had to use a woman in this piece because, as a gay man, the voracious sexuality was already all over on the street, and it was always an accepted given. And the woman, still, whether it's the 1800s or the 2000s, women have the onus of will.... So I thought it was very important that the woman use her sexuality as empowerment, aside from the men who already had that privilege. And also, for a woman to be sexualized onstage—she's always the whore, always cast away. So how to give a woman a sexual role that is about feeding society by moving forward—growing corn by being sexual." Petronio continued to evolve a savage choreography that mirrored what he heard in the score as brutally violent and erotic. Part One's opening "L'Adoration de la Terre" replicates what he created in *Wrong Wrong*—a naked, fully-endowed earth-woman, her long braid wrapped around her buttocks, performs a languorous solo, movements originating in the pelvis and traveling up the spine. "I was so much into pushing the pelvis forward, pulling back the curtains on it, trying to focus all the movements on that," says Petronio about a style that was to become signature—an impulse from the greater trochanter muscle, producing movement that seems to emanate from the sexual organs: "It became part of the language, like walking to me. It doesn't look like sex, but you get a very visceral feeling—moreover, because Stravinsky in the score is pounding his piano. That was the message we all had to keep pounding in *Rite of Spring*: Sex is here. It belongs here and is a part of who we are."

"Augurs of Spring," which followed, was designed as a series of playful trios, with two male-bodied dancers and one female-bodied dancer in teasing chases; the men framing, bookending, and encasing the female, only for her to slip away in outward-spiraling circles. There follows a lyrical quartet for two men and two women, all wearing identical leotards and tights that blur sexual difference. "Ritual of Abduction" is a male-female duet in which he chases and momentarily captures her, once again unable to contain her frenetic energy, and thus enabling her slippery release. Petronio flirts with bisexual attractions for the ensemble, males and females moving in synchrony or freely trading partners, but the omnificent power of the female continues to be foregrounded in "Spring Rounds," in which four female-bodied dancers lie on their backs along a diagonal, spreading their legs wide open, pelvises lifted to the sky. Their faces are turned downward, buried in earth, as they arch into catlike forms and slither into phallic extensions of the limbs that suggest a hungry thirst for penetration. Their "thirst" is answered, in the last section's "Ritual of the Rival Tribes," that begins with a male solo and expands into a frenzied sextet of male and female dancers that renders a social scene of sexually playful camaraderie and competition.

What was unusual in Petronio's *Half Wrong*, however, was that there was no "Danse Sacral," Virgin Sacrifice. "I erased the Virgin Sacrifice at the end," Petronio confessed. "I opted for a situationist version where no virgin shows up. I erased the Virgin Sacrifice. When the moment came in the score, I plunged the stage in darkness, save for the two grand pianos upstage, capturing the pianists' hands playing a four-handed version of the score." It was only in the next iteration of Petronio's *Rite* works that "Danse Sacral" would be restored.

Half Wrong Plus Laytext

While in Germany, in residence at the Deutsch Opera House in Berlin, Petronio began reading about the scandal that Nijinsky and Stravinsky had created in 1913 with *Le Sacre du Printemps*, looking to contemporary works that addressed the issue of scandal. Writings by such authors as Kathy Ackerman and William Burroughs, and Henry Miller's 1934 novel *Tropic of Cancer*, were initially deemed pornographic and challenged in court as obscene, but over time were deemed great works of literature. Inspired by the desire to further bring sexuality into his work, and "to bring into the dance world and into the movement vocabulary a fearlessness about sexuality," Petronio created a new set of verbal and projected aphorisms, called "Laytext" (the title a wordplay on "a layman's text, a text about getting laid, and rubber condoms"). Some were actual hand drawings of his own, erotic images of genitalia and breasts and other bodily parts that were inflated and deflated across the stage to "contextualize the dance in a very earthy and sexual way." *Half Wrong Plus Laytext* premiered at the Deutsch Opera House in Berlin, Germany and was subsequently presented on May 13, 1992, at The Kitchen in New York City.[16] "The first thing you notice about the Stephen Petronio Company is its astonishing dancers, and that is not because some are less than completely dressed, baring either breasts or bottoms," Anna Kisselgoff wrote about the New York premiere of *Half Wrong Plus Laytext*. "It is, rather, that these fearless performers always appear to be living on the edge. Whiplash is the word to describe the propulsive power on display. The dancers throw themselves into the air, to the floor, toward one another or, better yet, into stupefying flattened leaps, Frisbee style, across a cluster of curved bodies below."[17] The production maintained fragments of Stravinsky's *Le Sacre du Printemps*, along with Mitchell Lager's sound collage that fused the psychedelic music of Pink Floyd with middle-Eastern rhythms. Projections of the Laytext, in the second part, challenged the audience to simultaneously assimilate sound, movement, and text, instigating in the imagination a surreal orgiastic space devoid from rationality. The narration in the text pointedly described homosexual desire, with the voice of the female protagonist, the "Chosen One," remaining mute.

> Pictures of men and women...
> boys and girls...
> animals fish birds
> the copulating rhythms of the
> universe flows through the room
> vibrating hum of deep forest
> sudden quiet of cities
>
> Sacrifie a VIRGIN?
>
> We are dreaming of sex
> Huge thighs opening to us like
> The night
>
> Some folks like trains
> Some like ships
> I like the way you move your hips
> I like the taste of your lips...[18]
>
> Are you POSITIVE?
>
> It's unnatural to be sexless, eat alone.
> Don't you have to get laid?
> My bedroom's function is unclear.
> "It's a gathering place for men."
>
> MOTEL
> MOTEL
> MOTEL
>
> broken neon arabesque
> loneliness moans across the continent
> like fog horns over still oily water
> Death for dope fiends
> Death for sex queens (I mean fiends)
> Who offends the covered and
> Graceless flesh
> With broken animal innocence
> Of the movement
>
> Sacrifice a VIRGIN?
> Rite of transformation/phallocentric snuff
> Everything HANGS on Him

In the last section's "The Sacrifice," five women surround a woman in a gray long-sleeved leotard who shivers in coordination with the horns in the Stravinsky score. She is released into a solo of stalking madness that culminates in a coupling with a male-bodied dancer.

I asked Petronio about the line, "Sacrifice a VIRGIN?" He answered: "It got down to the whole thesis of the work. 'Sacrifice a Virgin?' is the author speaking. And I put it in the section when the communal group of dancers hits the floor, and they prostate themselves in unison. It is when all things are happening—when they hit the floor—that the text pops up. It happens between the actions, so there is no action going on. As they hit the floor, maybe she's walking on. And then the minute she starts moving, the text is gone. So I planted the question, *Sacrifice a Virgin?* in the audience's mind." "And how do you want the audience to respond to that question," I prodded. "I am asking the question, 'why aren't we questioning the fact, how can we sit in the twenty-first century and say, "Oh yes, Stravinsky was sacrificing the virgin, and it's great."' I just thought it was so weird. We would read about Pina Bausch's production and everybody else who made a *Rite of Spring*, but nobody is talking about the fact that they are killing a woman—to grow corn. To me, that sounds like that's the basic question."

Petronio's driving question is "Why are we killing a woman?" His answer is a complex circumspection that explores the mysteries of sexuality from his own point of view, and certainly not from the woman's experience, for the female voice in *Laytext* is alarmingly silenced. We see her only through the eyes of male perpetrators and a multiplicity of male personas. There is the voice of the romantic.

> *It was a summer night, the air is warm and electric, with the smell of incense, ozone, and the musky sweet, rotten smell of fever.*

Her copulation with the men reveals a sado-masochistic view of love-making:

> *She covered it in rubber...*
> *Strap it on*
> *Tie him up*
> *strip him with a razor*
> *so relieved I didn't castrate*
> *he came all over.*
>
> *Torn in two by a bull dyke*
> *most terrific vaginal grip I*
> *ever experienced*
>
> *Screams breaking glass, ripping cloth,*
> *a rising crescent of grunts, squeals,*
> *moans, whimpers, gasps*
> *the reek of semen and cunt and sweat.*

The text intimates that the seduction was not a rape, but instead an act instigated by the female:

> *Diamonds and fur pierces*
> *Evening dress and orchids and underwear*
> *Litter the floor*

In Part Two's final rape scene, Petronio waxes poetic on the thrill of male orgasm:

> The boys come at once
> Their young faces blaze
> Like a thousand S-H-O-O-T-I-N-G stars
>
> Some distance away
> a circle of cheering onlookers

Finally, there is Petronio's cynical and utterly frank perspective on the psyche of the male sex drive and the secret desire of "pussy envy":

> The male is completely incapable of love, friendship and tenderness. He is a half dead, unresponsive lump, at best an utter bore, an unresponsive viral blob. The male spends his life trying to become female. He is none the less obsessed with screwing. He'll swim a river of snot, wade nostril-deep through a mile of vomit if he thinks there will be a friendly pussy awaiting him. Friendship, and tenderness. He is a half dead unresponsive lump, at best an utter bore.[19]

> Women don't have penis envy
> Men have pussy envy
> SACRIFICE [six seconds]
> SACRE VICE (six seconds]
> NO SACRIFICE (eight seconds)

Full Half Wrong

The "laytext" in *Half Wrong Plus Laytext* did not survive in *Full Half Wrong*, which became the next iteration and blueprint for succeeding iterations of Petronio's *Rite* works.[20] The ballet, lit by Ken Tabachnick, had costumes by the Spanish fashion designer Manuel "*Manolo*" Blahnik Rodríguez, who reworked Lee Bowery's designs: one key element was a unitard that covered private parts, but with a strip cut out of the crotch, so that when the men or women opened their legs one would see the flesh of their inner thighs, which visualized the theme of virginity and sexual power. The first part of *Full Half Wrong* followed the Stravinsky score, with Petronio coming as close to a "straight" version of *The Rite of Spring*. In Part Two, however, we hear a sound collage by Mitchell Lager that fuses the psychedelic music of English rock band Pink Floyd with Middle Eastern tribal drums and chanting, as seven dancers support and maintain the female. A "mating ritual" follows with a male and female dancer that is manipulative and dispassionate, bodies lying prone, only to be prodded to uprightness. "That's when the whole world shifts, and it becomes crazily sexual and blatant," Petronio remarked. "The costumes turn rubberized and it gets into an abstract world of pelvic forward

Rebecca Hilton in *Full Half Wrong* (1992), Stephen Petronio Company (photograph courtesy Johan Elbers).

play, moving through that into surreality." Petronio reinforced this surreal scenography with projected epigrams onto the back wall of the stage—the projected text, deadpan observations on sex and sexual truisms, inspired by the aphorism-based installation art of Jenny Holzer, a key figure in the United States art scene of the late 1980s. One such projected aphorism was *Women don't have penis envy/men have Venus envy*.

Part Two begins with an overheard announcement:

*You are about to witness the ravishment
of a woman who has been abducted;
a woman whose initial fear and anxiety has mellowed into curious expectation.
Although at first, her reactions may lead you to believe that she is being tortured.
    Quite the contrary is true,
for no harm will come to those who are ravished.
In the morning she will be set free, unaware of anything.*

Petronio took the word "ravished," meaning "to seize and carry away by force; to force another to have sexual intercourse; to rape," literally. What followed in sound, movement, and text constituted a rhetorical debate about "what constitutes pornography?" It was a debate that took place through such juxtaposed projected texts as

> I'm trying to save you.
> Take me, take me back
> Oh, take me back,
> Earth Girl,

as this "Earth Girl" performs arabesques as a rapidly opening-and-closing scissoring of the legs, followed by spiral turns in the air activated by the initiation of the pelvis.

It is unlikely that Petronio, in the intertexuality of projected aphorisms while a (white) female dancer is "ravished" in an orgy set to Middle-Eastern rhythms, was aware of the danger of constructing an Orientalized subject as primitive, irrational, violent, despotic, and fanatic—essentially, an inferior "other." Orientalism is the exaggeration of difference, the application of clichéd analytical models for perceiving the "Oriental" world that results in a kind of cultural imperialism, and thus the source of inaccurate cultural representations that form the foundations of Western thought and perception of the Eastern world that is purely fictional.[21] Conceptions and myths about "the other" in dance have been part of the performative imagination in the West since early times, from the sari-and-point-shoe clad figure of the heroine of *La Bayadère*, and Nijinsky's role as the Hindu god *Krishna* in *Ballet Russes' Le Dieu Blue* and the golden slave in *Scheherazade*, to Ruth St. Denis' *Dance of the Red and Gold Sari*, to multiple renderings of the story of Salome, the Oriental virgin-whore by Loie Fuller (1905), Maud Allen (1907), Ada Overton Walker (1912), and Ruth St. Denis (1931).[22]

While Petronio's "primitivism" veers into a construction of the feminine as intuitively sexist, his dharmic path was to explore the fluidities of sexual behavior, and to deeply question his own sexual identity. "I have male and I have female in me and it is all energy, and sometimes I'm male and sometimes I'm very female," Petronio confessed. "I like being able to physicalize my body—my articulation is more in line with what women do, but I've got a fiery side as well that's bullish, and I think both are really important." That Petronio was trying to discern meaning in his sexuality through his choreography—and was clearly in new territory of exploration—is articulated in Anna Kisselgoff's 1993 review of *Full Half Wrong* in its premiere at New York's Joyce Theater: "Meaning itself is a concept that a new generation of choreographers is still trying to define," wrote Kisselgoff: "Mr. Petronio—brought up on formalists like Merce Cunningham and Trisha Brown, who tried to erase specific connotation from pure movement—is now attempting to convey specific dramatic images through form and energy. That Mr. Petronio's themes have something to do with a questioning of sexual identity is apparent, but what he has in mind does not necessarily come across onstage."[23]

The Rite Part

The Rite Part, Petronio's final Rite iteration, was premiered on April 18, 2006, at New York's Joyce Theater. Here was Petronio's *tour de force*, not only for the consecration of his craftsmanship as a choreographer with a distinct lexicon but for his ability to convey specific dramatic images through form and energy—the most important being the endowing of the female, not as a ravaged victim of sexual assault but an empowered, vital force of nature, a goddess in full control of her sexual and reproductive powers. Reviews brought to mind bits of past iterations of the *Rite* works:

> A deep male voice makes a loud intercom-like announcement to the effect that we are about to witness the ravishment of a woman who has allowed her sensual curiosity to overcome her fear. The sensual giving and taking of ravishment is well-shared among the dancers, regardless of gender. It is seductive rather than rapacious, and ultimately, the dancers become part of the set and watch as a Shila Turabassi, in a short white unitard appears among them and, strongly lit from above, interprets again the torso-centered body blooming, with limbs becoming petals and sepals.[24] Stephen Petronio's choreography braids together awkward and sophisticated movement. It can be kinetically exciting or only puzzling. His dancers move lower to the earth, savage. The opening solo recalled a dog hunching its back, or a baby crunching its toes. In the group work, heads heaved wildly off to the side, the rest of the body dragging along. The dancers bent forward, perhaps determined to plant their seeds, arms thrust behind them, moving within a stark architecture of bodies and light.[25]

Beginning and ending with the Stravinsky score, with a sound collage by Mitchell Lager, Simon Rattle, and the City of Birmingham Symphony Orchestra, costumes by Manolo, and lighting by Ken Tabatchnick, the look of *The Rite Part* was youthful and contemporary. It opens, in "Spring Rounds," not with a nymph with a braid entwined along her bare torso, but a strong and shapely female dancer—her hair in a bob and tinted red—in black tights and bright floral appliques. She walks downstage into a pool of light and smoothly descends to the ground. She leans back on her arms to feel the warmth of the sun; arches her back, pushing the pelvis to the sky. The movement is sensual, languorous, intimate—very much recalling Vaslav Nijinsky's performance in *L'Après-midi d'un faune*, first performed in Paris in 1912. She turns on the ground with rolls that extend the legs over the head, and cat-like arches that arc and curve the back. She stretches her legs into a full split—the movement suggesting a tantric ritual, the full somatic, the self, tasting the self.

The ballet continues to Part One's "Ritual of the Rival Tribes" with the same overhead announcement as in *Half Wrong*, "You are about to witness...." The most prominent change is the final "Sacrificial Dance," where five female dancers, with their hands tied behind their backs, surround the "Chosen One" and release her into a solo—kicking, rolling her hips, slashing her arms, and splaying open the legs to push the pelvis forward, giving birth to life. In the last beats of the finale, she rises from the earth to stand strong, gazing boldly at the audience. "What was striking about the final solo, the sacrifice, was how greatly it differed from the imagery that has come to be associated with this moment in Stravinsky's score," wrote Carl Kronin: "Rather than the quivering grasping and petrified stillness of Nijinsky's choreography, Petronio's movement, with its hyper-sensuality and physical prowess, depicts a woman who is in command of her sexuality and body. This is a far cry from a virgin sacrifice. He has chosen to show a woman who willfully works herself up into a state of rapturous transformation. Once again, Petronio's persistent curiosity of ecstatic sensuality."[26] "She was tough in that solo, taking in all of her powers ... and not apologizing about who she was at all," Petronio reflected. "The women's floor thing—very much opening the vagina and pushing the legs forward at the audience—that to me was a very big breakthrough. And the whole thing on the floor, like giving birth, pushing it out, was a vaginal awakening. Keeping it restricted to the floor was very exciting for me. It was one of the most exciting things I ever made."

Petronio's fascination with sex and violence should be viewed in the context of broader cultural currents in the 1980s and 1990s. At the same time that these subjects were being attacked by politicians waging a "family values" crusade, they were acquiring a new legitimacy as topics for serious academic study. In recent years, studies of eroticism and histories of the body have proliferated, heavily influenced by such philosophers as Georges Bataille and Michel Foucault. Petronio's best work has a Baudelairean quality in which the beautiful and the noisome inseparably intertwine. His dances are not quite "flowers of evil," but are often flowers of menace. The critic William Harris has aptly described the simultaneous discomfort and fascination provoked in watching Petronio's creations: "His dances are not easy to watch, and also not easy to forget."[27]

In my last conversation with Petronio, we spoke about the 1913 premiere of *Le Sacre du Printemps* and the riot that broke out at the Théâtre des Champs-Élysées, as Stravinsky's dissonant score pulsed, dancers darted and scuttered, and whispers in the audience gave way to agitated shouts and screams. At a restaurant afterward, celebrating with his collaborators, Stavinsky contentedly declared about the audience's provocation: "Exactly what I wanted." "Why do you like to provoke your audience," I asked Petronio. He answered: "Because I am provoked by things I don't understand."

NOTES

1. Stephen Petronio, *Confessions of a Motion Addict* (CreateSpace Independent, 2013), 119.

100 Part 1: Writing the Body

2. Igor Stravinsky and Robert Craft, *Expositions and Developments* (London: Faber & Faber, 1959), 143.

3. Unless otherwise noted, all quotes by Petronio are taken from interviews with author I conducted in January 2013 in Northampton and Amherst, Massachusetts, while Petronio was in a Five College choreographic residency.

4. Petronio, *Confessions of a Motion Addict*, 146.

5. Anna Kisselgoff, "Review/Dance; Petronio Company Explores Esthetics of a Post-Punk Era," *New York Times*, 19 January 1991.

6. Ibid.

7. Brandon Voss, "Stephen Petronio's Dance Revolution," *The Advocate*, 1 April 2010. http://brandonvoss.com/blog/stephen-petronio-stephen-petronios-dance-revolution.

8. Ibid.

9. Claudia La Rocco, "Upstart with the Marks of Experience," *New York Times*, 16 April 16, 2010.

10. Petronio, *Confessions of a Motion Addict*, 162.

11. The Joffrey Ballet's 1987 reconstruction of *Le Sacre du Printemps* resulted from years of research by Millicent Hodson, who pieced the choreography together from the original prompt books, contemporary sketches and photographs, and the recollections of Marie Rambert and other survivors.

12. Petronio, *Confessions of a Motion Addict*, 157.

13. David Gere, *How to Make Dances in an Epidemic: Tracking Choreography in the Age of AIDS* (Madison: University of Wisconsin Press, 2004), 12–14.

14. Petronio, *Confessions of a Motion Addict*, 120.

15. Petronio, *Confessions of a Motion Addict*, 107. Petronio was at the forefront of the AIDS crisis in 1981. Writes David Gere: "AIDS first entered public consciousness in the United States on 3 July 1981 when an obscure one-column article appeared on page A20 in the *New York Times*, reporting an explained cluster of cases of Kaposi's sarcoma, a rare cancer that until that time had almost exclusively affected older Italian men." Gere, *How to Make Dances in an Epidemic*, 28.

16. Dancers in the May 13, 1992, New York performance included Kristin Borg, Gerald Casel, Ori Flomin, Rebecca Hilton, Mia Lawrence, Ellis Wood, Rebecca Trower, Gordon Wright, and Petronio.

17. Anna Kisselgoff, "Review/Dance; Hurtling, Hurdling and Whirling Near the Edge," *New York Times*, 8 May 1992.

18. The quote about "lips and hips" is from Kathy Acker.

19. "The male is completely incapable of love…" is from Valerie Solaris, The S.C.U.M. Manifesto; all other parts of text from the writing of William Burroughs and Petronio.

20. *Half Wrong Plus Laytext* premiered in 1992 at the Deutsch Opera House in Germany and on May 13, 1992, in New York at The Kitchen; *Half Wrong* and *Full Half Wrong* would not be performed in New York under those titles until 1993.

21. Edward Said, *Orientalism* (New York: Random House, 1978), 2–3.

22. Ananya Chatterjea, *Butting Out: Reading Resistive Choreographies through Works by Jawole Willa Jo Zollar and Chandralekha* (Madison: Wesleyan University Press, 2004), 1–2.

23. Anna Kisselgoff, "Reviews/Dance; A Questioning of Sexual Identity," *New York Times*, 14 October 1993.

24. Richard Penberthy, "Ravishing Dance, just in time for spring! Stephen Petronio Company," ExploreDance.com, 23 April 2006.

25. Wendy Perron, "Stephen Petronio Company, The Joyce Theater, NYC, April 18–23, 2006," *Dance Magazine* 80, no. 7 (July 2006).

26. Carl Kronin, "The Rite Part," Ballet.dance.com, 21 March 2006, http://www.ballet-dance.com/forum/viewtopic.php?t=28978.

27. William Harris, "Dancing Out of Control," *Village Voice*, 31 January 1995.

Bibliography

Chatterjea, Ananya. *Butting Out: Reading Resistive Choreographies through Works by Jawole Willa Jo Zollar and Chandralekha*. Middletown, CT: Wesleyan University Press, 2004.

Gere, David. *How to Make Dances in an Epidemic: Tracking Choreography in the Age of AIDS.* Madison: University of Wisconsin Press, 2004.
Kronin, Carl. "The Rite Part." Ballet.dance.com, 21 March 2006. http://www.ballet-dance.com/forum/viewtopic.php?t=28978.
Petronio, Sephen. *Confessions of a Motion Addict.* New York: CreateSpace, 2013.
Kisselgoff, Anna. "Review/Dance; Petronio Company Explores Esthetics of a Post-Punk Era." *New York Times*, 19 January 1991.
_____. "Reviews/Dance; A Questioning of Sexual Identity." *New York Times*, 14 October 1993.
LaRocco, Claudia. "Upstart With the Marks of Experience." *New York Times*, 16 April 16, 2010.
Penberthy, Richard. "Ravishing Dance, just in time for spring! Stephen Petronio Company." ExploreDance.com, 23 April 2006.
Perron, Wendy. "Stephen Petronio Company, the Joyce Theater, NYC, April 18–23, 2006." *Dance Magazine* 80, no. 7, July 2006.
Said, Edward. *Orientalism.* New York: Random House, 1978.
Stravinsky, Igor, and Robert Craft. *Expositions and Developments.* London: Faber & Faber, 1959.
Voss, Brandon. "Stephen Petronio's Dance Revolution." *The Advocate*, 1 April 2010. http://brandonvoss.com/blog/stephen-petronio-stephen-petronios-dance-revolution.

Part 2

Transmissions and Traces

To dwell means to leave traces.
—Walter Benjamin

Othering the Religious Right
Ameritude, Whiteness and the USA Freedom Kids

Michelle T. Summers

On January 13, 2016, three young white girls in star spangled blue tops with red and white striped skirts took the stage to proudly perform their single, "Freedom's Call," to a large crowd in Pensacola, Florida. The trio, known as the USA Freedom Kids, patriotically sang, "stand up tall and answer freedom's call," as Donald Trump supporters clapped in time with the girls' hip bounces and *pas de bourrées*. Manager and dad, Jeff Popick, explains that the response from those at the rally was overwhelmingly positive with multiple requests for autographs and photo opportunities with the girls. However, within 48 hours, the song and dance had turned into a viral video sensation, prompting an outsized negative backlash to the girls' routine. Many major news media outlets reported on the performance, resulting in 30 million YouTube views and #Ameritude trending on Twitter. The song and dance were soon parodied by late night talk show hosts such as Jimmy Kimmel and Stephen Colbert. Many of the comments on social media, in particular, were brutal—calling the girls "brainwashed" and comparing them to Hitler's Youth or North Korean propaganda videos.[1] The video made the rounds on Facebook with viewers oscillating between amusement and disgust at what seemed to be just another one of the Trump campaign's circus tactics. But this video and the USA Freedom Kids take on a different significance when circulated in the wake of a realized Donald Trump presidency.

For this article, I am primarily interested in understanding how a Trump brand of fervor, as exemplified by the USA Freedom Kids' performance, was able to mobilize the Christian right, a conservative group whose emphasis on family values appears antithetical to a twice-divorced Donald Trump.

According to exit poll data, 81 percent of white evangelicals voted for Trump—a greater number than those who voted for born-again Christian George W. Bush.[2] In concurrence with recent scholarship that has been devoted to understanding the rise in populist candidates in Europe and the United States, I theorize this performance as embodying a form of cultural backlash rooted in liberalism's emphasis on identity politics.[3] However, this performance is not just about the values the USA Freedom Kids are enacting, but it is also about how these girls are framed as the Other within liberal discourses. If the USA Freedom Kids' danced Ameritude both reflects current cultural values and also shapes those values, then through performance analysis we can begin to understand how and why the Christian right might have been willing to invest in Donald Trump. This article is therefore framed around three song and dance routines performed by the USA Freedom Kids: "Freedom's Call," "American Mommies," and "National Anthem Part II." Each of these pieces contributes to an understanding of this term Ameritude by uniquely capturing the complexities of patriarchal strength, childhood futurity, whiteness as Americanness, and female empowerment integral to Trump's success.

I reached out to Jeff Popick, the manager of the USA Freedom Kids, in January of 2017, and he agreed to do a phone interview with me about the group and its experience with the Trump campaign. His daughter Alexis, the youngest girl USA Freedom Kid in pigtails, was the reason that Popick decided to form what was originally called the "Patriettes." When I asked Popick about the group's dance styles and training, he claimed "hip hop" dance as the basis for "Freedom's Call." While admittedly the group's music video "National Anthem Part II" is more explicitly presenting a choreo hip hop kind of style (complete with dancing WalMart employees and policemen), "Freedom's Call" really did not resonate with me as a "hip hop" dance, although it did remind me of another dance form imbued in African American vernacular dance traditions. I would argue that the movements are citing a very particular visual and embodied history of precision drill team dance that is widely popular in the American South.

Intimately intertwined with American football, military precision, beauty pageantry, and the image of the Southern "lady," many drill teams of this variety trace their origins to the formation of the Kilgore Rangerettes, a college dance group formed in Kilgore, Texas by Gussie Nell Davis in 1939. This group was the inspiration for my high school's dance team, the Southside Dixie Belles of Fort Smith, Arkansas, whose costumes are strikingly similar to those created for the USA Freedom Kids. And it was this costuming, paired with a comment by a YouTuber that lamented, "They're … so bad. So robotic. So graceless," that pointed me toward the militaristic and often patriotic impulses of the danced drill team.[4] The group's original name the "Patriettes" also continues a tradition of naming dance drill teams with the feminized suffix "ettes" (e.g., Rangerettes, Bearettes, and of course the Rockettes). These teams parallel and likely draw from African American female drill teams who, in response to the male dominated military drills, incorporate a history of parading and precision drilling into performances exhibiting African American dance vernacular traditions.[5] Steeped in a history of segregation and exclusion, the drill team's white counterpart usually features high kicking, militaristic conformity, patriotic motifs, and often tangentially "cowboy"-themed choreography. While both drill team forms have frequent associations with school spirit, football, and the suffix "ettes," white-dominated drill teams often invested in the maintenance of whiteness by invoking a similar racist rationale utilized around the segregated white *corps de ballet*. The Kilgore Rangerettes did not have an African American dance team member for 34 years. A 1987 *New York Times* article quoted the director of the Rockettes, Violet Holmes, as saying, "the dancers were supposed to be 'mirror images' of each other and ... 'one or two black girls in the line would definitely distract. You would lose the whole look of precision, which is the hallmark of the Rockettes.'"[6] Thus, while Popick's original association with hip hop may only be partially accurate, an analysis of this form of dance as drill team continues a history of "whitewashing" black dance forms to make them more mainstream and thus acceptable to white audiences.[7]

The military metaphor is central to the performance. Drill teams often have "officers" who are the leaders of the team, and they draw on military gestures such as the salute, stiff marching in formation, inexpressive faces, etc. This is reflected in the roboticism and over-played symbolic gestures that the USA Freedom Kids perform at the rally. At the same time, the performance is highly gendered. For example, the bevel bounce (with one knee popped) that the girls do throughout the rally is a chorus line tradition and beauty pageant secret meant to flatter the legs and show the body at a slight angle. The etymology of the suffix "ettes" also locates the group within a feminized and diminutive context—gesturing toward the imitative nature that these groups derive from the "real" military drill teams that perform dominance through enactments of war. These young girls are largely fulfilling the role of culture bearers of white metonymic power evidenced by their proximity to but inexact representation of military might.[8]

"Freedom's Call" is a glorification of the military mixed with feminine charm and youthful exuberance. The song itself is a reworking of the World War I propaganda tune "Over There" by George M. Cohan. In fact, Popick, who wrote the lyrics, originally created this song in honor of his personal hero, General George Patton,[9] but opportunistically shifted the lyrics to feature Donald Trump after his candidacy announcement. The choreography, in which Popick also had a hand, reflects hyper-visible symbols of U.S. patriotism (salutes, fist pumping, etc.), but Popick labels the movements mostly as "simple, silly fun." He identifies a disco-like John Travolta move and a "body builder" move that were meant to engage the audience and literally symbolize the words being sung. When asked about some of the more contentious phrases such as "Come on, boys, take 'em down!" and "Deal from strength or get crushed every time," Popick took an emotional pause and launched into an impassioned plea for Americans to not forget the fragile "gift we have ... freedom." While these are the phrases on which news outlets and social media commentators focused, the action-inducing lyrics "Our colors don't run, no-sirree" and the repeated call to "Stand up tall" are methods for thinking about how Ameritude is embodied. The slogan "Our colors don't run," a popular bumper sticker axiom post 9–11, conjures vague images of patriotism and bravery that are rooted in the play on words—both a reference to the running of colors on fabric, metaphorically implying the flag, and a reference to the belief that the U.S. military will not run from necessary battles. Thus, patriotism is predicated on a masculinist position of strength that does not allow one to back down from a fight, a reinforcement of Trump's then evolving America First strategy. Similarly, the call to "Stand up tall" takes on new meaning in light of NFL quarterback Colin Kaepernick's decision to kneel during the playing of the national anthem. Popick's lyrics represent those whose embodied ideal of patriotism is rooted in erect reverence rather than kneeling protest, a call to celebrate freedom while whitewashing who has access to perform that freedom and enjoy its privileges.[10] In fact, a reporter asked Popick directly if freedom was political, to which Popick responded, "Geez, I hope not."

Patriotism rooted in strength is also echoed in one of the key words that Popick invoked during our conversation—"empowerment." In particular, he spoke continually of the need to empower the young girls as they mature into adults. "Empowerment" frames the girls in an imagined language of feminism, appealing to a "liberal" sensibility by making statements difficult to contest. Why would someone want to criticize, or, as Popick exclaims, "bash," little girls who are working hard and trying to empower themselves through performance? Analogously, Popick previously reached out to the Hillary Clinton campaign, stating, "You're a very successful woman, and I would like to have these girls perform for you." He argues that it is not about liking Clinton, but rather about celebrating her as an "accomplished woman," even though she failed to respond to his performance request. The USA Freedom Kids seemingly herald the importance of powerful women, even going so far as to release a single titled "American Mommies," whose laudatory lyrics intone "Empowered Ladies, Enlightened Beauties, thanks for being you."

The reproductive relationship between parent and child is key to understanding the "American Mommies" music video. In her essay on white men and pregnancy, Peggy Phelan discusses the appropriation of feminist language and other grassroots movement rhetorics by the far-right group Operation Rescue.[11] As Phelan demonstrates, this rhetoric is often co-opted to reinforce whiteness and/or patriarchy in service of the future of our children (in the case of Operation Rescue, it is the unborn child that takes precedence). The lyrics of USA Freedom Kids' songs such as "American Mommies" creates an identificatory structure of feeling that is meant to bridge the political spectrum through the seemingly apolitical relationship of mother and child. The appearance of politics in this equation is an obvious point of tension for Popick as he continually asserts, "Forget Democrats, Republicans, liberals, conservatives—none of that means anything to me." But clearly the group has been strongly impacted by what Popick sees as a demonization project by the media and the left.[12] Female empowerment for young girls, like freedom, feels like a safe, agreed upon American norm that is above contestation. But the lyric "Empowered Ladies, Enlightened Beauties" speaks a different kind of narrative. "Ladies" is couched in a particular understanding of race, class, and gender, as Evelyn Higgenbotham reminds us in her analysis of the metalanguage of racial difference spoken to "black women" vs. "white ladies."[13] "Ladies" points to propriety, properly occupying one's place, and to social grace—this is staged in tension with the more strength-based (read masculine) term "empowered." Similarly, the aesthetic term "beauties" speaks to the circulation of women's bodies within a visual economy of taste, but this again is staged in tension with "enlightened," referencing intellectual knowledge over and against bodily attractiveness. So while it would seem that the USA Freedom Kids are utilizing language that positively connects young girls and their mothers with empowerment, they appear simultaneously to reinforce hegemonic structures of whiteness and patriarchy.

This language of empowerment is intimately intertwined with the politics of child futurity and its import to the "American" idealism wrapped up in family, nation, and race. Carol Mason's discussion of the 1974 textbook controversy over multiethnic curriculum in West Virginia points to the complex and sometimes dubious ways in which the souls of children come to represent the souls of the nation.[14] To this point, queer theorist Lee Edelman's work on the strategically apolitical appearance of the child within American public discourse also frames this viral video and its reception.[15] If the defense of the "child" is the defining element by which patriarchy becomes the structural, yet often unquestioned rational for America's political future, then the discussion surrounding the circulation of the USA Freedom Kids focuses on "protecting" these girls as a representative effort to protect the future of the United States itself. Popick claims he is only trying to empower the girls; yet, the YouTube and social media public are decrying this as "child abuse" precisely because the young girls are constructed as overtly political and thus disempowered.[16] So while as Mason, Edelman, and Phelan point out, children and their American (read white) futures are always political, the outsized reaction to the "Freedom's Call" video lies primarily in the extreme discomfort the U.S. public feels when this usually concealed fact is outed. This is evidenced in many YouTubers proclaiming this group to be "Hitler Youth 2.0" or part of the North Korean propaganda machine.[17]

The obvious connection to dictatorial regimes and their use of children to fuel political messages demonstrates this discomfort. Yet, ironically, while Popick's contact was with campaign manager Corey Lewandowski, a Trump endorsement of the group seems tangential at best. Popick says that after the media onslaught he was told by the "advance guy" for Trump that groups like his were billed as "throwaways" in the business—local groups meant to connect to community and raise morale, and little else. In the words of Popick, it "seems like they would just allow anybody" to perform regardless of talent. So the outsized response to this "Freedom's Call" as a political manipulation of children is complicated by the likelihood that the Trump campaign had little to no idea of the impact of this group, and also by the campaign's disassociation with the group after the media onslaught. Popick's group would go on to sue the campaign for not fulfilling its promises to the girls (such as not allowing them to have a table to sell CDs), a move which ironically made the media and U.S. public more empathetic rather than critical of the USA Freedom Kids.

"Freedom's Call" also allows us to think about these girls as a symbolic link between Donald Trump and the Christian right's family values platform. As reported by several news outlets, the election came down to a single voter issue for many on the religious right—would the nominee to the Supreme Court be pro-life? While initially many leaders of the religious right disavowed Trump, when he received the Republican nomination, he aligned with evangelical Vice-President Mike Pence in an attempt to woo the Christian right. Furthermore, in a February 2016 campaign speech, Trump appealed to the Christian right by declaring, "Christianity is under siege."[18] In this same speech, Trump calls televangelist Robert Jeffress onto the stage who immediately endorsed Trump because of his willingness to support pro-life policies and people. In June 2017, Trump would continue this rhetoric, this time identifying himself with this minoritization: "_We're_ under siege. You understand that.... But _we_ will come out bigger and better and stronger than ever."[19] The "we" is what mobilizes seemingly disparate groups, Donald Trump and the Christian right, under the banner of oppression, effectively co-opting an Othered identity. This also resonates as an appeal to save "our" children, effectively uniting the rhetoric of the far right, who operate through a language of marginalization of whites, with the Christian right, who feel as if their religion is marginalized. This fear, this ability to allow the election to boil down to a single issue, is rooted in the figure, and the future, of the child as imagined through persecution.

So perhaps the most compelling anxiety around these USA Freedom girls' dancing bodies is rooted in less obvious YouTube comments such as "Which One Is Honey Boo Boo?"[20] This question correlates these girls with what anthropologist Susan Harding has identified as the "Repugnant Cultural Other."[21] Honey Boo Boo is a young, southern, overweight, beauty pageant contestant whose show on TLC features her dancing and mugging for both the judges and the TV cameras. Her mother describes her and the family as "rednecks" at whom television audiences are invited to laugh. The YouTube comment, and many like it, does the work of aligning these girls' bodies within a modernist discourse that stigmatizes "rednecks" and allows working class whites to claim a marginalized position culturally and politically. But this idea of the "redneck" is misleading, as a recent *Atlantic* article by Ta-Nehisi Coates argues, for this emphasis on working class whites as the cultural Other fundamentally ignores the broad coalition of whites, rich and poor, who supported Donald Trump in his run for office.[22] Similarly, sociologist Arlie Hochschild's analysis of the Tea Party movement identifies the integral support of a handful of rich families to the appearance of a grassroots movement. So in the wake of the election and the Trump presidency, the tendency to blame the "liberal elites" inability to empathize with the white working class Other or the Christian Other or the "redneck" Other does not quite account for the complexities of the voting results. Yet, Hoschild's central argument focuses on the need to cross the empathy wall that has been constructed between conservatives and liberals, between red and blue regions of the country.[23] This echoes Joy Crosby's call for performance studies scholars to find methodologies for speaking with the Christian right, which has been made "other" both politically and academically. So there is a tension here as blame is often cast on the left's inability to reach the "ignorant" white working class, thus appearing to obscure perhaps the more important issue of understanding how whiteness and coalition building works across class, religion, and even gender in these contexts.

An analysis of the USA Freedom Kids' "National Anthem Part II" official dance remix video is helpful here to understand this strategic alignment with working class values that often consolidates imagined communities of whiteness. The video picks up where the YouTube viewer might have left off with a video snippet of the end of the USA Freedom Kids' performance at the Trump rally. After discussing how the group can top their recent media surge, the girls (including the two members omitted from the Trump Rally who, according to Popick, prefer to not be a part of live performances) land upon a new version of the national anthem as the key to "doing something better." The Star-Spangled banner lyrics are then intermixed with EDM music, as the five original group members in their red, white and blue costumes sing energetically, backed up by five other young girls in jeans, pink tops, and vests. Interposed throughout the video are shots of the girls with their hot pink, limo truck; a scene in front of and inside a Wal-Mart with dancing customers and workers; and policemen in uniform who wave in generic parking lots alongside the girls. Hyper-patriotic symbols also insert themselves including a recurring, large American flag, and a figure dressed as an eagle wearing a red, white and blue boxing-style robe. Singularly, each of the girls is featured in front of a giant Ameritude sign, saluting the viewer with a larger than life smile. Popick even makes a featured cameo as his daughter Alexis recites a portion of the National Anthem in front of him.

While people of color are included in this video (including one back up dancer who is black), the equation of Americanness with Walmart consumerism, police amicability, and suburban life (the girls often appear momentarily on hover boards or go carts in a typical suburban neighborhood) harken to a particular vision of Ameritude, as enunciated by one YouTube comment, "The real Americans are proud of you! :)"[24] What is insinuated is that this is the "real" America—the heartland that values trucks, Walmart, consumerism, suburban neighborhoods, and friendly policeman. The use of Walmart in particular summons associations with capitalist success, as Sam Walton's legend of the self-made billionaire aligns squarely with the U.S. mantra "pulling yourself up by your boot straps." Of course, there are criticisms of the company's economic and social impact, particularly in terms of its price-cutting structure that keeps employee wages low. There is also conflicting information about Walmart's role in the recurring problem of food deserts, areas of the country where poorer neighborhoods, often comprised of people of color, do not have access to fresh, healthy foods. While big box stores like Wal-Mart still largely sell packaged foods thus not necessarily leading to healthier habits, recent store closings by the world's largest retailer, often in these poorer neighborhoods, has led to the creation of new food deserts across the country. Still, the company image of regular people gaining access to consumer goods in the heartland of America is upon what this video trades.

Similarly, the image of friendly, dancing policemen ignores the politics of incarceration. As Michelle Alexander explains in her book *The New Jim Crow*, "The United States imprisons a larger percentage of its black population than South Africa did at the height of apartheid."[25] Alexander goes on to offer staggering statistics concerning the war on drugs and its creation of a new caste system within the United States, including a deconstruction of racial profiling and its persistent, quiet endorsement within the legal system despite claims of colorblind policies within police departments.[26] Finally, the tranquil images of suburbia presented in "National Anthem Part II" subtly point to the segregated housing practices of white flight resulting from school integration. As George Lipsitz recounts in his discussion of a possessive investment in whiteness in this country, FHA home loan discriminatory practices after World War II effectively segregated neighborhoods as a select few were awarded home loans that systematically whitened the suburbs.[27] Elements such as racial zoning, restrictive covenants, redlining, school remapping, and intergenerational transfers of wealth largely keep people of color out of white suburban neighborhoods. Thus, this video's imagination of America where lyrics like "USA, That's America, BABY, There's NO place better" ring true actually result from the obfuscation of histories of discriminatory housing practices, food deserts, and racial profiling. But the celebratory power of this narrative, of this America, is exactly the currency with which Ameritude trades.

White people benefit from identity politics.[28] This America imagined in "Ameritude" is figured in leftist discourse as the cultural Other that these girls represent, and the false creation of this Other, paired with the celebratory, homogenizing urge of the "real Americans," is what allows for white victimization to occur due to the limitations of identity politics. As Eva Cherniavksy argues, a primary component of white privilege is the ability to evacuate the center and claim racial marginalization, a seemingly paradoxical position of whiteness.[29] This in effect exposes the limits of identity politics for the political mobilization of subaltern peoples, and simultaneously, as Harding asserts, leads to a never-ending discussion of perceived oppression—who is more oppressed, who has the right to claim oppression, etc.[30] It is what enables the USA Freedom Kids to feel targeted by the liberal media for their "apolitical" representation of "freedom" without understanding how their "Ameritude" is filtered through a particular worldview that celebrates whiteness and patriarchy through the guise of patriotism.

Perceived oppression is also what allows the USA Freedom Kids to sue Trump for reneging on the performance and product contract that the campaign entered into with the group. Reported in September of 2016, the group cited that "the Trump campaign broke verbal agreements for performances at two events and refused to pay even a $2,500 stipend for the group's travel expenses."[31] Popick's logic in pursuing this suit brings the argument about empowerment full circle: "This is about empowering the girls. [The girls were] severely impacted by what had happened with the media and having no support from Donald Trump or his team. To me that just was not acceptable, so I felt like I had to show the girls that this is what we do [sue] in America…." Thus, as Anthea Kraut after Cheryl Harris has argued, the girls are made into both victim and inscribed into a discourse of whiteness as property, and this is all framed as contingent on their right to be empowered.[32] This is guaranteed by their whiteness, but threatened by both their femaleness and their youth. This assertion that suing others is the default action in making one American aligns the girls within the rhetoric of possessive individualism discussed at length by Eva Cherniavksy.[33] The privilege of white subjecthood works to ensure that the bodily labor of the girls, while circulating as a political commodity, was still maintained and possessed by the group itself. This is further complicated, however, by the girls' age since it is actually Popick as manager who claims this right on behalf of the girls, again invoking protection of childhood futurity through legal means. Popick ultimately decided to drop the case ("voluntarily dismissed with prejudice")—perhaps, the actuality of President Trump proved too difficult an opponent—but the urge to protect the girls' rights as performers frame them within this continuing discourse of whiteness as inalienable property that must always be negotiated.

To conclude, the USA Freedom Kids' performance is intrinsically linked with the American race as a euphemism for the white race. Powerfully resounding with the racial tension fraught within Trump's "Make America Great Again" slogan, the multifarious meanings of "American" are doing a particular kind of work in this USA Freedom Kids' term Ameritude. Building upon a militaristic American pride, Ameritude is a call to action built around Popick's continuous appeals to "freedom." The seeming apolitical nature of freedom is rooted in the contradictory rhetoric of American imperialism— as the song says, there is freedom and liberty everywhere *unless* you do not agree with us then you will be crushed. As scholars, it is important that we effectively understand the powerful force of Ameritude as packaged in the bodies of three little Southern white girls, in order to be able to converse with the culturally repugnant Other. As Joy Crosby argues, dialogue with the Christian right's religious other will only be possible when discourses privilege engagement over resistance.[34] The performance of the USA Freedom Kids is not transgressing or resisting or subverting the status quo—quite the opposite. It is important to create a language for understanding dances that maintain, inhabit, or even celebrate what seems to be reifying the "normal."[35] In doing so, only then can we begin to understand how Donald Trump's narratives of patriarchal strength, childhood futurity, whiteness as Americanness, and female empowerment captivate the Christian right, and thereby theorize how such engagement might mobilize a different kind of racial and political future for the United States.

Notes

1. *Northwest Florida Daily News*, "Trump's 'USA Freedom Kids sing at rally," *YouTube*, Jan. 14, 2016, accessed Oct. 11, 2017, https://www.youtube.com/watch?v=KT2oAYGkB3c.

2. Gregory A. Smith and Jessica Martinez, "How the faithful voted: A preliminary 2016 analysis," *Pew Research Center*, Nov. 9, 2016, accessed Oct. 11, 2017, http://www.pewresearch.org/fact-tank/2016/11/09/how-the-faithful-voted-a-preliminary-2016-analysis/.

3. Much of this work is emerging from the newly created Center for Right Wing Studies at UC Berkeley headed by **Lawrence Rosenthal and Christine Trost.**

4. Pausen Think, *Northwest Florida Daily News*, "Trump's 'USA Freedom Kids sing at

rally," *YouTube*, 2016, accessed Oct. 11, 2017, https://www.youtube.com/watch?v=KT2oAYGkB3c.

5. Rooted in minstrelsy and traced through African American parading traditions, military drilling has a long history as part of African American vernacular dance. Often mock productions involving parody, drilling is regularly incorporated into African-American dance performances, competitions, and events. See Jacqui Malone, *Steppin' on the Blues: The Visible Rhythms of African American Dance* (Champaign: University of Illinois Press, 1996).

6. Bruce Lambert, "Rockettes and Race: Barrier Slips," *New York Times*, Dec. 26, 1987, accessed Oct. 11, 2017, http://www.nytimes.com/1987/12/26/nyregion/rockettes-and-race-barrier-slips.html?mcubz=0.

7. Many scholars from Jane Desmond to Thomas DeFrantz have identified this cultural appropriation or borrowing from U.S. black dance forms and subsequent sanitization of perceived sexuality or aggression within the forms in order to make them more acceptable within a mainstream, read white, context. See Jane Desmond, "Embodying Difference: Issues in Dance and Cultural Studies," in *Meaning in Motion: New Cultural Studies of Dance*, ed. Jane Desmond (Durham: Duke University Press, 1997), 29–54, and Thomas DeFrantz, "The Black Beat Made Visible: Hip Hop Dance and Body Power," in *Of the Presence of the Body: Essays on Dance and Performance Theory*, ed. André Lepecki (Middletown, CT: Wesleyan University Press, 2004), 64–81.

8. Framing my understanding of gendered nationalism is Anne McClintock, *Imperial Leather: Race, Gender, and Sexuality in the Colonial Conquest* (New York: Routledge, 1995).

9. Other iterations including a tribute to Fox News anchor Megyn Kelly.

10. As one commenter at a recent DSA conference in 2017 noted, the irony of this symbolic kneeling also lies in the fact that kneeling is often one of the most devout and humbling enactments within religious or political action.

11. Peggy *Phelan*, "*White Men and Pregnancy*: Discovering the Body to be Rescued," in *Unmarked: The Politics of Performance* (New York: Routledge, 1993).

12. Popick told me a rather long story about a kid on a Nickelodeon show he had seen recently who was wearing an "I Love Michelle Obama" shirt, pointing out that no one criticized this child for his politics. He also gave examples of kids who rap and say four-letter words and do not receive attention form the media. Popick states: "It's really interesting that the media will bash kids singing about freedom and find a way to put a pejorative spin on it when they don't do the same thing on the other side of the political spectrum." Jeff Popick in discussion with the author, January 2016.

13. Evelyn Higginbotham, "African-American Women's History and the Metalanguage of Race," *Signs* 17, no. 2 (Winter 1992): 254.

14. Carol Mason, "Reproducing the Souls of White Folk," *Hypatia* 22, no. 2 (Spring 2007): 100.

15. Lee Edelman, *No Future: Queer Theory and the Death Drive* (Durham: Duke University Press, 2004).

16. Toli Bera, *Northwest Florida Daily News*, "Trump's 'USA Freedom Kids' sing at rally," *YouTube*, 2016, accessed Oct. 11, 2017, https://www.youtube.com/watch?v=KT2oAYGkB3c.

17. Will Barnerd and Sneaky Deej, *Northwest Florida Daily News*, "Trump's 'USA Freedom Kids' sing at rally," *YouTube*, 2016, accessed Oct. 11, 2017, https://www.youtube.com/watch?v=KT2oAYGkB3c.

18. CNN Newsroom, "Trump Speech; Christie Endorses Trump," *CNN*, Feb. 26, 2016, accessed Oct. 11, 2017, http://www.cnn.com/TRANSCRIPTS/1602/26/cnr.05.html.

19. Tim Alberta, "Trump and the Religious Right: A Match Made in Heaven," *Politico Magazine*, June 13, 2017, accessed Oct. 11, 2017, http://www.politico.com/magazine/story/2017/06/13/trump-and-the-religious-right-a-match-made-in-heaven-215251.

20. fa de, Northwest Florida Daily News, "Trump's 'USA Freedom Kids' sing at rally," *YouTube*, 2016, Accessed Oct. 11, 2017, https://www.youtube.com/watch?v=KT2oAYGkB3c.

21. Susan Harding, "Representing Fundamentalism: The Problem of the Repugnant Cultural Other," *Social Research* 58, no. 2 (Summer 1991): 373–393.

22. Ta-Nehisi Coates, "The First White President," *The Atlantic*, Oct. 2017, Accessed Oct.

11, 2017, https://www.theatlantic.com/magazine/archive/2017/10/the-first-white-president-ta-nehisi-coates/537909/.
23. Arlie Hochschild, *Strangers in their Own Land: Anger and Mourning on the American Right* (New York: The New Press, 2016).
24. The USA Freedom Kids, "USA FREEDOM KIDS dance remix [OFFICIAL MUSIC VIDEO]—National Anthem Part 2," *YouTube*, June 25, 2016, https://www.youtube.com/watch?v=qLHEJQ6NEhA.
25. Michelle Alexander, *New Jim Crow: Mass Incarceration in the Age of Colorblindness* (New York: The New Press, 2010), 6.
26. *Ibid.*, 129.
27. George Lipsitz, *The Possessive Investment in Whiteness: How White People Profit from Identity Politics* (Philadelphia: Temple University Press, 2006), 7–14.
28. *Ibid.*
29. Eva Cherniavsky, *Incorporations: Race, Nation, and the Body Politics of Capital* (Minneapolis: University of Minnesota Press, 2006), 62.
30. Harding, "Representing Fundamentalism," 392.
31. Kelly Weill, "USA Freedom Kids Sue Trump Campaign for Stiffing Them," *The Daily Beast*, Sept. 6, 2016, Accessed Oct. 11, 2017, https://www.thedailybeast.com/usa-freedom-girls-sue-trump-campaign-for-stiffing-them.
32. See Anthea Kraut, "White Womanhood, Property Rights, and the Campaign for Choreographic Copyright: Loie Fuller's 'Serpentine Dance,'" *Dance Research Journal* 43, no. 1 (Summer 2011): 2–26, and Cheryl L. Harris, "Whiteness as Property," *Harvard Law Review* 106, no. 8 (1993): 1710–1791.
33. Cherniavksy, *Incorporations*, xv.
34. See Joy Crosby, "Liminality and the Sacred: Discipline Building and Speaking with the Other," *Liminalities: A Journal of Performance Studies* 5, no. 1 (April 2009): 1–19.
35. I am thinking here in particular of Saba Mahmood's work on inhabitation as a theory of alternative agency in her ethnography of the women's piety movement within the contemporary Islamic revival. See Saba Mahmood, *The Politics of Piety: The Islamic Revival and the Feminist Subject* (Princeton: Princeton University Press: 2011).

BIBLIOGRAPHY

Alberta, Tim. "Trump and the Religious Right: A Match Made in Heaven." *Politico Magazine*. June 13, 2017. Accessed Oct. 11, 2017. http://www.politico.com/magazine/story/2017/06/13/trump-and-the-religious-right-a-match-made-in-heaven-215251.
Alexander, Michelle. *The New Jim Crow: Mass Incarceration in the Age of Colorblindness*. New York: The New Press, 2010.
Cherniavsky, Eva. *Incorporations: Race, Nation, and the Body Politics of Capital*. Minneapolis: University of Minnesota Press, 2006.
CNN Newsroom. "Trump Speech; Christie Endorses Trump." *CNN*. Feb. 26, 2016. Accessed Oct. 11, 2017. http://www.cnn.com/TRANSCRIPTS/1602/26/cnr.05.html.
Coates, Ta-Nehisi. "The First White President." *The Atlantic*, Oct. 2017. Accessed Oct. 11, 2017. https://www.theatlantic.com/magazine/archive/2017/10/the-first-white-president-ta-nehisi-coates/537909/.
Crosby, Joy. "Liminality and the Sacred: Discipline Building and Speaking with the Other." *Liminalities: A Journal of Performance Studies* 5, no. 1 (April 2009): 1–19.
DeFrantz, Thomas. "The Black Beat Made Visible: Hip Hop Dance and Body Power." In *Of the Presence of the Body: Essays on Dance and Performance Theory*, edited by Andre Lepecki. Middeltown, CT: Wesleyan University Press, 2004: 64–81
Desmond, Jane. "Embodying Difference: Issues in Dance and Cultural Studies." In *Meaning in Motion: New Cultural Studies of Dance*, edited by Jane Desmond. Durham: Duke University Press, 1997, 29–54.
Edelman, Lee. *No Future: Queer Theory and the Death Drive*. Durham: Duke University Press, 2004.
Harding, Susan. "Representing Fundamentalism: The Problem of the Repugnant Cultural Other." *Social Research* 58, no. 2 (Summer 1991): 373–393.

Harris, Cheryl L. "Whiteness as Property." *Harvard Law Review* 106, no. 8 (1993): 1710–1791.

Higginbotham, Evelyn. "African-American Women's History and the Metalanguage of Race." *Signs* 17, no. 2 (Winter 1992): 251–274.

Hochschild, Arlie. *Strangers in their Own Land: Anger and Mourning on the American Right.* New York: The New Press, 2016.

Kraut, Anthea. "White Womanhood, Property Rights, and the Campaign for Choreographic Copyright: Loie Fuller's 'Serpentine Dance.'" *Dance Research Journal* 43, no. 1 (Summer 2011): 2–26

Lambert, Bruce. "Rockettes and Race: Barrier Slips." *New York Times*, Dec. 26, 1987. Accessed Oct. 11, 2017. http://www.nytimes.com/1987/12/26/nyregion/rockettes-and-race-barrier-slips.html?mcubz=0.

Lipsitz, George. *The Possessive Investment in Whiteness: How White People Profit from Identity Politics.* Philadelphia: Temple University Press, 2006.

Mahmood, Saba. *The Politics of Piety: The Islamic Revival and the Feminist Subject.* Princeton: Princeton University Press, 2011.

Malone, Jacqui. *Steppin' on the Blues: The Visible Rhythms of African American Dance.* Champaign: University of Illinois Press, 1996.

Mason, Carol. "Reproducing the Souls of White Folk." *Hypatia* 22, no. 2 (Spring 2007): 98–121.

McClintock, Anne. *Imperial Leather: Race, Gender, and Sexuality in the Colonial Conquest.* New York: Routledge, 1995.

Northwest Florida Daily News. "Trump's 'USA Freedom Kids' sing at rally." *YouTube*, Jan. 14, 2016. Accessed Oct. 11, 2017. https://www.youtube.com/watch?v= KT2oAYGkB3c.

Phelan, Peggy. "White Men and Pregnancy: Discovering the Body to Be Rescued." *Unmarked: The Politics of Performance.* New York: Routledge, 1993.

Smith, Gregory A., and Jessica Martinez. "How the faithful voted: A preliminary 2016 Analysis." *Pew Research Center*, Nov. 9, 2016. Accessed Oct. 11, 2017. http://www.pewresearch.org/fact-tank/2016/11/09/how-the-faithful-voted-a-preliminary-2016-analysis/.

Weill, Kelly. "USA Freedom Kids Sue Trump Campaign for Stiffing Them." *The Daily Beast*, September 6, 2016. Accessed Oct. 11, 2017. https://www.thedailybeast.com/usa-freedom-girls-sue-trump-campaign-for-stiffing-them.

Escape Routes and Roots
Rewriting the Narrative of the Vulgar Body
A'KEITHA CAREY

> **Bazodi**[1]:
> *As a Caribbean woman, I observed embodied practices of transgression and resistance particularly in various Caribbean performances such as Carnival, Dancehall, and Junkanoo. In these settings, the hip wine is a central theme. Growing up in the Bahamas, I observed family and friends, young and old, moving their hips to soca and reggae music at these events. The hip wine took on many iterations and the levels of rotation and articulation varied communicating divergent life stories and histories. Winding hips illustrate multifarious narratives that reflect the many dimensions of freedom, power, and censorship.*[2]

After over a decade-long hiatus, I attended the Miami Broward Carnival 2017, in hopes that the event would provide rich research for this essay. The event took place on Sunday, October 8, at the Youth Fair Grounds at Florida International University. Excitement built and wavered after an hour-long fight with traffic. Hundreds of cars filed in from every direction. After finding parking at the local Publix grocery store and sampling some red wine on sale while engaging in the obligatory bathroom break, my friend and I finally walked across the street and entered the party zone. I decided that I wouldn't be playing mas'[3] in this year's parade but I would enter the space as a participant-observer.

As we chipped[4] down the walkway to the sweet sounds of soca music towards the energy of the crowd, I immediately perked up. My eyes were drawn to bodies winin' up performed as solos, duets (in variations of female/female and male/female) and also ensembles. I was also attentive to the various countries represented by flags in hands, on heads, tied around waists or draped across the body. I tucked my Bahamian flag in my back pocket and joined the masses exhibiting national pride. Vivid and gorgeous costumes were on display worn by women and men of all nationalities, race, and sizes.

My senses were being aroused and teased in dynamic ways. As we sojourned through the crowd, the smell of weed, food, and liquor was overwhelming in some areas, signifying an authentic island vibe. I was particularly cognizant of this essay and through such a lens, I attempted to wear the hat of both dance scholar and Caribbean-winer woman. I immediately took out my camera, whistle and started to document and dance along the parade route to the song "Full Extreme" by the Ultimate Rejects. It was pumping an infectious riddim'[5] which commanded and demanded the hips, waist, and torso to wuk up[6] and express unadulterated freedom. This welcome song served as a processional for the bands that were competing in the parade, the revelers[7] and the participant/observers. The drum roll at the beginning of the track required a prance reminiscent of a drum major of an HBCU (Historically Black College University) marching band drum line. This salute indicated it was time to get in formation and get ready. My hips respond in a figure 8, rolling from east to west, as I made way through the crowd. The singer chanted:

> *Tell dem ah feeling good*
> like a new machine
> like morning dew fresh on the scene
> and we go party to the full extreme
> and light it up
> with gasoline
> oh lord the city could burn down
> we jammin still we jamming still
> the building could fall down
> we jammin still we jammin
> just hold them and wuk them[8]

These lyrics served as an offering of freedom and liberation from societal constraints and oppressive occurrences.

The ritual of winin' was an offering to this invocation presented by the Ultimate Rejects, producing a series of gyrations, spirals, accents, pulsations, counter riddims, and polyriddims from the masses. As I analyzed the movers, I acknowledged that many of them initiated this movement from the hips performing hip-mancipation—an embodied practice of freedom. In the first stanza of "Full Extreme," the lead singer informs his patrons that his body is renewed like a high performing apparatus, ready to take on what the fete[9] has to offer. He is ready to participate in this experience to the ultimate, no holds barred. His level of commitment is so profound, takes on a spiritual occurrence. He becomes so entranced in his worship and wukin' up, that if the city were to become an inferno with superstructures collapsing, it would not stop him from reverencing the sweet sounds of soca and holding his partner and gettin' on bad.[10] This notion of celebration as a religious practice is rooted in the premise of Carnival.

In this essay, I will attempt to rewrite the narrative of the vulgar body discussing how the misunderstood dance movement "the hip wine" is a subversive practice that is rooted in the spiritual and sublime, provides redemption and empowerment, contributes to potent levels of self-confidence and esteem for many women, and is a major practice of Caribbean identity. I will also explore new interventions of the hip wine in the Americas, analyzing how this movement gesture is located in the dance technique CaribFunk[11] and Caribbean cultural performances. I will conclude with a discussion of how this technology is an example of Afrofuturism, providing a divine space for women of color to re-write their histories.

As I communicate my experience of Carnival and my theoretical claims concerning the vulgar body, it is imperative that I engage in the language that is germane to my identity as a Caribbean woman. As a Caribbean scholar, I continuously search for terminology that translates my experience and also respects my culture; I realize that this language needs to articulate and represent all facets of the people, history, politics, and social and cultural nuances of Caribbean people; therefore, I engage with Caribbean phraseology throughout the text to maintain a sense of authenticity to my experience.

Ca'nival Back Den[12]

The history of Carnival is inclusive of African and European traditions demonstrating a synchronized jubilee associated with the Lent festivities rooted in the Catholic traditions in Italy. The exegesis of the word Carnival is the giving up, or "farewell," to meat. Its Latin translation is: carni (meat) and vale (farewell). In anticipation for this festival, observers of Easter fasted. This European tradition was imposed upon the slaves in the Americas when the colonizers invaded their territories. Africans fused this custom with their own traditions, developing their own unique interpretation of their slave owner's festival.

Carnival is a ritual that is transformative. Many participants often describe it as spiritual, which expresses how they believe that the pelvis is connected to the soul. In this context, the pelvis is the preacher or spiritual deity mandating the hips to witness, signify, and enunciate a consciousness that is metaphysical. Through this multilayered articulation of power and spirituality, women are able to redefine and confront expectations and stereotypes that limit who they are—controlling how and what they do with their bodies. These women audaciously perform vilified movements, rebelling against systems that challenge a "normalized" culture, one that frowns on such behavior. Women who perform these hip rotations negotiate their power, reclaim their citizenship, thereby valorizing the "disembodied body."[13] Their embodied knowledge and corpo-reality is a recognition of "potentiality and prowess,"[14] challenging Western standards. This winin' body postures a transformative liberating power, acting as a free agent of resistance.[15]

> **Bruk Out**[16]
> *I present the winin' Afro-Caribbean female body as a border body that constantly negotiates the clashing politics of decolonization (read: a body constructed by the complexities of carnival, dancehall, and histories of resistance within the Caribbean) and colonization (read: a body seized and organized by gendered constructs of heteropatriarchy, imperialism, and racism).*[17]

In this essay, I attempt to answer the question with which my students and I grapple: how can a winin' body that performs this level of intense and virtuosic physicality, that is rooted in the spiritual and sublime, provide redemption? Furthermore, how can the winin' body which represents potent levels of self-confidence and esteem for many women, and is a major practice of Caribbean identity be deemed as vulgar?

In Caribbean societies, the hip wine is central to identity politics; it provides a sense of community and citizenship particularly in the United States (as exhibited at Carnival). This practice of erotic agency and pleasure has been associated with deviant behavior in black communities for centuries. This has also been a topic of discussion in regards to the violence inflicted on women winers, often suggesting that women who perform the wine are soliciting violence upon themselves. Through my investigation, I also attempt to reframe the existing narrative of black female sexuality by "positioning desire, agency and black women's engagements with pleasure as a viable theoretical paradigm."[18] This essay provides an inquiry on the arduous history of hip wine (rotation of the pelvis), the black female body, and a strategy to circumvent and transgress the negative stereotypes associated with bodies of color.

Within my effort to reframe the vulgar body, I define the neologism "hip-mancipation" with the intent of providing a term that serves to make sense of not only the aesthetic value of the wine but also its history and its purpose in our current social sphere. Hip-mancipation is an "embodied freedom and erotic agency in wider contemporary contexts of the neocolonial restructuring of citizenship, sovereignty, and power across both national and transnational terrains."[19] The term "hip-mancipation" is based on redemption from the negative stereotypes associated with ways of moving the pelvis—particularly the rotation of the pelvis termed 'the hip wine' in Afro Diasporic societies. Caribbean performance, specifically the hip wine, speaks to the embodied citizenship about which sociologist Mimi Sheller theorizes about, introducing the lower region and movements associated with this zone as emancipatory in the face of marginalization. These biases are often associated with race, class, gender, and culture.

Hip-mancipation as a theoretical claim addresses the prohibitive practices and institutionalized interests with embodied liberty and utterances that explore erotic power.[20] Performers of this agency demonstrate subversive practices that refute colonial ideologies, this includes, according to Sheller, "infractions of policing, [which are] reserved primarily for bodies of color, and the stipulation of respectability and Puritanism that was expected of women of a particular social status. Women that performed erotic power without shame, not succumbing to bodily censorship were expressing erotic agency."[21]

Within hip-mancipation, what is considered taboo (the winding and pulsations of the hip) by those who have chosen to engage in Puritan ideology surrounding the body is celebrated by subversive bodies. Within the lens of empowerment, the profane and indecent is deemed spiritual and powerful by winers; the subaltern express "citizenship from below" through the awareness, articulation, and consciousness of the pelvis. Sheller states, "Citizenship from below addresses the deeper constitutive struggles over embodied freedoms and embodied constraints within unequal interpersonal and international relations."[22] Citizenship was denied to those who performed such vulgarities (bodies of color) and also included women and those who were disenfranchised; they were considered citizens from below. Those who engage with hip-mancipation seek transgression from negative stereotypes associated with bodies of color.

History: Deh Black Women Dem Body[23]

Historically, the black woman's body has been policed in Dancehall and Carnival culture due to colonial ideals concerning respectability. Women who performed winin' were labeled indecent and immoral by some; I assert that women who engaged in winin' were "aware of how the body could be used as a 'form of protest'"[24] and resistance by creating new models and definitions of femaleness, respectability and agency—the "erotic as power."

Jamaican scholar Carolyn Cooper discusses this idea of the erotic as a performed identity in the Dancehall, participants (women) are able to express themselves in contrast to the restrictive roles that are assigned to them due to their status[25] and race. In this social sphere, women can "embrace their sexual selves and appreciate the erotic power of their bodies without 'feeling shame.'"[26] This power is an insurgency against disenfranchisement and oppression entrenched in colonialism; this agency encompasses the rebellious energy that is an expression of an Afro-Diasporic reading of 'citizenship from below'; "this embodied protest went against everything the female colonial subject symbolized to the colonialist"[27] providing an integrity associated with Afro Diasporic corporeal expressions.[28]

Women's bodies, specifically the bodies of women of color, have been ridiculed, marginalized, oppressed, and victimized for centuries by colonization, patriarchy, racism, and by those who found no value in beauty, power, and brilliance of black women. A major character in the affront on the black female body is Sara Baartman—the Hottentot Venus. Baartman serves as the iconic image of "racial and sexual alterity."[29] This South African woman's body was placed on display in Europe where her buttocks was stigmatized, criticized, ridiculed, and pathologized: "cutting across continents and culture—subjugated under slavery and colonization."[30] Upon her death, she continued to be victimized undergoing inhumane examination of her genitalia. The dissection of her body played an immense role in the scientific racism that constructed and stereotyped black women's sexuality and bodies, reducing the black female to body parts. These fallacies and myths concerning the black female body continue to impact women of color in the twenty-first century. Popular culture theorist Janell Hobson has written extensively about blackness and the black female body in her text *Venus in the Dark: Blackness and Beauty in Popular Culture* (2005) stating that Baartman has been used to typify the entire black/African race and that her derriere has been used to stereotype black women's bodies, affirming that black women's bodies have been isolated to body parts paying particular attention to their buttocks; "Baartman inspire[d] a whole stereotype in which black women, en masse, are 'known to have big behinds.'"[31] This has impacted the ways in which women of color express their desire, sexuality, and sensuality, and in how performances of the hip are read and critiqued.

The policing of women of color and their sexuality is a colonial transgression, one which bears witness to the shaping and demonstration of corporeal expressions and practices of rebellion, liberation, and agency throughout the African Diaspora. Slavery and colonial ideologies concerning respectability and purity contributed to the constrained practices of embodied freedom as illustrated by bodies of color. This policing limited and silenced the ways in which they exhibited their sexuality in public and in private; "to truly write a history of embodied freedom, and to understand its contemporary limitations and possibilities, we need better accounts of the emergence of inter-bodily relations in the aftermath of slavery."[32] Hobson refers to this as a "culture of dissemblance." This ideology prohibits any expression of the black female body; she is expected to mute her body, disallowing any publicity or exhibition. Under this ideology, the vulgar body is of particular interest. This includes a body that isolates the lower region, articulating the pelvis in a contract and release action, delighting in sinuous revolutions hastening in matrimony with the percussion of the drum fracturing "Victorian ideological and highly conservative tenets of respectability."[33] Bodies that participated in this dissent from normative social standards of morality were stigmatized, sexualized, and demonized. The raced, classed, and sexualized bodies of the subaltern were disruptive to the bourgeoisie and seen as rebellious souls who fractured colonial order. These disorderly patrons of immorality were disallowed any opportunity to advertise any sense of humanity and respectability. They were dehumanized, pushed to the margins, and oftentimes implicated in prostitution, needing to be controlled. Performances of embodied agency and freedom conflicted with governing agencies who were sanctioned to regulate sexuality. Embodied agency and fractured colonial order were pervasive at the Miami Broward Carnival; patrons were seemingly unconcerned with notions of Victorian ideology concerning respectability.

The carnival participants did "bruk out," to the sweet soca and the blazing reggae tunes demonstrating a freedom that did not produce shock or distress. Hip-mancipation was exhibited in fierce and ferocious doses. Bodies of color freely wined and got down low, saluting the earth, massaging the pelvic girdle, and exposing their pride in their culture, bodies, and winin' prowess. We, the community, the revelers were enlightened, participating in a bacchanal[34] that was pleasurable: spiritual and sensual. I did not observe anyone struggling with respectability politics who may be fearful of colonial infractions that included policing or shaming. I did not observe anyone who was adhering to the politics surrounding bodily censorship. The women that were dressed in Carnival regalia were expressing a sense of erotic agency, an altered sensibility about female power, strength, and sensuality.

Black feminist writer, Audre Lorde provides an acute and poignant term for this defiant expression of embodied freedom and consciousness. She defines this interpretation the 'erotic as power.' This term opposes the Western masculine patriarchal definition which is concerned with the sexual (read: pornographic)[35]; "Lorde's understanding of the erotic moves beyond the sexual as a purely physical relationship to encompass a wider realm of feeling and the sensual."[36] Sheller's theory on erotic agency purports that it serves as a recovery project from the mental and physical entrapment of slavery. Addressing the whole body in a collective effort, the spiritual and sensual serve as the conduit to achieve embodied empowerment and liberation[37]; this is exhibited through the pelvis. A woman who is conscious of her power and performs this virtue is potent. As I have written in "Junkanoo and Carnival as Theoretical Frameworks: A Pedagogical Narrative," "this embodied freedom is pleasurable, natural, and inviting—drawing those in who are seeking the same type of liberation."[38] A woman who recognizes this endowment is powerful and a threat to some; she through her subversiveness performs biomythography.

Biomythography

Biomythography, a transgressive literary mechanism developed by black feminist scholar Audre Lorde fuses myth, history, and biography to illustrate the ways women want to be identified and understood. I identify how the hip wine, an Afro-Caribbean performative technology and corporeal illustration of biomythography functions as a vehicle for transformation and re-interpretation—establishing a new truth. The goal of engaging with biomythography as a corporeal expression serves to challenge both the dancer and the audience; in the sense that there is a duality which is a simultaneous reflection that is external and internal. This involves bodies articulating, expressing, miming, celebrating and/or abhorring signifiers of race, sex, and class. Movers are on a journey, their task is to bring awareness of oppressive systems that consume our everyday lives, dance their stories of the past and present while actively thinking of interventions that can also transcend those narratives.

> **Hip wine**
> *A rolling hip dance that includes, but is not limited to dexterous and vigorous rolls, gyrations, thrusts, and shakes of the hip, pelvis, and buttocks....*[39]

This embodied freedom is rooted in the sublime, virtuosic, and lyrical articulation of the pelvis performed by bodies of color who embrace and celebrate their ancestral landscape and memories. It is a technology rooted in controversy associated with the politics of respectability—these are the social, political, and sexual narratives that identify the performer as less than or "Other." Performers of the hip wine have historically been deemed as citizens from below—the second-class citizen, enslaved, abject, low class, vulgar, and grotesque by those who do not and cannot celebrate black bodies, culture, and experiences. Under the scrutiny of the colonial gaze, those who chose to perform this vulgarity were considered deviant, solely performing a sexual act removed from the metaphysical, political, and historical. This "entrap[ment] of the black dancing body within the negative spaces of primitivism"[40] functioned to erase the black identity of the performer, inciting a sense of shame and neurosis where the performer is left to question their own positionality within an expression that served multiple purposes, primarily serving as a redemptive practice and subversive act of protest. In my own experience performing the hip wine or movement that articulate the pelvic region, I have been shamed. In those moments, I questioned my participation in this ridicule and how I could be verbally and visually punished for performing a movement that provided such pleasure and satisfaction for me. Many of my students shared similar encounters in which they were demonized by family and in social spheres for performing this expression of freedom. Later in my career, I was afforded a sense of emancipation from my complicated history with the hip wine when I met Caribbean artist/scholar Cynthia Oliver in the last year in my MFA program at Florida State University; this encounter changed my life.

Cynthia Oliver and *Rigidigidim De Bamba De*

See How It Go[41]

"Yet no matter how deep the dig, I return to the beginning. To de riddim. To hip shake, hip swill."[42]

Artist/scholar Cynthia Oliver speaks of her relationship with calypso music and dance and how the experience is somewhat cyclical—similar to how the hips wine and roll in a circle to tell her story. In the *Making Caribbean Dance: Continuity and Creativity in Island Cultures* anthology, Oliver will, "inevitably circle round, circle back, and begin again"[43] in order to articulate the power of the pelvis. This description captures the kinesthetic articulation of the hip wine, a movement which grounds this piece. *Rigidigidim De Bamba De* recalls the stories of women and their connections to and experiences with the wine. This transnational piece features the stories of six women from all over the African Diaspora with roots and family ties in Canada, Trinidad, Jamaica, St. Croix, the Bahamas, Brooklyn, the Gambia, Liverpool and St. Lucia discussing race, sexuality, colonialism, patriarchy and much more. I had the wonderful opportunity of performing the work, which premiered in 2009.

Oliver expressed her interest in the wine and how it functions as a communicator of lived experiences and national identity and how this transnational movement resonates across coasts, hemispheres and continents. Referring to the audition process, she said: "[I] endeavored to make this coming together a space where we were sharing information, locating ourselves in the city [New York], but carving out a private place for Caribbean gyuls to be themselves safely, unabashedly, without censor, and this is exactly what I got. And I was moved."[44]

Oliver refers to the "politic of pleasure" that Morgan references, which enables women of color to perform their corporeal knowledge, undoing the censorship in which they have been conditioned to participate, subduing their voice. The piece advocates for the power of the pelvis, allowing us to become citizens of our societies. As a participant in the audition process and later performing the work, I can attest to the embodied freedom and corporeal knowledge that I experienced.

> The hip wine, though considered taboo in many Caribbean (i.e., colonial) societies, provided a great source of "power and prowess" for me. I found strength in the circling of my waist and the undulation of my pelvis. I delighted in the opportunity to carve out space for my identity, race, gender, and culture with my hip/butt in a performance piece that was scheduled to be performed nationally—showing the world how we speak, activate/deactivate, and represent with our pelvis. This was a fairytale for me. Performing in this work instilled not only a sense of national pride, but it allowed me to acknowledge my history, the history of my black body. I was able to take ownership of the area that has been ridiculed and scorned for decades, allowing me to discuss my complex history with race, culture, and the body, elaborating on my positionality as an insider/outsider in my community, place of employment, and even in my family. As I performed the wine, I reflected on moments in my adolescent years of shame, ridicule, and the fear of being caught moving that way, but through this endowed expression of power and agency, I was able to locate my voice—my erotic power, regaining my citizenship.[45]

I suggest that Rigidigidim De Bamba De is an illustration of hip-mancipation which allows for a reading of "Caribbean forms of erotic agency [that] address[es] fundamental issues regarding the praxis of embodied freedom,"[46] specifically through the hip wine, which enables many women to own their power and locate their voice.

CaribFunk Technique Is a Woman Ting[47]

As a dance educator whose focus is on citizenship, identity, female power, and transgressive performances of the hip, I am on a continuous journey to discover methods that increase my students' awareness of not only Diasporic communities and also constantly trying to provide them with organic entry points on the journey. I engage with my educational system titled C.E.L. to support our explorations of womanhood, agency, and pleasure.

I developed CaribFunk technique as a system of movement that is centered on the idea of using the body as a form of protest particularly the hip wine, reclaiming the dejected and sexualized body part(s) (pelvis, hip, derriere) that produce resistance in some women. In my class, these women are encouraged to Ci-P(Her)/De-CiP(Her) their positionality inside and outside of the classroom/studio. CaribFunk encourages the exploration of self while investigating identity, citizenship, and culture through a kinesthetic expression of the rhythmic gestures of the hips—we are signifyin' with the pelvis. The technique samples and remixes Afro-Caribbean (traditional and social dance forms), classical ballet, modern (Lester Horton, Martha Graham, and José Limón) and fitness elements.

CaribFunk is grounded in "the erotic as power." Lorde defines "the erotic" as the connection to our sense of self and our internal satisfaction; this power resides in the female's spiritual realm of consciousness.[48] Participants engage with the hip wine, female strength, liberation, sensuality and dynamism. In this context, I affirm that the erotic as power sanctions women to restore their lives, thoughts, expressions, and bodies from oppressive constructs in society. I theorize that the hip wine is the conduit of this power, affirming that there is power, vitality, and discourse in the pelvis—a discourse that is subversive; one that rejects colonial notions of respectability. For some, this power is the source of conflict.

I have searched for terminology and phraseology that nestled my interest in multiple genres and styles, one that allowed me to address political and social themes, providing voice to those that felt invisible. I was in pursuit of a technique that was culturally responsive, one that recognized identity politics, cultural variances, and the African and European epistemologies in which I had been trained in. I valued both my Afro-Caribbean and American identity and desired to construct a movement technology that was reflective of my multifarious history, interests, and background.

CaribFunk, an Afro-Caribbean feminist praxis responds to the role of the oppression and subjugation of the female body. Participants are encouraged to not only speak their truth concerning body politics but to (re) envision the role of women, how they traverse from object to subject, and the variegated ways they can perform these identities (biomythography). CaribFunk encourages students to (re) position their lens, one which may be trained to portray the female body as sexualized; a phenomenon on exhibition. Rather, I propose that the technique attempts to provoke the dancer to "confront the issues of spectacle that surround the representation of the female body."[49] I suggest that the technique builds community through the performance of bodily histories, constructing and restoring one's existence and purpose, which may have been destroyed or ruptured through various forms of oppression. CaribFunk "interrogates the limitations and possibilities of hip hop, [dance], feminism, and pedagogy, and is therefore self-adjusting; and, stages the political through performance-based cultural criticism."[50]

All'yuh Overstand[51]

> **C.E.L.: An Educational System**
> **CaribFunk** is a transnational experience, one that is supported with a creolized physical language and methodology, one that informs the dancer of the cultural experience: the **history, music, and dance**. C.E.L. is predicated on three principles 1. preserving the Culture 2. embarking on a new Experience and 3. speaking a new Language.[52]

There are several elements that I focus on within my technique: visualization, self-love, self-reflection, personal narrative, creative writing, passion/compassion, and the respect of history and culture. These principles are further discussed through my educational system entitled C.E.L. C.E.L. is an acronym for preserving the CULTURE; embarking on a new EXPERIENCE; and learning a new LANGUAGE.

Visualization: Is used to incite power, freedom, and ownership over one's body when discussing the various systems of oppression.

Self-love: As Oshun (the Yoruba deity of love) adorns her body with gold jewelry, embellishing her beauty, moving through the space with power and confidence, students are taught to accept who they are. They are encouraged to embrace who they are (height, race, sex, color, religion, sexuality, size, class, ability), own it unapologetically, and know that they are enough.

Self-reflection: Through this new ownership, one must acknowledge and come to terms with how they achieved this sense of empowerment and the journey they took.

Personal Narrative: Students are encouraged to tell their (own) stories. There is something about transparency that moves the reader or the listener to enter your zone and embrace and respect your truth.

Creative Writing: This is my absolute favorite way of expressing my thoughts. I encourage students to participate in performative writing to share their stories.

Passion and love for oneself (self-confidence).

Compassion for humanity. The teachings, discourse, personal stories, research and scholarship encourages respect for humanity.

History and culture are vital to one's growth and development

On Di Road[53]: Exercise 1

> *I visualize soca artist Destra Garcia playing mas at Carnival, her body delightfully embellished wearing a bikini that is radiant with sparking glitter and gorgeous feathers chippin' down the road with music blazin' featuring the hottest tunes of the season from*

the big truck. I examine possible movement and gestures that can be utilized from this rich example of Caribbean cultural performance. I also take note of the relations and examples of community that are present. How do I now transcribe and interpret this visualization and biomythography?

The beginning of all of my CaribFunk classes begins with a processional. I give my directives such as: 1. See you neighbor and community of participants; 2. Imagine that you are at Carnival in Trinidad or Crop Over in Barbados or in the Dancehall in Jamaica; think about the sensory/sensual experience; 3. What are you seeing? Beautiful colors of extravagant costumes and beautiful women and men; 4. What do you smell? Food? Alcohol? Ganja; 5. What do you taste; 6. What do you feel? The sweat dripping down your back? Someone's skin pressed up against yours? I am interested in how these various sensory experiences influences their movement and what narratives are created.

I then press play on my IPhone, "Trailor Load" is blaring out of the speakers. The processional begins. Dancers enter the space. It is an exercise designed to warm up the body in the sense that the mind, body, and spirit are asked to be transposed to a celebratory environment such as Carnival, Junkanoo, or the Dancehall and to truly allow themselves to authentically perform gestures and recreate experiences and relationships that could function in these very improvisational, multidimensional, and expressive cultural communities. The dancers focus on the internal and external, while allowing their entire body to participate in the experience. I observe the various articulations of the bodies of my students. I am particularly interested in their hip, waist, and pelvis (dis)connections. I am looking for freedom, familiarity, trust, censorship, excitement, frustration, community, power, and pain.

Afrofuturism: Ci-P-Her/De-Ci-P-Her-Ing the Hip Wine and She/Her Power

Ci-P(Her)/De-CiP(Her)Ing is an expression of the constant flow of riddim'—bodies performing stories—signifyin'. I engage with this term to discuss the role of the body in the educational process focusing on the Afro-Caribbean feminist pedagogical practice of CaribFunk. This embodied theory is a sampling and remixing of traditional and popular dance styles and aesthetics that interrogates and repositions the "other" (whether that is articulated as a physical body or body of knowledge), challenging and critiquing politics through performance. CaribFunk technique is a "call to action" advocating the body as a site for knowledge that encourages the reimagining of life experiences, allowing students to explore their authentic selves through an empowering performance of the pelvis. This emerging praxis that is geared towards the twenty-first-century student identifies methods which serve as unorthodox interventions that can possibly "interrogate, subvert, and transform socially mediated practices and ways of seeing that contain, dehumanize, and exploit ... culture, politics, and bodies—especially black bodies,"[54] specifically in the Americas.

It was also imperative that I created a thread that distinctly recognized and illustrated that my discourse was centered on Caribbean feminism and existentialism while supplying current examples of the philosophy and praxis I explore. Within the praxis, I am concerned with having a discussion that portrays women of color as whole beings, proud, exuberant artists and conquerors—women who win and do not waver in the face of adversity, despair, and even death. It is my hope that my courses express the luxurious and multifarious ways black women perform their strength, power, and identity while proffering an elaborate sometimes mangled expression of Hip Hop culture.[55]

Ci-P(Her)/De-CiP(Her)Ing is Afrofuturistic expression of the constant flow of riddim'—bodies performing stories—signifyin'. For many of us who are involved in social justice work, we are Ci-p(her)/De-cip(her)Ing, fighting, maneuvering, and performing our activism on multiple levels. Our strategies are cyclical, inclusive of the continuous rotation of involvement in coalitions and support groups. Our energy is exerted in unique ways, which includes letter- writing, organizing and attending meetings, marching, creating artistic work, and performing that work, etc. We as activists traverse in and out of the Ci-p(her), expressing our voices and demonstrating unity for social justice.

Party Dun: Conclusion[56]

The party dun (at the end of a semester) but it is really never the end for my students (in higher education) as many reflect on their journey and transformation, particularly what happens for them in their day-to-day lives outside of my tutelage. I encourage them to reflect on the class discussions, journal entries, and the goals they set for themselves. My goal as the facilitator is to provide them with the information that provides insight (culturally) concerning the history of the wine but also the politics surrounding women who express themselves in this manner and what it means for women to express themselves "freely." My overall objective is to convey that CaribFunk technique is a performative movement that is signifyin', demanding change in the way the black winin' body is perceived and described. The technique serves as a catalyst for permutation, one which is embedded in media and popular culture, rooted in civil rights but radicalized as a contemporary and modernized representation of freedom fighting and empowerment. Through this awareness and articulation, I encourage my students to negotiate their identity while at the same time exposing systems of power that have historically disenfranchised women, blacks and other people of color. Through this intervention, I encourage my students to use multiple platforms to eradicate systems of oppression, machinating "bring[ing] about freedom for all of us once and for all"[57] while rewriting the narrative of the vulgar body.

Students in my CaribFunk class are provided tools to assist in controlling and re-writing their own narrative: (1) Women are able to Ci-P(Her)/De-CiP(Her) their positionality inside and outside of the classroom/studio while demonstrating an acute awareness of their bodies in systems and constructs that are designed to abhor, condemn and shame winin' bodies; (2) the performative technology and corporeal illustration of biomythography introduced in the course circumvents the castigation of signifyin' hips allowing curious and subversive women alike to speak their own truths without censorship or tempering their voice while repositioning the lens often times used to sexualize and sensationalize winin' bodies; and (3) students are introduced to the C. E. L. Educational System (Visualizing, Self-Love, Self-Reflection, Personal Narrative, Creative Writing, Passion, Compassion and History and Culture) which is an attempt to celebrate the bodies and the voices behind those stories revealing truth, hip-mancipation and erotic agency that are most commonly celebrated in Caribbean spaces.

Notes

1. Afro-Trini creole (from French word *abasourdir*, meaning to daze) that describes being in a state of lightheadedness or shock as if in a daze or if stunned.
2. Personal reflection written by A'Keitha Carey in 2017.
3. A common Trinidadian term for the word masquerade.
4. A Trinidadian term for a walking step in the middle of Carnival. The dragging of the feet with bent knees provide a form of rest from the exuberant dancing.
5. Shortened for rhythm.
6. A Barbadian term synonymous for winin.'
7. Those who perform boisterous enjoyment of Carnival.
8. An excerpt from the song "Full Extreme" performed by Ultimate Rejects.
9. Creolized Trinidadian term for party.
10. Afro Caribbean term for behaving in a wild and rambunctious manner. It also refers to a person's winin' skill.
11. A fusion dance technique that incorporates Afro Caribbean (traditional and social) dance, modern, ballet, and fitness principles was developed as a system of movement that is centered on the idea of using the body as a form of protest particularly the hip wine, reclaiming the dejected and sexualized body part(s) (pelvis, hip, derriere).
12. Ca'naval is the Anglophone pronunciation of Carnival and den translates to 'then.'
13. Niaah 2010, 136.
14. Borelli 1998, 129.
15. Niaah 2010.
16. To break away from social codes and standards.
17. Jones 2016, 2.
18. Morgan 2015, 36.
19. Sheller 2012, "Imagining Citizenship from Below."
20. *Ibid.*
21. Carey 2016, 23.
22. Sheller 2012, "Chapter One: Histories from the Bottom(s) Up."
23. Translates to the history of black women's bodies.
24. Noel 2010, 66.
25. Springer 2007
26. *Ibid.*, 114.
27. Noel 2010, 67.
28. Noel 2010

29. Hobson 2005, 57.
30. Ibid.
31. Ibid.
32. Sheller, 2012, Ch. 1: History from the Bottom(s) Up, 1st paragraph.
33. Noel 2010, 65.
34. Traditionally used to refer to Bacchus, the god of "wine." In Trinidadian culture, it often refers to wild parties, aggressive winin,' or fights.
35. Lorde 1984.
36. Sheller 2012, "Theorizing Erotic Agency."
37. Ibid.
38. Carey 2016, 23.
39. Jones 2016, 1.
40. Ibid.
41. Translates to "look at this example."
42. Oliver 2010, 4.
43. Ibid.
44. Ibid., 6.
45. Personal reflection written by A'Keitha Carey in 2017.
46. Sheller 2012.
47. Translates as CaribFunk technique is a woman thing (is centered on the experiences of women).
48. Lorde 1984
49. Albright 2013, 64.
50. Brown and Kwakye 2012, 4
51. All'yuh is a Trinidadian term for "everyone." Overstand: In Rastafari philosophy, it is understood that one should not "understand" or stand underneath an idea (negative connotation); when one correctly discovers or discerns an idea or concept, they "overstand" it.
52. Carey 2016, 285.
53. Translates to "on the road." This reflection was written by A'Keitha Carey in 2017.
54. Isoke 2012, 37.
55. Morgan 1999.
56. Translates to "the party is over."
57. Guynn 2015.

Bibliography

Albright, Ann C. 2013. *Engaging Bodies: The Politics and Poetics of Corporeality*. Middletown, CT: Wesleyan University Press.
Borelli, Melissa B. 1998. "Hips, Hip-notism, Hip(g)nosis: The Mulata Performances of Ninón Sevilla." In *the Routledge Dance Studies Reader*, eds. Alexandra Carter and Janet O'Shea, 122–132. New York: Routledge.
Brown, Ruth N., and Chamara J. Kwakye. 2012. *Wish to Live: The Hip Hop Pedagogy Reader*. Eds. R. N. Brown and C. J. Kwakye. New York: Peter Lang.
Carey, A'Keitha. 2016. "Junkanoo and Carnival as Theoretical Frameworks: A Pedagogical Narrative." *Society of Dance History Scholars* 36:21–27.
_____. 2016. "Visualizing Caribbean Performance (Jamaican Dancehall and Trinidadian Carnival) as Praxis: An Autohistoria of KiKi's Journey." *Journal of Dance Education* 17.4:129–138.
Guynn, Jessica. 2015. "Meet the Woman Who Coined #BlackLivesMatter." *USA Today*, March 4.
Hobson, Janell. 2005. *Venus in the Dark: Blackness and Beauty in Popular Culture*. New York: Routledge.
Isoke, Z. 2012. "Lighting the Fire: Hip Hop Feminism in a Midwestern Classroom." In *Wish to Live: A Hip Hop Feminist Reader*, eds. R. N. Brown and C. J. Kwakye, 33–45. New York: Peter Lang.
Jones, Adanna. 2016. "Can Rihanna Have Her Cake and Eat It Too? A Schizophrenic Search

for Resistance within the Screened Spectacles of a Winin' Fatale." *In the Oxford Handbook of Screendance*. Ed. Douglas Rosenberg. Oxford: Oxford University Press.
Lorde, A. 1984. "Uses of the Erotic: The Erotic as Power." In *Sister Outsider*, 53–59. Berkley: Crossing Press.
Morgan, Joan. 2015. "Why We Get Off: Moving Towards a Black Feminist Politics of Pleasure." *The Black Scholar: Journal of Black Studies and Research* 45(4):36–46.
_____. 1999. *When Chickenheads Come Home to Roost: A Hip-Hop Feminist Breaks it Down*. New York. Simon & Schuster.
Niaah, S. S. 2010. *DanceHall: From Slave Ship to Ghetto*. Ottawa: University of Ottawa Press.
Noel, S. A. 2010. "De Jamette in We: Redefining Performance in Contemporary Trinidad Carnival." *Small Axe* 14(1):60–78.
Oliver, Cynthia. 2010. "Rigidigidim De Bamba De: A Calypso Journey from Start to…" in *Making Caribbean Dance: Continuity and Creativity in Island Cultures*, ed. Susanna Sloat, 3–10. Gainesville: University Press of Florida.
Sheller, Mimi. 2012. *Citizenship From Below: Erotic Agency and Caribbean Freedom*. Durham: Duke University Press. Kindle edition.
Springer, J. T. 2007. "Roll It Gal: Alison Hinds, Female Empowerment, and Calypso." *Meridians: Feminism, Race, Transnationalism* 8(1):93–129.
Ultimate Rejects. *Full Extreme*. 2017.

Lo Que Queda/ That Which Remains

Dancing Bodies, Historical Erasure and Cultural Transmission

Michelle Heffner Hayes

Histories and Resonances

"*Pero que las alas arraiguen y sus raíces vuelen*/Our wings ground us, and our roots make us fly."[1] Flamencos often invoke the words of the Nobel Prize–winning Spanish poet Juan Ramón Jiménez in the discussion of tradition. The reference has become shorthand, appearing in descriptions of music by *cantaora*/singer Carmen Linares (for her album *Alas y Raices*) and, most recently, in reflections on *bailaora*/dancer Belén Maya's solo work, *Romnia*, by Roma activist Joaquín López Bustamante.[2,3] The metaphor captures the continuous counter-tension of "cultural legacy and experimental imperative" in the practice of tradition.[4] For me, the words capture the distinct physical experience of performing flamenco, with its grounded stance (*raíces*/ roots) and arms that extend into space (*alas*/wings). I use these images as I negotiate the physical experience of dancing traditional forms with their oral, written, filmic, musical or photographic accounts. In the process of analysis, I consider the many filters that illuminate or obscure the transmission of cultural practices "exactly" as they once were performed.

Dance historians grapple with the difference between the attempt to reconstruct a historical event and the desire to reinvent it.[5] From period-specific reconstructions of Baroque dances using Feuillet notation to Katherine Dunham's restaging of Afro-Caribbean rituals, choreographers consider the intention, the evidence and their effects in performance. Even when we think we "know" the choreography, what does it take to maintain *Revelations* in repertory for the Alvin Ailey American Dance Theater? How do we view new performances of postmodern choreography like Yvonne Rainer's *Trio A*? While we can never fully recapture the "original moment" of a performance, the informed investigation of the conditions that produced the event, and the tensions surrounding its contemporary reinvention, fascinate me. Certainly, mindfulness and rigorous scholarly research play a role in my own investigation of flamenco as a tradition. However, I recognize that movement traditions form a complex language and a multilayered system of meaning. As discourses, they hold vital information for research. Knowledge production in dance can be difficult to document, so embodied "evidence" may seem too ephemeral to be of use. As scholars, we are only beginning to explore the value of movement as a text in scholarly settings.

Bodies and Cultural Transmission

Several contemporary artist scholars advocate for the knowledge accrued and expressed through movement. Flamenco choreographer and performer Juan Carlos Lérida founded an international, experimental collective called Flamenco Empírico, marked by the first festival by that name in Barcelona with Mercat de les Flors in 2009. The empiricism in the name of the aesthetic movement refers to the practices of the body.[6] Dance studies scholar Claudia Jeschke weighs "performative knowledge" in dance historiography.[7] Her work in reconstructing Spanish dances suggests that choreographic practices "place dance identities in the gaps between ethnography, artistic demands, and strategies of embodiment."[8] Yvonne Daniel refers to "embodied knowledge" in her analysis of ritual dances of the African diaspora: "While dance and music behaviors are not immediately associated with religious, sociopolitical or politico-economic issues, they can reveal intriguing nonaesthetic information."[9]

Moving bodies contain part of what is recovered through documentation, *and also* what is constructed in the process of choreography in traditional forms. Teachers, parents, elders and communities impart this information to practitioners. My focus on the training, choreography and performance of dancing bodies is, in itself, a pedagogical legacy. Susan Leigh Foster, one of my teachers, linked "how the processes of making a dance and making a dancer are bound together" in her pivotal book *Reading Dancing: Bodies and Subjects in Contemporary American Dance* (University of California Press, 1986).[10] Further, she analyzed, in *Choreographing Empathy: Kinesthesia in Performance* (Routledge, 2011): "The dancer's performance draws upon and engages with prevailing senses of the body and of subjectivity in a given historical moment. Likewise, the viewer's rapport is shaped by common and prevailing senses of the body and of subjectivity in a given social moment as well as by the unique circumstances of watching a particular dance."[11] Not only does the embodied knowledge of the performer produce meaning, the danced information also registers with its audience. That is, the intelligibility of movement depends upon a familiarity with how we produce knowledge within the tradition. Knowledge production constitutes bodies and bodies participate in this discursive function of making meaning.

Consider this example: in 2015, a gifted performer and teacher from Spain, Juan Paredes, taught a *bulerías* workshop to students in Mission, Kansas, part of the Kansas City Metropolitan Area.[12] Paredes is a figure who specializes in teaching *bulerías* to the uninitiated, with comedy and skill. He teaches people in locations like large city plazas, train stations, and studios all over the world. In Kansas City, experienced adult flamenco students attended his workshop. They wanted to learn how and when to enter with a singer.

The relationship between the singer, guitarist and dancer in flamenco demands that each participant understand the conventions of when to "play." Paredes answered, "Cuando entro yo? ¡Cuando yo quiero!/When do I enter? *Whenever I want!*"[13] In response, I doubled over in laughter. Several students stared in disbelief. Paredes was making a joke, but also telling the truth, another convention of flamenco. Only an advanced performer, steeped in the innumerable decisions and techniques occurring in that moment, would make that comment. *Bulerías* is a flamenco *palo*/song form in 6/8 time, but it is counted in phrases of twelve. In fact, the first "count" is twelve, not one. To make things more confusing to an outsider, the "twelve" is often counted in Spanish as a "two," because a closing phrase ends on "ten," and "*once*/eleven, *doce*/ twelve" take too long to say if you want to maintain the rhythm. That is, unless you are counting it in six, or in three, according to regional variations (like *bulerías de Jerez*). Many practitioners do not "count" at all. Traditional *letras*/lyrics have a specific phrase length, and gaps between the verses, but the length of time in those phrases and gaps depends on the decisions of the singer, or on the unspoken dialogue between the dancer and musicians. Coded steps, vocal cues and musical figures announce the entrance of the guitarist, singer or dancer. So, yes, Paredes can do *whatever he wants* because he can assess his various choices and decide in a millisecond. The rest of us, when learning *bulerías*, walk around with halting steps, muttering *DOS. Un dos TRES. Cuatro, cinco, SEIS, siete, OCHO, nueve DIEZ, un DOS. Un dos TRES* … TWO. One, two, THREE. Four, five SIX, seven, EIGHT, nine, TEN, One, TWO. One, two, THREE … with a sense of anxiety and terrible purpose. Dancers who do not grow up in the tradition learn sequences of steps, but not necessarily the values surrounding the decisions that give those steps coherence as choreography. Teachers like Paredes model or explain the conditions for making choices within a system of meaning.

That moment of knowledge production by Paredes, the ability to enter "whenever I want!" (and also to joke about it) involves years of practice, and not just in the studio. The knowledge depends upon information from the environment of the dancing: spoken conversations, jokes, and social conventions. All of these layers of discourse produce decisions within the dancing body. In my own practice, I think of how I hear Latin popular music and distinguish between *merengue* or *cumbias* or *son* by moving *first* and *then* naming the form. When I explain specific flamenco *palos*, I fall into different postures because each *palo* has a distinct vocabulary, rhythm pattern, melodic structure and emotional tone. Of course, any number of cultural discourses shape my body and subjectivity in space and time, but there is information in the bodily practice. I think perhaps I am attentive to these details because I started dance training late, at seventeen. I did not take flamenco or Latin popular dance classes until I was in my 20s. I was often "behind the curve," struck by what was never explained, but was clearly reflected in the bodies around me. So, often, I listen and move, then translate that experience into written or verbal statements.

Razón de Son

In 2014, in a moment of listening and moving, I discovered an exquisite album, *Razón de Son*, by Raúl Rodríguez.[14] Rodríguez, a musician and anthropologist, describes the album as "an imaginary folklore" that incorporates the varied intercultural dialogue that eventually became flamenco.[15] In my research, words like "imaginary" and "folklore" resonate deeply with the tensions of exoticism and nationalism, but the music was so persuasive, I suspended my skepticism. Flamenco's antecedents come from Andalucía (itself a rich mix of cultures: African, Arab, Sephardic, Roma, and regional folklore), Latin America, the Caribbean and the American South.[16] From the fourteenth to the nineteenth centuries, the slave trade and colonialism displaced and forced the decimation and migration of people and industry back and forth from Spain to the "New World." The dance forms that grew from this period of violent contact also reflect myriad intersectional identities, particularly with respect to race and social class. Although scholars of Caribbean dance recount the Creolization of European, African and indigenous forms, few researchers discuss in detail the influence of African and indigenous forms in Spanish dance.[17]

At the same time that I was listening obsessively to the album, new scholarship in flamenco studies drew attention to the Africanist influences in flamenco from scholars like K. Meira Goldberg, and others.[18] In 2016, M. Angel Rosales directed *Gurumbé: Canciónes de tu memoria negra* (2016), and in

2017, the Association for the Study of the Worldwide African Diaspora hosted their conference in Seville. In my ostensibly isolated small community in Kansas, a new wave of social justice activists made their voices heard in response to anti-black violence. Protestors decried the violence, physical and symbolic, from racist graffiti in dorm rooms to the murder of Michael Brown in Ferguson, Missouri, and other fatal shootings of black Americans by police officers.

As I listened to these songs by a flamenco musician, I read and moved, and found the rhythms and movements of *son, rumba cubana, punto cubano, rumba flamenca, sevillanas, bulerías* and more. Just as the rhythms and structures of the past are still intelligible in the myriad forms of Spain, Latin America and the Caribbean, the traces of the bodies that danced are also present. The Africanist elements in flamenco studies seemed, until recently, to be designated to a footnote or completely erased and obscured by the arguments about the Roma versus "Spanish" origin of flamenco. That symbolic violence of erasure and absence—in a daily news feed that attests to material violence, the continuing legacy of systemic oppression that persists in our cultures—compelled me to explore and recognize the multiple influences in the dance vocabulary.

What struck me about *Razón de Son* was not just its AntropoMúsica (to borrow the name of a conference in Barcelona in 2015 where Rodríguez discussed the album), the anthropological investigation of music, but the framing of the work as an imaginary folklore, an act of personal creation as well as archival recuperation. "Imaginary folklore" emerged as a jazz movement in Europe in the 1970s. Jazz specialists Joachim-Ernst Berendt, and Günther Huesmann describe the music of this movement less as reconstruction than a reinvention: "[T]he 'imaginary folklore' musicians reflected their own backgrounds in the telling of personal jazz stories. Spanish pianist Chano Domingo combined the jazz tradition with the melodies and rhythms of flamenco; kaval flutist Theodosii Spassov brought Bulgarian hora melodies into contemporary jazz; Armenian-Turkish percussionist Arto Tuncboyacian used the sounds of the Caucasus and Anatolia in Western European improvisations."[19] Rodríguez, distinct from the "imaginary folklore" jazz artists, excavates the musical traditions present in Latin America and Caribbean from the perspective of a contemporary flamenco artist. He plays an instrument he invented, the flamenco *tres*, a modified version of the Cuban instrument.[20] As a choreographer, I was intrigued by the phantom dances that would accompany Rodríguez's "imaginary folklore." I wanted to reinvent them, not as they originally occurred, but as a contemporary expression informed by scholarly research, the movement vocabularies, and the "performative knowledge" of what remains.

I contacted Raúl Rodríguez. We had met briefly in Miami in 2005, when

I was co-presenting (on behalf of Miami Dade College, Cultural Affairs) a festival with Bailes Ferrer and FUNDarte, called Flamenco in the Sun. Then, he was performing with the group Son de la Frontera, a group dedicated to preserving the tradition of Diego del Gastor. I sent him samples of my previous work with music from the Cuban artists Gema y Pavel (Gema Corredera and Pavel Urkiza), who had also given me permission to choreograph to their music. The choreography was a mix of contemporary dance theatre and *son* movements. The music has been characterized as *filin progresivo* (inspired by the "feeling" genre in Cuban music of the 40s and 50s).[21] As it turns out, I had chosen one of Rodríguez's favorite songs from the duo, and we discovered we had several people in common. He very generously agreed to allow me to choreograph to his music for performance.[22]

Still, I was hesitant. I choreograph for dance students at a Midwestern university in a place my friend flamenco guitarist José Luis de la Paz affectionately refers to as "*el culo del mundo.*"[23] I was concerned that the reinvention of dances once performed by Africans, Afro-Caribbeans, Spaniards, and possibly indigenous Caribbeans like the Arawak/Taíno, by a white professor in a predominantly white institution (PWI) would represent an act of objectionable cultural appropriation. I am conscious of my role in cultural transmission and how any choreography that results is marked by the "empiricism" of the bodies who perform it. Those bodies reflect and construct their positions in complex, intersecting hierarchies of power.

Dance and Cultural Appropriation

Scholarship about dance and cultural appropriation considers context, marginalization, cultural notions of property, hegemony, commodification, exploitation, erasure, and more.[24] In *Flamenco: Conflicting Histories of the Dance* (McFarland, 2009), I write about the difficulty of defining cultures of origin in flamenco. I analyze the ways in which claims of "authenticity" have led to the commodification and exploitation of the people who perform it, usually by hegemonic discourses, like the Franco dictatorship or the tourist industry. In my analysis, any representation of a culture by an "outsider" is a cultural appropriation. Whether or not it is objectionable depends on the people, the relationship and the context. It can be difficult to define where the line of "insider" and "outsider" is drawn without resorting to essentializing important concepts like race and identity. Flamenco scholars trace its antecedents to the cultures that were oppressed and expelled from Spain: the Muslims, the Sephardic Jews and the Roma, in addition to Andalucían folk practices. Due to the presence of these many cultures in Andalucía for centuries, including Afro-Andalucíans, it becomes difficult to assign authorship to a specific race or ethnicity, or even to define what is Andalucían culture.[25] Until recently, the material violence of slavery in Spanish history disappears from the flamenco narrative of how these cultures encountered one another, subsumed under the classification of the songs and dances *de ida y vuelta*/of coming and going.

The role of cultural appropriation in the development of African American and Caribbean dance has been well documented. Any inquiry into the Africanist elements of flamenco stands on the shoulders of scholars such as Robert Farris Thompson, Kariamu Welsh Asante, Brenda Dixon Gottschild, Thomas F. DeFrantz, Constance Valis Hill, Nadine Georges-Graves, and others.[26] Yvonne Daniel and Juliet McMains contribute important scholarship documenting the Africanist presence in the dances of the Caribbean.[27] In these works, cultural appropriation occurs when the dances of the African diaspora are taken by white dancers and audiences, without attribution or compensation for their labor. Further, despite the popularity and commercial success of African diaspora dances, black people face continued oppression. Just as Nadine George-Graves discusses in her analysis of ragtime dance in the United States, "[I]t is less the mixing of cultures that is problematic than the subsequent disenfranchisement, exploitative commercialization, and disavowal of African American culture by hegemonic forces."[28] My choreography, a personal creative expression, brought to bear the privilege that foregrounds cultural appropriation.

Following some reflective reading, I listened to the dialogue occurring around issues of social justice in my own community. Many activists use the "Yes, and …" model of dialogue, borrowed from theatrical improvisation.[29] This model requires that the improviser respond to statements with the phrase "Yes, and …" The response recognizes the value of the previous statement and continues the conversation. "Yes, and …" allows multiple voices to speak and deepens the dialogue. More than one thing can be true at one time. *Yes*, my unearned, unasked-for white privilege has allowed me to engage with master teachers and performers in traditions outside my immediate cultural sphere for decades. I recognize that my privilege has enabled me to travel, to study critically, and to form important relationships with artists. White privilege grants me access to a teaching job at an institution of higher education.[30] *And …* my privilege comes with a responsibility: if I do not do this critical work that I receive from my privilege, then those same narratives of exclusion remain uncontested in the classroom, in the scholarly dialogue, and in our cultures. There is no space of innocence here; I do not deny that the very systems of oppression that created my privilege also perpetuated these biases. But by remaining silent, I do more harm. Moreover, the way I work should include proper attribution to the work of marginalized people, and payment for their labor. If I cannot pay someone directly, I try to advocate for them to receive a paid performance or publication opportunity. *And*, I have an obligation to extend social justice work beyond scholarship, to classroom practices, studio etiquette, curriculum, department leadership, university administration, our community and further.

Practice as Research

Critical dance and race studies, as well as the desire to address the production of knowledge in the studio and on the concert stage, led me to an exploration of Practice as Research (PaR).[31] Estelle Barrett's *Introduction to Practice as Research: Approaches to Creative Arts Enquiry* recognizes artistic research as a "mode of knowledge production.... The innovative and critical potential of practice-based research lies in its capacity to generate personally situated knowledge while at the same time, revealing philosophical, social and cultural contexts for the critical intervention and application of knowledge outcomes."[32] One of the outcomes described by Barrett is "the potential of studio-based research to demonstrate how knowledge is revealed and how we come to acquire knowledge."[33] By analyzing the way we teach these dances, we learn about how bodies are constructed and erased in the process.

In the rehearsal studio, I focused on different movement motifs that appear in Afro-Caribbean and flamenco forms, based on decades of study with different teachers.[34] These Creolized Africanist and European influences, beautifully identified and explored in the context of American concert dance by Brenda Dixon Gottschild in her precedent-setting *Digging the Africanist Presence in American Performance*,[35] include the flexion of knees, displacement of hips, mobile torsos, polyrhythms, polycentrism, and "high-affect juxtaposition as Africanist elements."[36] Afro-Caribbean-Andalucían dances are replete with these movements, locating counter-rhythms in hips and feet, or competing areas of emphasis in spiraling torsos and circling wrists. I focused on partner dancing, the *sevillanas* step and *pasada*, and isolated moments of footwork from the fandango traditions of Spain, Latin America and the Caribbean.[37] Also, I incorporated specific *rumba* steps that appear in Cuban folkloric performances, such as the "basic," or the virtuoso display of the solo *rumba colombia*.[38]

Combining these various shapes and steps into sequences and patterns that might logically be danced together—creating a syntax for the vocabulary—was far more difficult. How does this vocabulary create intelligible "sentences," or even poetry? Random words do not create a sentence that can be understood in a language. Choreographers consider phrase lengths, rhythm, shift of weight, accents, textures, lines, and the aggregate movement of bodies through space to construct ideas, among many other decisions. Fortunately,

University Dance Company members perform in *Lo Que Queda/That Which Remains*, Lied Center of Kansas, November 18, 2016, at the Lied Center of Kansas. In the foreground are Alyssa Rivera (left) and Kayla Wegley (photograph by Brett Pruitt, used with permission of the Department of Dance at the University of Kansas).

Rodríguez's *Razón de Son* ("Intro" and the song by the same name as the album) provided an overarching structure in 6/8 time, and accent patterns that allowed for combinations from the different vocabularies. Within that musical structure, I based many decisions on the embodied knowledge of sequencing from my years of study in flamenco and African diaspora dances. I soon recognized that the students did not have access to this "empiricism."

The contemporary dancers in *Lo Que Queda/That Which Remains* bear very little relationship to the social dancers who practiced popular dances of the past, or even contemporary practitioners of flamenco or Latin popular dances. I cast eight aspiring professionals, with regular training in ballet, concert modern dance, and jazz. One of those dancers identifies as Latinx, and one as Afro-Latinx. The rest of the dancers identify as white.[39] Among their electives, students may take a single semester of flamenco technique with me as part of their training. Also, they have the option to take a semester of African dance and a semester of classical Indian dances. Recently, we added two levels of training in hip hop and a course in rhythm tap to our curriculum offerings. But those experiences provide very little preparation to perform flamenco or Afro-Caribbean forms with mastery. Again, I wondered if I was simply perpetuating cultural appropriation by creating a dance with students who were not steeped in these traditions. And again, I realized that by not teaching these forms to the best of my ability, I was complicit in their exclusion from curriculum. Then, all of my students, including students of color, would lose the opportunity to learn about the vibrant people and dances of the African diaspora, and their role in shaping the cultures of the Americas. So, in addition to three hours of rehearsal per week for three months, I asked students to watch videos and documentaries to understand the context of the forms, read an English translation of the lyrics for the songs, and listen to other cuts from the album.

This method represents a departure from established studio practice. It is unusual for our dancers to discuss the context and history of dance forms in rehearsal. Generally, choreographers demonstrate and assume that dancers are able to reproduce the movement. In ballet, a choreographer can gesture and indicate steps simply by using the terminology. These assumptions makes sense when the core training in ballet, jazz or classical modern dance matches the works performed by the students. By contrast, information about cultural studies in dance is presented to students in lecture courses, although we in higher education are hybridizing more lecture and studio as techniques like hip hop, rhythm tap, African, Indian classical dance and flamenco have been added to the curriculum. In rehearsal, the dancers and I discussed the relationship between African dances and the Caribbean, Spain as a colonial power, the slave trade, and the near-genocide of the Arawak/Taíno people. We also talked about the social status of itinerant workers, landowners, and slaves, the complex and changing definitions of race, coded expressions of gender and sexuality, and the mechanics of power that govern these categories in different cultural contexts. And we danced.

Outcomes: Knowledge Production

Physically, we discovered the conflicts between the training of the dancers and the demands of the choreography. These dancers excel at explosive, athletic movements, like leaps and jumps, which are completely absent from *Lo Que Queda/That Which Remains*. They also pick up movement sequences very quickly, which almost works to their disadvantage, because they do not need to repeat movement several times to reproduce the sequence. However, to change the body's organization and movement logic—to learn a new technique—a dancer needs to repeat sequences and cultivate new tendencies.[40]

Specific moments of effort and disjuncture emerged in the execution of the choreography:

   1. The micro-bend of the knees in flamenco and African diaspora dance involves the flexion and displacement of the hips. This bodily organization presented a major challenge, because it contradicts the squared-off, lifted pelvis in the dancers' previous training in ballet and often in concert modern/contemporary dance.
   2. Even though ballet allows for *cambré* in the torso, it was very difficult to coordinate an arching upper torso on a moving body, what Dixon Gottschild calls "polycentrism" and "high affect juxtaposition."[41]
   3. The coordination of the arms/*braceo* and hands/*floreo* with the rest of the body frustrated the dancers. Dancers perform *port au bras* as part of their daily training, but the arms move in opposition to a stable torso. These dances demanded the spiraling action of the flamenco torso or the integration of mobile, plastic or percussive, torso in African diaspora forms. The circling wrists and spiraling fingers brought attention to the ways in which the hands are fixed, and largely ignored, in forms that are more familiar.
   4. Moving in sixes, with changing accents, nearly drove the dancers to despair. A basic *bulerías* pattern consists of two phrases of three and three phrases of two. By contrast, the dominant training of the dancers places the accent on the first beat in 3/4 or 4/4 time. It took weeks to master a basic walking pattern. Then, when we added the movements of hips, torso, arms and hands on top of the polyrhythms, they faced the challenge of coordinating polycentrism.
   5. *Lo Que Queda/That Which Remains* features few, short sequences of *zapateado*/footwork. Many of the dancers have several years of experience in tap. The ways that the feet strike against the floor in tap differ from the way that flamenco dancers direct their feet down and through the floor. At the end of the piece, dancers gather in a half-circle around a soloist (an experienced rhythm tapper) for a more extended section of footwork. The dancers play *palmas*/hand-clapping as the soloist builds a section of footwork. *Palmas* should support the dancer by maintaining the rhythm, and adjust to the speed of the dancer. Frequently, the *palmas* would speed up, lose the accent pattern, and "pull" the soloist out of the *compás*/rhythmic structure.

During periods of self-reflection in rehearsal, these accomplished dancers revealed they felt vulnerable in forms that were not considered "technically difficult" by their peers. This new awareness allowed us an opening to discuss cultural assumptions about the invisibilization of labor in forms outside "classical" dance, specifically ballet and concert modern dance. If we examine the requirements of the dance degrees at our institution, ballet, modern/contemporary, and jazz techniques are the "core" competencies required for completion.[42] Other techniques are classified among dance studies courses, suggesting that they are not "techniques." In our studio environment, we temporarily "decentered" the discussion from concert to popular dance as a locus of mastery. We emphasized the aesthetics of social dance rather than "high art dance," and celebrated a history of the lower classes instead of the courts or the upper classes. We set aside the normalized white, Euro-centric identity to explore expressions of Creolization.

I find it interesting that these Africanist elements of flamenco and Latin popular dances presented such a challenge to dancers who have decades of performance experience, and in forms like tap, jazz and hip hop, which are African diaspora forms. My sense is that we (and here, I refer to our program, but also many private studios and other universities) are teaching these Africanist movement traditions as ornaments on a white body, regardless of the race of the dancer. That may not be our intention, but the majority of the training and the weight of courses to fulfill degree requirements perpetuate

Soloist Brennon Madrid, with *jaleo*/shouts of encouragement and *palmas* by (left to right) Briana Herrington, Christine Bessey, Ana Glocker, Kayla Wegley, Alyssa Rivera, Jillian Armstrong, and Tori Hilger, November 18, 2016, at the Lied Center of Kansas (photograph by Brett Pruitt, used with permission of the Department of Dance at the University of Kansas).

the values of the white body. This body possesses an orientation that is centered and upright. It moves with straight legs that occasionally bend. The limbs operate around a stable torso, to music that falls on the downbeat, mostly in 4/4 time. The assumptions that underpin what constitutes training in higher education privileges a Euro-centric notion of virtuosity. These discursive practices naturalize or normalize an ableist, racist, sizeist, cisgendered, heterosexist white body, even when we strive to be more "inclusive" in our curriculum.[43]

The discomfort among the dancers in an immersive Creolized movement environment stems from the embodied realization that (temporarily) their highly-trained bodies are *other*. Bodies are not "neutral," and dance training is normative. This knowledge is powerful, even if it is for a fleeting moment, when we consider what social justice advocates tell us about marginalized bodies. The white corporeal identity is a construction predicated by violence, symbolic, material and systemic. In a critique of how higher education fails to deliver on its promises of diversity, equity and inclusion, Cody Charles, a social justice educator, says, "We have no capacity to create spaces that hold the identities that are at the frontline of violence. We have no capacity to create systemic and sustainable change—informed by challenging the status quo, thinking creatively, and taking intentional and strategic risk unremittingly. Moreover, I'm noticing that we lack the capacity to simply experience empathy for the bodies that are constantly under attack."[44] Judith Butler, in *Bodies That Matter*, reminds us that the "exclusionary matrix by which subjects are formed thus requires the simultaneous production of a domain of abject beings, those who are not yet 'subjects,' but who form the constitutive outside to the domain of the subject."[45] Through our studio practices, our syllabi, our degree programs, our recruitment practices, our scholarship, and our hiring and promotion choices, university dance departments create such an apparatus. If we (at our institution, but this is a question for others in academia) wonder why we cannot recruit and retain students and faculty of color, we must recognize how cultural biases are perpetuated and assess how our "normal" practices exclude bodies and subjects.

While expanding our curriculum is a start, we have not de-centered the narrative. The rehearsal process for *Lo Que Queda/That Which Remains* allowed us to create a temporary dialogic space of "imaginary folklore." We were able to reconsider the context of the original moment of performance of the dances that may have preceded flamenco. Students gave me feedback that the history and context of the movements allowed them to physically connect to elements like the micro-bend in the knees more effectively than the correction to "bend the knees." During the process, we temporarily re-centered the dialogue to highlight the presence of Africanist influences and the production of knowledge in the histories of flamenco. We discovered how our training shapes systems of organizing the body and its vocabulary. We recognized the normalization of practices that render invisible hierarchies of power across centuries and how that process plays out in the studios of academia. We related the construction of bodies in dance to other discursive practices, like systems of government and the policing of its citizenry, not only in the colonization of the Caribbean, but also our daily lives in Kansas.

Lo Que Queda/That Which Remains is in dialogue with the archive of written histories, dance manuals and music notation, recordings, images, oral traditions and the transmission of knowledge from teachers to students over generations. As with any performance, the dance is reflective of and constitutive of the conditions that produce/are produced by it. While we may never be able to represent the performative knowledge of the past, the reimagining of these roots has value in the present. We honor those bodies erased by history and recognize how we choreograph culture. Our practice may transform our performance. *Y que las raíces vuelen/*And our roots make us fly...

NOTES

1. From "*Hacía el mar*/To the Sea," in *Diario de un poeta recién casado/Diary of a Recently-Married Poet* (Madrid: Calleja, 1916), 22.
2. Fermín Lobotón, "Alas y Raices: la inspiración andaluza de Juan Ramón Jiménez," *Diario de Cadiz 150*, December 15, 2008, accessed February 16, 2017. http://www.diariode cadiz.es/ocio/Raices-Alas-Juan-Ramon-Jimenez_0_214178874.html.
3. "*Romnia*, About the Project," website for Belén Maya, accessed May 31, 2017, http://belenmaya.com/?p=1656&lang=en
4. "Choreographing Contemporaneity: Cultural Legacy and Experimental Imperative," in *Flamenco on the Global Stage: Historical, Critical and Theoretical Perspectives*, ed. K. Meira Goldber, Ninotchka Bennahum and Michelle Heffner Hayes (Jefferson, NC: McFarland, 2015), 280–291.
5. In the interest of plotting histories and their resonances in the current moment, I think the definitive article on this topic, frequently referenced or refuted, is Mark Franko's "Repeatability, Reconstruction and Beyond," *Theatre Journal* 41.1 (March 1989): 56–74.
6. In an interview with Victor Ginesta in *Barcelones*, Lérida, who is the founder of a movement called Flamenco Empírico (Empirical Flamenco), describes, in the training of the body, "an empirical component in daily practices." "Juan Carlos Lérida: El aura de autenticidad en flamenco es falsa," February 12, 2016, accessed May 31, 2017. http://barcelones.com/cultura/juan-carlos-lerida-el-aura-de-autenticidad-del-flamenco-es-falsa/2016/02/.
7. Claudia Jeschke, "Notation Systems as Texts of Performative Knowledge," *Dance Research Journal* 31.1 (Spring 1999): 4–7.
8. "Hispanomania in Nineteenth Century Dance Theory and Choreography," in *Flamenco on the Global Stage: Historical, Critical and Theoretical Perspectives*, ed. K. Meira Goldberg, Ninotchka Devorah Bennahum, and Michelle Heffner Hayes (Jefferson, NC: McFarland, 2015), 100–101.
9. Yvonne Daniel, *Dancing Wisdom: Embodied Knowledge: in Haitian Voudou, Cuban Yoruba, and Bahian Candomblé* (Urbana: University of Illinois Press, 2005), 1.
10. Susan Foster, *Reading Dancing: Bodies and Subjects in Contemporary American Dance* (Berkeley: University of California Press, 1986), 3.
11. Susan Foster, *Choreographing Empathy: Kinesthesia in Performance* (New York: Routledge, 2011), 2.
12. *Bulerías* is a fast-paced festive flamenco form often performed improvisationally

in party settings or as part of an encore for a professional performance. It is one of the core song forms in the flamenco repertory.

13. *Bulerías* workshop with Juan Paredes, Viva Social Dance Studio: KC's Premier Salsa/Latin Dance School, Mission, KS, May 30, 2015.

14. Raúl Rodríguez, *Razón de Son*, Mp3 (Madrid: Fol Música, 2014).

15. Rafael Manjavacas, "Interview with Raúl Rodriguez, 'Razón de Son,'" February 9, 2015, accessed February 16, 2017, https://www.deflamenco.com/revista/entrevistas/interview-with-raul-rodriguez-razon-de-son-1.html

16. Michelle Heffner Hayes, "Desiring Narratives: Flamenco in History and Film," in *Flamenco: Conflicting Histories of the Dance* (Jefferson, NC: McFarland, 2009), 29–52.

17. K. Meira Goldberg, "Sonidos Negros: On the Blackness of Flamenco," *Dance Chronicle* 37, no. 1 (January 2014): 85–113. Also listed in the "Corrigendum," *Dance Chronicle* 38, no. 1 (January 2015): 125: Kathy Milazzo, "The Tango de Negros in Spain's Romantic Age: Lost in Translation," in Proceedings from Society of Dance History Scholars Conference (Guildford, UK, July 2010); Kathy Milazzo, "The Negro in Flamenco Dance," Congress on Research in Dance Conference, Albuquerque, NM, 2012; Santiago Auserón, *El ritmo perdido* (Barcelona: Ediciones Peninsula, 2012); Alberto del Campo Tejedor and Rafael Cáceres Feria, *Historia cultural del flamenco (1546–1910): El barbero y la guitarra* (Cordoba: Almuzara, 2013).

18. Goldberg "Sonidos Negros"; Milazzo "The Black African in Spain's Romantic Age," "The Negro in Flamenco Dance."

19. Joachim-Ernst Berendt, and Günther Huesmann, *The Jazz Book: From Ragtime to the 21st Century* (Chicago: Chicago Review Press, 2009), 64, ProQuest Ebook Central.

20. Rodríguez discusses his early influences and the development of the flamenco *tres* in an interview with the Festival Etnosur, July 3, 2014, accessed July 8, 2017, https://www.youtube.com/watch?v=7-vbtzzJjQg.

21. Julienne Gage, "Gema y Pavel in Miami Beach," *Miami New Times*, January 17, 2008, accessed July 5, 2017, http://www.miaminewtimes.com/music/gema-and-pavel-in-miami-beach-6332821

22. After the dance was performed, I sent Rodríguez a video link to the final choreography. We maintained a correspondence, and were able to meet again in person at the Spaniards, Natives, Africans, and Gypsies: Transatlantic Malagueñas and Zapateados in Music, Song, and Dance conference, co-sponsored by Center for Iberian and Latin American Music (CILAM), UCR Department of Music, and the Foundation for Iberian Music at the Barry S. Brook Center for Music Research and Documentation, The Graduate Center, The City University of New York, in Riverside, CA, April 6–7, 2017. He has seen the presentation about the work that preceded this chapter.

23. Literally, "the ass of the world." Personal conversation, February 9, 2007, author's translation.

24. Some excellent sources in this debate include recent examples like Anthea Kraut's *Choreographing Copyright: Race, Gender and Intellectual Property Rights in American Dance* (Oxford: Oxford University Press, 2016), *The Oxford Handbook of Dance and Theatre* (Oxford: Oxford University Press, 2015), edited by Nadine George-Graves, which has several chapters that deal with the politics of cultural appropriation, Cindy Garcia's *Salsa Crossings: Dancing Latinidad in Los Angeles* (Durham: Duke University Press, 2013) and Priya Srinivasan's *Sweating Saris: Indian Dance as Transnational Labor* (Philadelphia: Temple University Press, 2011).

25. Hayes, "Desiring Narratives," 29–52, Please note that my own discussion mentions, but does not comment extensively, on the Africanist presence in flamenco. Also, the use of terminology to refer to the Roma people has changed in recent years. At the time of publication, I translated *gitano* directly to English as "gypsy," and used the term interchangeably with Roma. Currently, scholars use the terms Roma and Gitanos to refer to the Spanish Roma.

26. Thomas F. DeFrantz did the world of dance studies a great service in editing *Dancing Many Drums: Excavations in African American Dance, Studies in Dance History Series* (Madison: University of Wisconsin Press, 2002). His introduction, "African American Dance: A Complex History," provides a valuable historiography of scholarship in race and dance studies, 3–35.

27. Yvonne Daniel, *Caribbean and Altantic Diaspora Dance: Igniting Citizenship* (Urbana, Chicago and Springfield, IL: University of Illinois Press, 2011); Juliet McMains *Spinning Mambo into Salsa: Caribbean Dance in Global Commerce* (Oxford: Oxford University Press, 2015).

28. "'Just Like Being at the Zoo': Primitivity and Ragtime Dance" in *Ballroom, Boogie, Shimmy, Sham, Shake: A Social and Popular Dance Reader*, ed. Julie Malig (Urbana: University of Illinois Press, 2009), 63.

29. For an example of how this model works in activist theatre, see Jessy Ardern's "Forum Theatre and the Power of Yes, and..." in *Staging Social Justice: Collaborating to Create Activist Theatre*, ed. Norma Bowles and Daniel-Raymond Nadon (Carbondale: Southern Illinois University Press, 2013), 219–224. I want to thank my colleagues and teachers, Emily Gullickson and Cody Charles, for their labor in modeling this form in their Social Justice Fellows seminar at the University of Kansas.

30. This is not an exhaustive list.

31. I am indebted to Niurca E. Márquez for introducing me to the methodology of Practice as Research (PaR). Her master's thesis, "The Collage as Cartography: A Methodology for Choreographic Composition in a Practice as Research Framework," (Jacksonville University, April 2017) educated me as her advisor.

32. Estelle Barrett, *Practice as Research: Approaches to Creative Arts Inquiry*, ed. Estelle Barrett and Barbara Bolt (London: I.B.Tauris, 2010), 2.

33. Barrett, *Practice as Research*, 2.

34. In order to give credit to my teachers of African diaspora forms, including *baile popular*/Latin popular dances, I list them here: Willie Lenoir, Bernard Johnson, Liliana Valle and members of Curubande, Carmen Amalia Reina de Nelson and members of Grupo Macondo, Amaniyea Payne, Rosângela Silvestre, Richard Rodriguez, Blanche Brown, Baba Chuck Davis, Rennie Harris, Jawole Willa Jo Zollar, Charmaine Warren, and Niurca Márquez.

35. Brenda Dixon Gottschild, *Digging the Africanist Presence in American Performance: Dance and Other Contexts* (Westport, CT: Greenwood Press, 1996).

36. Robert Farris Thompson originally identified the "get down" posture, as well as several Africanist elements in the arts in "An Aesthetic of the Cool II," *African Arts* 7.1 (Autumn 1973), pp. 40–43, 64–67, 89–91. Kariamu Welsh Asante further developed these Africanist dance elements in "Commonalities in African Dance: an Aesthetic Foundation," in *African Cultures: the Rhythms of Unity*, ed. Molefi Kete Asante and Kariamu Welsh Asante, 71–82 (Westport, CT: Greenwood Press, 1985). Gottschild extended this analysis to American concert dance in 1996. Brenda Dixon Gottschild, *Digging the Africanist Presence in American Performance: Dance and Other Contexts* (Westport, CT: Greenwood Press, 1996).

37. For theories and historical mapping of the migrations of the fandango, please see the essays in *The Global Reach of the Fandango in Music, Song and Dance: Spaniards, Indians, Africans and Gypsies*, ed. K. Meira Goldberg and Antoni Pizá (Newcastle upon Tyne: Cambridge Scholars, 2017).

38. Daniel, *Caribbean and Atlantic Diasporic Dance*, 98.

39. I am deeply grateful for the bravery of these dancers in accepting the challenge of learning, in trusting me with their vulnerability, and for performing the work. They are Jillian Armstrong, Christine Bessey, Ana Glocker, Briana Herrington, Tori Hilger, Brennon Madrid, Alyssa Rivera, and Kayla Wegley.

40. I should recognize that there are several arguments about repetition and dancing, but here I am referring to establishing what dancers call "muscle memory" as explained by Sally Sevey Fitt in *Dance Kinesiology* (Schirmer Books: New York, 1996), p. 134. I'm not claiming that "rote" repetition is the key to mastery, but good practice is essential to learning. I would argue that also there are several other cognitive factors in place, including the cultural context of various dance forms.

41. Gottschild, *Digging the Africanist Presence in American Performance*, 11.

42. The Department of Dance at the University of Kansas now accepts hip hop and rhythm tap as technique courses for the degrees in dance.

43. Since the premiere of the work, our department has revised its technical requirements, the language we use to describe etiquette, and recruitment efforts. These changes involve

risk to a small program in a university. Fewer people are taking ballet classes when hip hop "counts" as a technique course. Dancers who graduate will have a different training profile than in previous generations. The faculty have committed to an ongoing effort to de-center the values of the program.

44. Cody Charles, "#WhyYouAlwaysLyin," *Reclaiming Anger: I Am the Rage That Baldwin Speaks of* (blog), September 29, 2016, accessed June 29, 2017, https://medium.com/reclaiming-anger/tagged/other

45. Judith Butler, *Bodies That Matter: On the Discursive Limits of "Sex"* (New York: Routledge, 1993), 3.

BIBLIOGRAPHY

Ardern, Jessy. "Forum Theatre and the Power of Yes, and…." In *Staging Social Justice: Collaborating to Create Activist Theatre*, eds. Norma Bowles and Daniel-Raymond Nadon, 219–224. Carbondale: Southern Illinois University Press, 2013.

Asante, Kariamu Welsh. "Commonalities in African Dance: an Aesthetic Foundation." In *African Cultures: The Rhythms of Unity*, eds. Molefi Kete Asante and Kariamu Welsh Asante, 71–82.Westport, CT: Greenwood Press, 1985.

Auserón, Santiago. *El ritmo perdido*. Barcelona: Ediciones Peninsula, 2012.

Barrett, Estelle. Introduction to *Practice as Research: Approaches to Creative Arts Inquiry*, eds. Estelle Barrett and Barbara Bolt, 1–14. London: I.B. Tauris, 2010.

Berendt, Joachim-Ernst, and Günther Huesmann. *The Jazz Book: From Ragtime to the 21st Century*. Chicago: Chicago Review Press, 2009. ProQuest Ebook Central.

Butler, Judith. *Bodies That Matter: On the Discursive Limits of "Sex."* New York: Routledge, 1993.

Charles, Cody. "#WhyYouAlwaysLyin." *Reclaiming Anger: I Am the Rage That Baldwin Speaks of* Blog. September 29, 2016. Accessed June 29, 2017. https://medium.com/reclaiming-anger/tagged/other.

Daniel, Yvonne. *Caribbean and Altantic Diaspora Dance: Igniting Citizenship*. Urbana: University of Illinois Press, 2011.

_____. *Dancing Wisdom: Embodied Knowledge: in Haitian Voudou, Cuban Yoruba, and Bahian Candomblé*. Urbana: University of Illinois Press, 2005.

DeFrantz, Thomas F. "African American Dance: A Complex History." In *Dancing Many Drums: Excavations in African American Dance, Studies in Dance History Series*, ed. Thomas F. DeFrantz, 3–35. Madison: University of Wisconsin Press, 2002.

Fitt, Sally Sevey. *Dance Kinesiology*. New York: Schirmer Books, 1996.

Foster, Susan. *Reading Dancing: Bodies and Subjects in Contemporary American Dance* Berkeley: University of California Press, 1986.

_____. *Choreographing Empathy: Kinesthesia in Performance*. New York: Routledge, 2011.

Franko, Mark. "Repeatability, Reconstruction and Beyond." *Theatre Journal* 41.1 (March 1989): 56–74.

Gage, Julienne. "Gema y Pavel in Miami Beach." *Miami New Times*, January 17, 2008. Accessed July 5, 2017. http://www.miaminewtimes.com/music/gema-and-pavel-in-miami-beach-6332821.

George-Graves, Nadine. "'Just Like Being at gthe Zoo': Primitivity and Ragtime Dance." In *Ballroom, Boogie, Shimmy, Sham, Shake: A Social and Popular Dance Reader*, ed. Julie Malnig, 55–71. Urbana: University of Illinois Press, 2009.

_____, ed. *Oxford Handbook of Dance and Theatre*. Oxford: Oxford University Press, 2015.

Ginesta, Victor. "Juan Carlos Lérida: El aura de autenticided de flamenco es falsa." *Barcelones*. February 12, 2016. Accessed May 31, 2017. http://barcelones.com/cultura/juan-carlos-lerida-el-aura-de-autenticidad-del-flamenco-es-falsa/2016/02/.

Goldberg, K. Meira. "Corrigendum." *Dance Chronicle* 38, no. 1 (January 2015): 125.

_____. "Sonidos Negros: On the Blackness of Flamenco." *Dance Chronicle* 37, no. 1 (January 2014): 85–113.

_____, and Antoni Pizá, eds. *The Global Reach of the Fandango in Music, Song and Dance: Spaniards, Indians, Africans and Gypsies*. Newcastle upon Tyne: Cambridge Scholars, 2017.

Gottschild, Brenda Dixon. *Digging the Africanist Presence in American Performance: Dance and Other Contexts*. Westport, CT: Greenwood Press, 1996.
Heffner Hayes, Michelle. "Choreographing Contemporaneity: Cultural Legacy and Experimental Imperative." In *Flamenco on the Global Stage: Historical, Critical and Theoretical Perspectives*, eds. K. Meira Goldberg, Ninotchka Bennahum and Michelle Heffner Hayes. Jefferson, NC: McFarland, 2015.
_____. *Flamenco: Conflicting Histories of the Dance*. Jefferson, NC: McFarland, 2009.
Jeschke, Claudia, with Robert Atwood. "Hispanomania in Nineteenth Century Dance Theory and Choreography." In *Flamenco on the Global Stage: Historical, Critical and Theoretical Perspectives*, eds. K. Meira Goldberg, Ninotchka Devorah Bennahum, and Michelle Heffner Hayes, 95–102. Jefferson, NC: McFarland, 2015.
_____. "Notation Systems as Texts of Performative Knowledge," *Dance Research Journal* 31.1 (Spring 1999): 4–7.
Jiménez, Juan Ramón. "Hacía el mar/To the Sea." *Diario de un poeta recién casado/Diary of a Recently-Married Poet*. Madrid: Calleja, 1917.
Kraut, Anthea. *Choreographing Copyright: Race, Gender and Intellectual Property Rights in American Dance*. Oxford: Oxford University Press, 2016.
Lobotón, Fermín. "Alas y Raices: la inspiración andaluza de Juan Ramón Jiménez." Diario de Cadiz. December 15, 2008. Accessed February 16, 2017. http://www.diariodecadiz.es/ocio/Raices-Alas-Juan-Ramon-Jimenez_0_214178874.html.
López Bustamante, Joaquín. "Romnia, About the Project." Website for Belén May. Accessed May 31, 2017. http://belenmaya.com/?p=1656&lang=en.
Manjavacas, Rafael. "Interview with Raúl Rodriguez, 'Razón de Son.'" February 9, 2015. Accessed February 16, 2017. https://www.deflamenco.com/revista/entrevistas/interview-with-raul-rodriguez-razon-de-son-1.html.
Márquez, Niurca E. "The Collage as Cartography: A Methodology for Choreographic Composition in a Practice as Research Framework." Master's thesis, Jacksonville University, April 2017.
Milazzo, Kathy. "The Black African in Spain's Romantic Age: Negotiations of Identity." 2010. Proceedings of the Society of Dance History Scholars Conference, Guildford, UK, July 9–11, 2010.
_____. "The Negro in Flamenco Dance." Paper presented at the Congress on Research in Dance Conference, Albuquerque, NM, November 8–11, 2012.
Paredes, Juan. *Bulerías* workshop. Viva Social Dance Studio, KC's Premier Salsa/Latin Dance School, Mission, KS, May 30, 2015.
Paz, José Luis de la. Personal conversation, February 9, 2007.
Rodríguez, Raúl. *Razón de Son*. Mp3. Madrid: Fol Música, 2014.
_____.YouTube interview with Festival Etnosur, July 3, 2014. Accessed July 8, 2017. https://www.youtube.com/watch?v=7-vbtzzJjQg.
Srinivasan, Priya. *Sweating Saris: Indian Dance as Transnational Labor*. Philadelphia: Temple University Press, 2011.
Tejedor, Alberto del Campo, and Rafael Cáceres Feria. *Historia cultural del flamenco (1546–1910): El barbero y la guitarra*. Cordoba: Almuzara, 2013.
Thompson, Robert Farris. "An Aesthetic of the Cool II." *African Arts* 7:1 (Autumn 1973), 40–43, 64–67, 89–91.

Screaming Soundscapes
The Sounds of Puerto Rican Contemporary Performance in the Work of Teresa Hernández and Ivette Román

Lydia Platón Lázaro

> Writing should erase itself before the plenitude of living speech, perfectly represented in the transparency of its notation, immediately present for the subject who speaks it, and for the subject who receives its meaning content, value.
> —Derrida, "Semiology and Grammatology"

"*Siento una voz que me dice agúzate...*" (I hear/feel a voice that says: watch out!) warns the classic salsa song by Richie Ray and Bobby Cruz, alluding to the sense or maybe even the intuition of the presence of spirits, or something unexplainable. How can we best make sense of sensations, intuitions, even premonitions that exist both as body and as language? Must that inner voice, indeed a part of our bodies, partake in logocentric reasoning? Can it only be accounted for using the logic of the written word, inherited from the European colonizers of the Americas? Can language accurately represent sensation, feeling, and experience, or is its logic necessarily one step removed from embodied experience? This voice that is "felt" in the example of "*Agúzate*," part of Puerto Rican popular music culture, salsa, which itself is the result of multiple diasporic relations between voice, body, and language.[1] *Agúzate*—watch out!

In this essay, I listen to and observe soundscapes that attempt to escape the fixed regimes of cultural representation that have negatively marked Caribbean subjects through analyses of two experimental contemporary performers: Ivette Román and Teresa Hernández.[2] In a myriad of ways, both these artists "voice" these obscure silences of oppression on stage. Román creates her own register of sounds and an original use of the vocal apparatus, the actual organs involved to emit sound, while Hernández creates characters whose voices may find themselves in the register of the scream as well as in the discourse of post-colonialism. In considering both artists' deformations and transformations of sound, I will examine intersections of voice and body, voice and language, and language and word to illuminate issues related to "staging" the other from the point of view of the "othered." Both artists have been active in Puerto Rico and, to a limited extent, internationally since the 1980s, and continue to explore uses of the voice and the body as means of articulating freedom from "othering" dynamics that are the result of the double colonialism of Puerto Rico: first by the Europeans, and second by American imperialists.

These two Puerto Rican performance artists create soundscapes of their island realities as they express their passion for formal experimentation within the world of independent stage practice. This voice-centered work results, in the case of Ivette Román, in an original genre that is in dialogue with "singing" and "performance art," but that cannot be easily categorized. It is also close to long-practiced Caribbean improvisational intonations of sound, legacies of the tonalities brought from Africa in the voices of enslaved people. Román also experiments with the sonic representation of psychologically impacted sound, which renders a soundscape that embodies a politics of resistance long expressed in Caribbean popular musical and theater forms. Teresa Hernández, on the other hand, uses a more traditional style of theatrical experimentation to create characters defined by their vocal registers. Thus we can *hear* imperialism, in the form of Puerto Rican accents and tonalities, including the voices of migration. Hernández combines these characters' utterances with movement, drawing from her long career as a contemporary dance artist.

Puerto Rico, the Caribbean and the Sound of Foundational Paradigms

Traditionally, logocentrism as a "higher" form of knowledge tends to place the embodied voice as a lower form in terms of hierarchical language relations, where the body and voice need to be separated in order to account for the sensation in language. Sensation is being used here as equal to feeling, not necessarily easily accounted for in speech. How do we word what our senses perceive, the hairs standing on the back of our necks, or the chills that announce a coming illness? Now, imagine not being accorded the humanity to perceive the insight of your senses, because your humanity itself is in question. As ethnomusicologist Ana María Ochoa points out: "The relation between the voice and the ear then implies a *zoé*, a particular notion of life that involves addressing different conceptions of the human and the boundaries between the human and nonhuman. In the colonial context of the Americas, where peoples from different places came together, such a definition of life through the voice was certainly a contested political issue."[3] Because Caribbean discourse was first an oral discourse, it was considered "lesser" by Europeans, who privileged writing over oral expression in their valuation of cultural production. Many Anglophone and Francophone Caribbean scholars and artists have developed groundbreaking theories exploring the effects of this forced primacy of the written word, which had a traumatic effect on Caribbean cultures,[4] especially as it attempted to silence the embodied word, for example in attempts to suppress both the use and the value of the creoles spoken on many islands, as well as the cultural performances of dance, singing, and music, to name only a few of the Caribbean performative arts that exist within the context of a system of double-standards. On one hand, many Caribbean performative expressions of sound and movement are praised by non–Caribbean audiences and critics, even revered, for their originality and virtuosity, but on the other, they are considered raw, uncultivated, and accidental, and hence are under-represented as forms of vital art production in the islands of the archipelago when it comes to the Caribbean arts being part of a wider artistic scene outside of the Caribbean region.

The word borne of embodied experience, likened to the use of the voice in performance, is in fact the "witness-cry," the screaming of the voices that came together[5] mostly in brutal form, in this geography of islands that "repeat themselves"[6] in their histories of conquest, slavery, and forced diaspora, as well as carnival. The texts bearing witnesses to these vocal encounters were not originally written by those conquered and enslaved, but rather conjured by the conqueror as travel narratives, conquest chronicles, clinical observations, and even the written law that would govern and enforce an order of submission, a petition for "silence."[7] In these pages, Caribbean peoples appear romanticized, infantilized, and/or grossly deformed.[8] Such representations, with their agenda of cultural domination, are still visible in many of the ways that countries like the United States enact capitalist imperialism today, and are especially recognizable in media coverage of invasions, hurricanes, and migration policies, as well as in the discussion of allies and enemies. Such accounts are often told from the perspective of only one voice, and hence cement the representation of Caribbean realities to such an extent that resistance necessitates screaming over the noise of a domination that fans the flames of racism, classicism, and lack of opportunity.

Although the language spoken in Puerto Rico is Spanish, with aspirations from political leaders for full English/Spanish bilingualism, language politics have changed intermittently depending on the governments in power dating back to the 1898 invasion of Puerto Rico by the United States.[9] These politics have ranged from English as the language of instruction in the first half of the twentieth century, to the public policy of Spanish as the official language in the early 90s,[10] to the promulgation of both. Issues of language and language use are still subjects of ideological debate in contemporary Puerto Rico, while younger generations, depending on their social class and connections with family members living in the United States, glide freely between globalized Internet- and cable TV-acquired English and Puerto Rican Spanish. The dynamics of the oral and the written do not take shape around Spanish and/or English versus a creolized tongue, as in many of our neighboring islands, but rather in relation to the esteem awarded to form and language in literature, theater, and performance. It is artistically both a weapon and a tool to be able to render the local context through a critical use of sound and speech in performance. Although it is not within the scope of this study, there are also interesting linguistic experiments visible in the work of Puerto Ricans living in the diaspora, especially in the form created by Nuyorican Poerty, and later in other cities with rich cultural venues, such as Chicago. These word performers play with the notions of sounds and meaning born from their Spanglish imaginaries.[11]

To further contextualize the concept of sound in a contemporary performance framework in the Caribbean, it is useful to understand the complexity of our geographical region, as Edouard Glissant explains it: "The Caribbean, the Other America. Banging away incessantly at the main ideas will perhaps lead to exposing the space they occupy in us. Repetition of these ideas does not clarify their expression; on the contrary, it perhaps leads to obscurity. We need those stubborn shadows where repetition leads to perpetual concealment, which is our form of resistance."[12] These ideas have helped to create an interdisciplinary vision of artistic and cultural practice that has accounted for a particularly embodied relationship to history and memory emerging from the realities of slavery, migration, and colonization, which have, together, generated a psychic master of souls,[13] and hence, of sounds.

One such canonical example can be found in Antonio Benítez Rojo's *The Repeating Island*, specifically referring to the sound of these Caribbean-based dynamics: "In any case, the impossibility of being able to assume a stable identity, not even the color marked on one's skin, can only be reconstructed in being a 'certain way,' in the midst of the noise and the fury of the chaos."[14] The Caribbean soundscape as counter rhythm[15] is something at which contemporary performance excels, while existing within chaos, stridency, and noise. Benítez Rojo adds: "el texto caribeño muestra los rasgos de la cultura supersincrética de donde emerge. Es sin duda un consumado *per-*

former que acude a las más aventuradas improvisaciones para no dejarse atrapar por su propia textualidad"[16] (the Caribbean text shows the traces of the supersynchretic culture from whence it emerges. Without a doubt, a consummated performer who uses the most adventurous improvisations in order to not be trapped by his own textuality). In that cross between voice and body, voice and language, and language and word, multiple paths of interpretation become available for the field of performance art. It is possible to consider this work from the perspective of cultural legacy, but also, and perhaps less obviously, there exists the possibility of creating art forms that serve the individual artistic imagination without regard for disciplinary rules, resulting in art with a profound political engagement that is based in embodied Caribbean histories. While trained in music or dance, for example, the artist may flow between genres, focusing on the soundscape as a raw articulation (closer to the body) as well as a complex literary play on words (closer to writing).

The performances created by both Román and Hernández propose critically engaged content in experimental form to address inequities of gender, political corruption in Puerto Rico, and personal situations that arise out of being artist-subjects on this island. They touch on themes such as personal relationships, class inequities, and the impossibility for artists to find paid work, among other ruminations. Sound and vocal experimentation feed off the minimalist exploration of sound itself, in the case of Ivette Román, and between fragmentation, interruption, and silence. Teresa Hernández creates a theatrical framing of the polarities of sound and language, as described by Caribbean performance scholar and critic Lowell Fiet: "While exploring the anxieties of Puertorrican daily life in general, her actions and movements theatrically inquire about the transversality of issues, like gender, class, and race within the colonial condition, social and domestic violence, ethnic, linguistic, and cultural identity, and even post-modern esthetics and intellectual utterances as elitist and assimiliationist."[17] There is a conversation taking place in these artists' work between Benítez Rojo's concept of "a certain way" and the choice of the universe of sound as a sonic reiteration of the distorted soundscapes issuing from and within Puerto Rico.

Another approach to maximizing the potential of the topic of the sound universe of performance as a possibility for liberating strategies is proposed by José Esteban Muñoz as "disidentification"[18] Although Muñoz's work is not specifically centered on the sound universe, he analyzes artists who perform personally engaged monologues, such as Cuban-American artist Carmelita Tropicana. Muñoz proposes the potential of disidentification to destabilize dominant discourses in a variety of ways, including the transformation of the public sphere itself in what he calls "sub-cultural circuits"[19] In this way, power-voice is granted back to the artist in the creation of a "counter-public" sphere. The work of Román and Hernández, using a polyphony of texts and movements in a soundscape that is recognizably from the Spanish-speaking Caribbean, is also a reflection of disidentification from its very source. The Puerto Rican soundscapes of Román and Hernández are intentionally presented where small audiences convene to see live art, venues usually defined by the audience's acceptance of experimentation. This reality also aspires to be a route for connectivity beyond the island geography, informed by many sources that combine high-culture and popular-culture approaches to sound, theatricality, and movement simultaneously.

The relationship between voice and language lends itself to infinite explorations. Artists like Román and Hernández, who choose to privilege vocal experimentation as an essential part of their message, allow spectators to listen to words with which we are familiar as part of colonial discourse while freeing us from the supposed supremacy of this discourse via the way these words sound, are played with, or are communicated through the body. The voice I reference here, the one that I imagine as a "scream," stems from the sound of something that is meant to be heard as raw sound unmediated by technique. The best description of such a sound can be found in Roland Barthes' notion of the "grain" of the voice, a word that conjures materiality and texture. Moreover, African American poet Nathaniel Mackey develops the useful concept of *"el duende"* in his analysis of Anglo-Caribbean writers Wilson Harris and Kamau Brathwaite, used originally by Spanish playwright and poet Federico García Lorca to describe the voice of the Flamenco singer, when a focus on technical skill is released in order to give texture to the voice: "The duende is both an omen and a goad. It insists upon the insufficiency, the essential silence of mere technical eloquence, stretching the singer's voice to the breaking point. The pursuit of a meta-voice, of an acknowledged and thus more authentic 'silence' beyond where conventional elocution leaves off, this impoverishment or tearing of the voice, corresponds to what Harris, quoting the Barbadian poet Kamau Brathwait, refers to as 'tunelessness,' the essential condition of the Caribbean's orchestra of deprivation."[20] This particular form of embodiment, then, is what I am proposing as a Caribbean strategy when joined with Edouard Glissant's theories of relation: "What is Caribbean in fact? A series of relationships. We all feel it, we express it in all kinds of hidden or twisted ways, or we fiercely deny it. But we sense that this sea exists within us with its weight of now revealed islands. The Caribbean Sea is not an American Lake. It is the estuary of the Americas."[21] What Glissant refers to as "hidden or twisted ways" relates both to the "certain way" exposed by Antonio Benítez Rojo and to García Lorca's *"duende,"* while also reflecting a strategy of disidentification.

Debate around orality also "gives faith" to other kinds of embodiments that are silenced, very similar to the way in which creoles and theories of creolité give faith to screams and the chiasms of the voice, especially prevalent in Caribbean poetry. The chiasm represents a space of possibility, in the way that it has been conceived by Glissant according to J. Micahel Dash: "In contrast to the cataloged, monolingual, monochrome world that Glissant identifies with Europe, New World landscape offers the creative imagination a kind of metalanguage in which a new grammar of feeling and sensation is externalized."[22] Therefore, voices that are working through words work also with tone, stridency, and emotional expression in order to present a Caribbean subject whose voice contains the colonial trace that documents language which has been overloaded with experience and memory. Despite being conscious of this fact, however, audiences still seem unaccustomed to receiving sound in itself as meaning, continuing to look, rather, for the established codes of the written word and of European music. This issue relates to the problematic question of "staging" the other and the desire for success that artists need in order for their work to be shared widely across cultures. It is both liberating and limiting.

Creole languages, even in my metaphorical use of the term, are considered one of the major witnesses to the silencing of the Caribbean body. The relationship between voice and body in the example of creole reveals the conceptual abyss that still exists today in the ongoing insistence on a voice that is separated from the body, in a privileging of written over oral cultural and ephemeral forms such as performance, which results inevitably in the stigmatization of Caribbean populations and their expressions. Quoting Ochoa:

> The politics of regimentation of the voice are also multiple and often show us how the body and the voice do not necessarily coincide (Connor 2000; Weidman 2006). To the contrary, voices have the potential to disembody themselves into objects as in ventriloquism (Connor 2000), to travel between human and nonhuman entities as when animals teach humans songs (Seeger 1987), to incarnate other worldly beings in a body of this world as in rites of "possession" (Matory 2005), or are presumed to represent an autonomous or unique individual, as in the predominant Western philosophical political tradition (Weidman 2011). Hearing voices thus frequently invokes the need to ontologically address implied questions about the cosmologies (Schmidt 2000) or the ear (Steege 2012) and the definition of life they bring forth.[23]

In the case of Puerto Rico, the adoption of Spanish as a main language inserted us heavily within a literary tradition of the word (logos), and not so obviously within the traditions of orality, of creolized languages, evident on other islands. We express our relationship to (creolized and imperial) language through the way we use tone, volume, and word play, through the powerful embodiment of sound exemplified in the work of the artists I am discussing here, as well as through our African-diasporic music. Literary critic Juan Gelpí points out that the literary vocation in Puerto Rico also exhibits a paternalist and masculine predominance, and that this attitude still permeates our reception of colonial traces in our literature, theater, and performance. It is also, then, part of the vocation of Román and Hernández to read and interpret differently in their particular soundscapes.

Embodying Sound: Vocal Technique and Freedom in the Work of Ivette Román

Ivette Román recognizes the use of all the possible references within her reach to deconstruct them, note by note. This approach results in use of alliteration, rhythm patterns that insist on repetition and counter-rhythm. In the following example, rhyme schemes popular in Puerto Rican nursery rhymes set the tone for an apocalyptic message dedicated to the island of Vieques, where the U.S. Navy held military exercises for more than 60 years, until they were evicted by civilian protest around the year 2000:

> La tierra guerra y yo paz paz pus gas rasga la cara tez te tú tití se fue a la
> Que dolor que dolor que pena doña Ana acusa la musa qué excusa excuseme señor que está en lo cierto Celestes pestes el lente observador del cosmos y el colmo cólmame de tu espíritu ceño irritable
> Estables estables estallidos idos y yo paz paz pus gas y tú...[24]
> (The earth, war and I, peace peace pus, gas tear the face of your auntie tantie who went to war, pain, pain, and sadness, so sad Doña Ana who accuses the muse that excuses, excuse me sir you are right celestial pestilences of the observing lens of the cosmos culminating in me with your spirit, marked irritably in your frown. Stable, stable, exploding explosions and I peace, pus, gas and you...)

In the performance of this text, Román used her voice to emulate the sound of war machines, characteristic of the military exercises performed in Vieques, which have been held accountable for the alarming number of cancer cases on the island. Again, a polyphony of text and sound creates parallel universes for the imagination that are at once highly political and aesthetically experimental.

At other times, Román returns to singing, in performances that can be considered concerts. In the case of *El hombre San Juan, Puerto Rico* (*The Man San Juan, Puerto Rico*) (1995), she engages in feminist protest of the sexist practices of many men in their attitudes towards women. This piece was performed as a blues along with other pieces that revolved around nostalgic storytelling, for example about childhood homes as a metaphor for that which has been lost to hurricanes.[25] In *Voces del Maleficio* (*Cursed Voices*) (1995), the venue was a formal theater space in San Juan where she also performed abstract experimentation using her voice as a cello, among many other sounds evocative of the contradictions of colonial experience, including the example of the mastery of technique and the breaking of the voice as raw emotion.

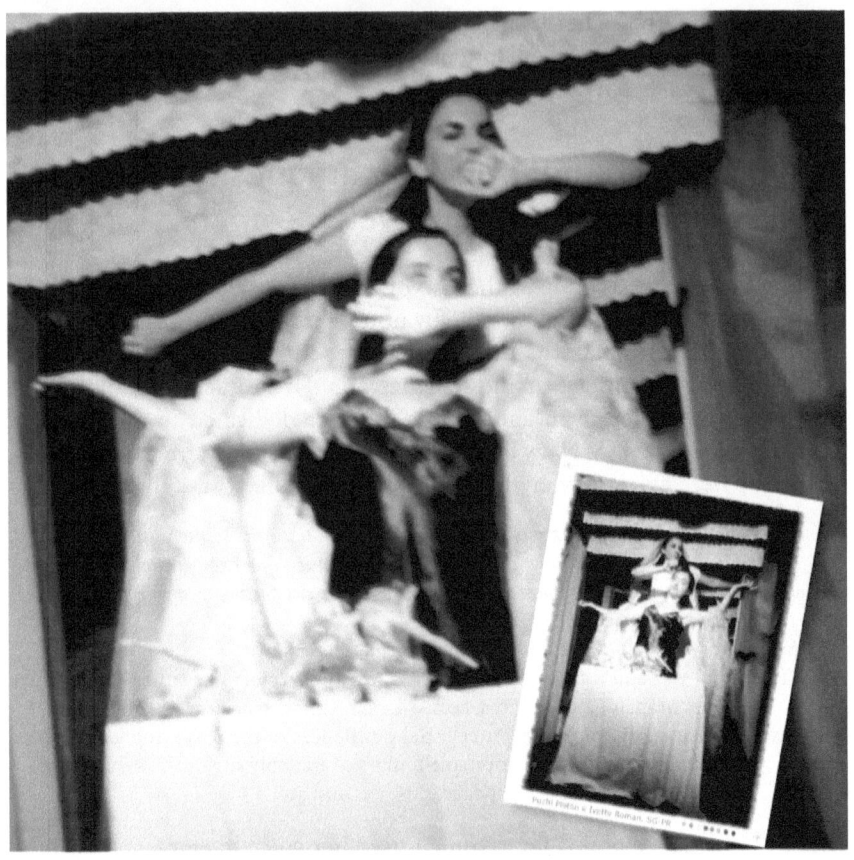

Ivette Román (top) and Lydia Platón in *Matropofagia*, a play written by Mayra Santos Febres (photograph courtesy Ana Rosa Rivera).

During this period, her signature sound was to regress from word play and storytelling in song to silence and pure sound. Here the silences scream of injustice, as a mirror of silenced resistance. When considering Román's work, as I do as representative of disidentified Caribbean performance, I am invoking the ideas of the aural so well explained by Benítez-Rojo as chaos, and by Glissant as the sound of Antillanité, as an exploded discourse: "The natural reaction of the freed body of the slave is the explosive scream, the excited gesture."[26] This aural sense of injustice accompanies Román's sound of the colonial contradictions of Puerto Rican politics.

Toma de Posesión (1997) is an example of political content that performs the work that protest songs have in the past, in our Latin American musical heritage. This piece is a critical comment on the second term of governor Pedro Rosselló (father of current governor Ricardo Rosselló), known for the illegal appropriation of government funds during his first term. Here, Román used a combination of chanting, singing, and an evocation of the collective memory of protest songs. Screaming, she presents desperation, along with the New Age philosophy that serves to further induce denial and resistance to action: she screams while saying, "Control yourself, Ivette." The vocalized internal voice is what makes this piece both hilariously funny and profoundly tragic at the same time. It refers clearly to the lack of control we have as Puerto Ricans in the face of the corruption we have experienced, which today has led to the humanitarian crisis of an unpayable public debt among many other socio-economic consequences.[27] Román chants, *"Nirvana, Osho, Yoga, Kundalini, creative visualization, candidatura independiente, Spanish Only, uye! Uye! Uye! de San Juan, Puerto Rico...."*

Román plays here with the use of Spanglish, as well as with the idea that the primal scream serves as a consciousness-building tool. The last refrain is taken from a protest song of the 1970s by Puerto Rican singer/songwriter Roy Brown, *"Mister con Macana"* ("Mister with a Club"), originally written to protest police violence. *"Huye"* literally means to run away, to escape. Román's use of the word, spelled without an h phonetically and orthographically, is a call to flee that takes the form of a battle cry.

This early work of Román evokes the rhythmic power of Caribbean poetry, where rhythm, counter rhythm, and word play join to form a unique expression of embodied sound. The artist says that her work stems in part from an interest in the voice as a musical instrument and all the possibilities thereof. Her ideas about existence, specifically regarding a power struggle within a power struggle, become the basis for a message about emptiness that emerges in pieces in which silences between noise and rhythmic segments become loud, uncomfortable reminders of her message.[28]

After 1999, under the influence of the Latin American political cabaret, there was a major shift in Román's work. Her sound began to come through characters, as an homage to performers like the recently deceased Antonio Pantoja, the first and most powerful queer icon of cabaret, song, and comedy in Puerto Rico. Román says the following about her cabaret work:

> Here, the word returns and so do explicit political comments. I am also innovating with Caribbean humor (before, the humor in my work was "under the table"), the use of costumes (this is the first time I am not working as Ivette, but as a character: La Diva PosApocalíptica (The PostApocalyptic Diva) and another more recent one, La Calandria Guaraguá de Manatí (The lark-hawk of the town of Manatí). It is also the first time, since my LAPD years that I collaborate with other colleagues (Amed Irizarri and Freddie Mercado),[29] and the first time that I interpret other people's songs (performers Ivy Andino and Eduardo Alegría, and writer Mayra Santos, among others).[30]

Román's cabaret mode of creating and performing her own texts as well as others' continued into the 2000s.

As a result of the economic reality of the island and the poor cultural infrastructure that exists there, Román was forced to move to the mainland United States in 2005, where she continued to perform her solo work in collaboration with experimental music artists in Houston, Texas. This new influence resulted in an even greater tendency towards the abstract, and a minimized focus on words. Almost as if left without voice, the body insists on articulating sound to evoke emotion, nature, and the lack of a reasonable approach to its pain. Despite the great dramatism that can be achieved using this approach, the work is also a peaceful exploration of breath, the key mode of connection between body, voice, and language.

Román returned to Puerto Rico ten years later and founded a sound collective, *La Sociedad Sonora del Amor* (*The Sound Society of Love*). She did not abandon her highly political performances, but rather attempted a more personal connection to loss and emptiness. She began to focus on training others, while also encouraging improvisation sessions in front of audiences. It was a way to keep this art form vital in times of even greater austerity on the island. As I write, she has had to leave Puerto Rico once again, and continues to collaborate with Houston-based artists with the intention of helping to keep the Puerto Rico project alive.

The Character Voice: Teresa Hernández's Throat as Medium

Teresa Hernández began her artistic career as a contemporary dancer. She was one of the original members of the collective formed by choreographer and movement artist Viveca Vázquez, *Taller de Otra Cosa*, literally translated as *The Workshop of Something Other*. She co-created pieces that in Puerto Rico were groundbreaking in the realm of dance-theater, including the use of video as a narrative force in choreographic pieces. This group was highly influenced by Vázquez's participation in the downtown dance scene of the 1970s and 80s.[31] Hernández has been central to the work of *Taller de Otra Cosa* in her writing and contributions to theatrical and conceptual processes that have appeared in their shows over the past 25 years. She began her solo performance career simultaneously and became an established figure in the experimental and academic arts scene. One of the most important characters she created around that time portrays how accents and tonalities shape the sound universe of the critique Hernández strived to make. The character of Isabella, a postmodern literary critic, was featured as part of a dance concert of *Taller de Otra Cosa* entitled *Kan't Translate: Tradúcelo* (1992), an exploration of the relationship between speaking bodies and the concepts of understanding and communicating via the moving body. The show highlighted language, the use of words and movement, and playing with silences to explore the absence and presence of language in its relationship to the body. The show also explored imperial and colonial connections embedded in language in Puerto Rico. The character of Isabella as a public intellectual rooted in the Hispanic critical traditions of Puerto Rican scholars was a parody of the way critics receive and perceive experimental work dedicated to deconstructing movement and its relationship to the body, language, and power.

Isabella's voice and use of body and language is a very powerful example of the way humor can also be used critically to address the particularities of Puerto Rican identity issues, while exploring the aesthetic limits of the voice and body. Hernández followed up with a site-specific show that featured new characters, as well as new video and dance pieces, titled *Acceso Controlado (Controlled Access)*. The name refers to the changes that swept over Puerto Rico in 1995[32] to address crime rates. Security guards popped up everywhere at the entrances to neighborhoods in a measure that contributed to perpetuating racism and class differences in neighboring communities that had previously interacted normally. Even housing projects became gated communities, shutting in poor populations, while in middle-class and affluent neighborhoods, the issue became one of who decided access.[33] This initiative occurred under Governor Pedro Rosselló, whom Román had criticized in *Toma de Posesión*, mentioned earlier.

Acceso Controlado was created and performed at Casa Aboy, a cultural center in an old art-deco house that used to belong to a prominent family. The show opened with audience interaction guided by an Afro-Puerto Rican police officer, in uniform but wearing rubber slippers as shoes, conjuring poverty and tropical informality; a European nanny, dressed in a suit from the fifties; and an androgynous secret service security guard whose face was covered with pantyhose. These characters, although very funny, were also very scary as they instructed the audience not to breathe, think, or move except when prompted. Once the audience had been corralled in the parking lot, the upstairs window of the house opened and down came a plastic blue skirt of the dimensions of the building. Atop that huge installation, the color of the ocean, was *"La Reina"* (The Queen), complete with an eighteenth-century-style bright yellow wig made of Styrofoam cups. The voice of the Queen was monstrous, a combination of accents and imperial languages somewhere between Spanish, English, French, and German, and her speech was an invented mixture of famous historical phrases of imperial and colonial times in the Caribbean and around the world. This animal-like representation of the powerful decreed "Death to the African" and an end to the "Little People" of the region, while proclaiming herself as part of the "Select countries of the world with a bomb." This powerful monologue came to life again in 2005, this time in a Spanish-era historical building, the Arsenal de la Puntilla, also a museum in Old San Juan. Each time the Queen has reappeared in academic or performance settings, the current political and social circumstances make her criticism just as powerful as it was when she appeared in 1995.

Other iconic characters in *Acceso Controlado* had particular voice personas that played ironically with the violence that is embedded in the sound of our cultural memory. After the Queen's appearance, a private security guard escorted the audience into the house for the rest of the show. The character of *Teniente Cortés* (*The Courteous Lieutenant*), speaking in a familiar Puerto Rican female voice, gave life to the personal narratives of mothers and fathers in communities dealing with the criminalization of youth and the unfair justice system that decides their destinies. Private security guard is often the job of choice for women who are forced to hold two jobs to survive economically. This lieutenant responds to an ethnographic rendering of the women who still "guard" access to many buildings and gated communities.

The complete vocal and social class opposite of Lieutenant Cortés was *La Primera Dama en Solo Operático en Tiempos Desafortunados* (*The First Lady in an Operatic Solo for Desperate Times*), a characterization that criticized first lady Maga Rosselló and her questionable campaigns for children and youth, which included a curfew for the young. Here, the voice work, as in Román's case, takes advantage of existing cultural traditions that come with musical memory, this time by singing a *zarzuela*, a Spanish form of theater that was very popular throughout the Caribbean during the nineteenth century. It is also identified with whiteness in general, due to the fact that Spanish blood legacies are still brought up to establish racial relations in Puerto Rico based on pigmentocracy. The song refers to the island as being isolated and petrified: "la isla aislada, petrificada de nuestro señor padre el creador...." The character appeared again in 2005, in El Regreso de la Reina (*The Return of the Queen*), this time emulating Argentinian icon Evita, as if the character had acquired some kind of global awareness beyond the island.

In *La Nostalgia del quinqué: una huida* (*Nostalgia for the Gas Lamp: An Escape*)[34] (1999), we met *Licenciada Perdóname* (*Attorney Forgive Me*), a very large and asthmatic district representative; *Pragma*, an athletic motivational speaker; and a global singer, *Perpetua*. Each one of these characters had a totally different vocal characteristic. *Perdóname*, for example, inhaled her words in a sickly sounding, troubled voice, a metaphor for the colonial legislation that oversees the laws and represents traumatized discourse, now reduced to respiratory ailments. She is literally unable to breathe while trying to present her proposals. *Pragma* is the voice of progress, the flattening discourse of accelerated construction, the false promise of development that isn't planned or organized. *Perpetua*, the most elusive of the three "sisters," speaks in a language invented by Hernández called "la lingua globale," which sounds like Spanish mixed with Italian and Portuguese. It is a sound-invention of accents to highlight "globalization" as an unfair imperialist influence over territories whose access to modernity has been at the very least chaotic, but who live as if full modernity were the reality.

The nostalgia referred to ironically by these dysfunctional characters, whose voices represent Puerto Rican approaches to identity, were accompanied by a series of skits of Puerto Rican "*jibaros*," peasants who once populated the pages of classic Puerto Rican literature. Now considered emblematic folkloric images of the past, their appearance in literature heralds a "jibarismo literario"[35] the defining force of national literature throughout the twentieth century. Hernández problematizes this unilateral view of identity, which silences and obscures African diasporic culture as well as the influences of the Puerto Rican diaspora.[36]

After presenting these successful pieces, Hernández went on to create other solo performance pieces in which the voices of the characters continued to play with ideas of storytelling from the perspective of the voiceless other, or through presenting the sound of the powerful as deformed and monstrous, using the sound wave as well as hurtful language to show how those in power perpetuate inequality. Her most recent creation, *Privada* (a play on the word "private," which in Spanish also means "to be deprived of"), featured a monologue by a Cuban character, Lazara, who has never lived outside of Puerto Rico. A 52-year-old woman like the performer, Lazara tells

Teresa Hernández in *Privada* (photograph courtesy Antonio Ramírez Aponte).

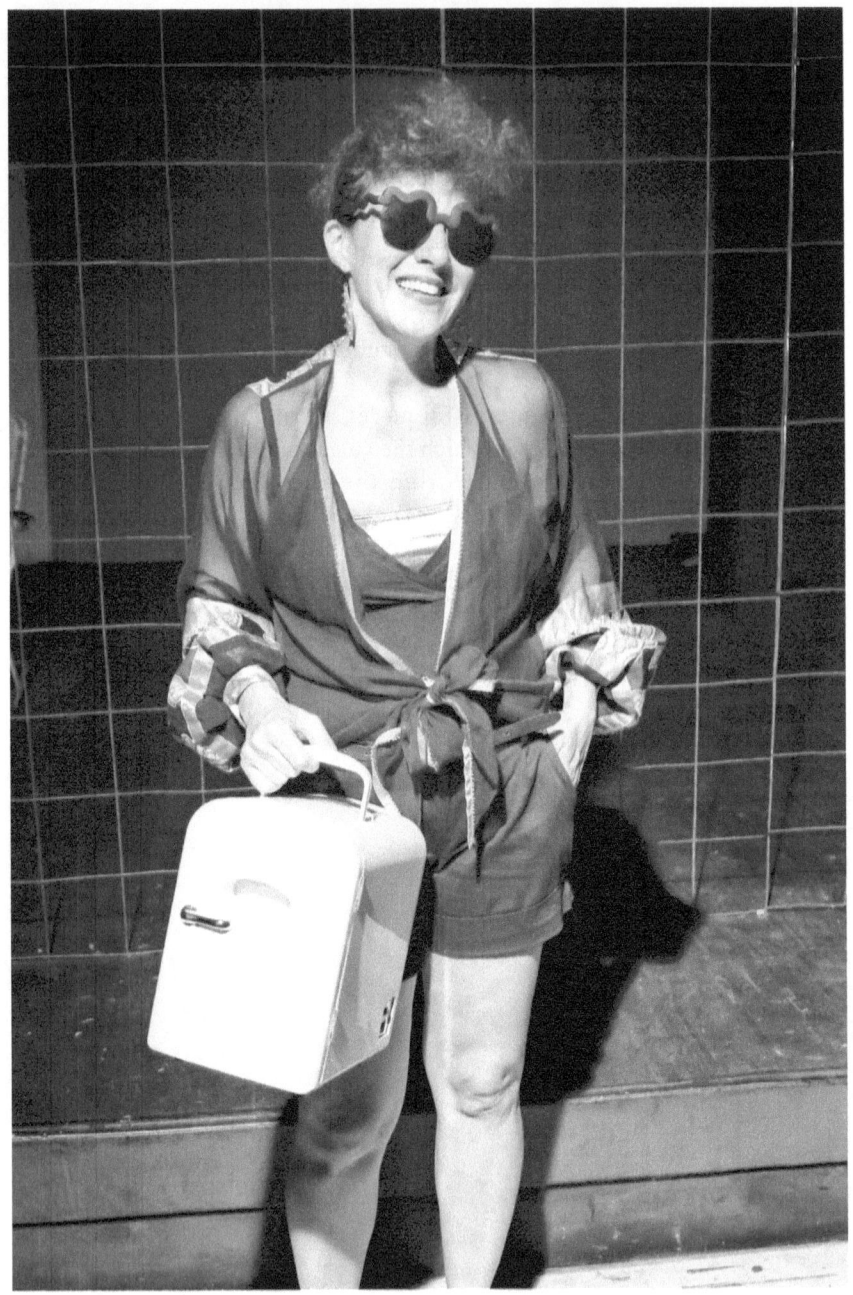

Teresa Hernández in *Privada* (photograph courtesy Antonio Ramírez Aponte).

the audience about her "private" life, using this narrative as a form of disidentification. This personal, politically engaged narrative is both a reminder of pressing topics related to the arts as well as to the economically threatened public university system, while also representing deep compassion and empathy through the private narrative of a woman who takes care of her elderly grandmother—another common reality of the island, where caretaking is often done by family members rather than by institutions. The "Cuban" accent and expressions help the audience to usefully distance itself, enabling Puerto Rican spectators to be able to engage in a forthright critique of their own society, while at the same time making us more aware of Hernández's vocal-local identity.

Stormy Seas: The Unpredictable Wave of Experimental Communication

The soundscapes of creole keep coming back as poetic metaphor in the vocal explorations I have explored here. Glissant refers to creole language as a "forced poetics," a constructive way to refer to the use of the voice and its relation to the Puerto Rican soundscape by the artists I have studied in this essay. When addressing the scream, the "grain"—the texture of the voice, which may be located in a place that is not comfortable with "technique"—there is a window into the aesthetics of the broken, damaged chord that still reverberates in Puerto Rican society as a result of colonialism. The experimental stage art forms of Román and Hernández, which purposefully engage the broken voice, the monstrous and the melodic, present new possibilities for identity as they play with and reinvent received notions of form. Great.

The confrontation between language, voice, and body—or rather, the energy released in finally accepting the voice—the Caribbean (woman's) voice—as an integral part of the body, and establishing the value of that embodied voice as equal to the value of the written word, is not only found in specifically Puerto Rican soundscapes, or in the imaginations of these particular artists. But one of the critical things that the work of these artists reveals, however, is the importance of connecting such performative political practices across geographies and across languages in order to strengthen an anticolonial movement and an anti-imperialist agenda.

I want to end with a graffiti I encountered in a bathroom at the University of Puerto Rico: "*Aquí sólo saben escribir basura, hablen y griten...*" (Here they only know how to write garbage, speak, and scream...). I interpret these words as an invitation to continue this conversation in whatever tone of voice we wish.

NOTES

1. Salsa is an African diasporic form born in New York, where migrant Puerto Ricans made music with other New World African descendants, such as Cubans, Venezuelans, Colombians, Panamians, to name only a few.

2. Most of my ideas about Ivette Román and Teresa Hernández come from my participation as a performer in their work, as a producer, and as an active audience member for the last 25 years. Their work may be found as part of the digital archive compiled by the Hemispheric Institute of Performance and Politics, at www.hemisphericinstitute.org.

3. Ana María Ochoa, *Aurality: Listening and Knowledge in Nineteenth-Century Colombia (Sign, Storage, Transmission)* (Durham: Duke University Press, 2014), 9.

4. This is the main discussion of *In Praise of Creoleness* by Martinicans Patrice Chamoiseau, Rapahel Confiant, and Jean Bernabé (1989); the driving argument in Caribbean Discourse, one of Edouard Glissant's seminal texts (1981), and the base for literary studies focused on Caribbean poetry, drama, short stories, and novels.

5. Édouard Glissant, *Caribbean Discourse* (Charlottesville: University of Virginia Press, 1989).

6. Antonio Benítez Rojo, *The Repeating Island* (Barcelona: Editorial Casiopea,1992).

7. Gayatri Spivak, "Can the Subaltern Speak?" (1988).

8. See the work of Ian Gregory Strachan, *Paradise and Plantation* (2002), Mary Louise Pratt, *Imperial Eyes* (1992), and Peter Hulme, *Colonial Encounters: Europe and the Native Caribbean 1492–1797* (1986), among many other explorations of this argument written during the post-colonial critique of the 90s.

9. Close to the end of the Spanish-American War, U.S. troops invaded Puerto Rico on July 25, 1898. The United States acquired the island as part of the Paris Treaty later that year. Since then Puerto Rico has been under the U.S. territories law of the Congress.

10. See González Torres Roamé, *Idioma, bilinguismo y nacionalidad: la presencia del Inglés en Puerto Rico* (Rio Piedras: Editorial de la Universidad de Puerto Rico, 2002).

11. Urayoan Noel, *In Visible Movement: Nuyorican Poetry form the Sixties to Slam* (Iowa City: University of Iowa Press, 2014).

12. Glissant, *Caribbean*, 4.

13. In *Black Skin, White Masks* (1952), Frantz Fanon discusses the psychological repercussions of colonialism as they play out in the field of racial relations.

14. Benítez Rojo, *Repeating*, 44.

15. See Juan Otero Garabis, *Nación y ritmo: descargas del Caribe* (2000) for the discussion of National ideology and salsa music.

16. Benítez Rojo, *Repeating*, 45.

17. Lowell Fiet, "Teresa Hernández: la nueva acción y escritura teatral puertorriquño," Hemispheric Institute of Performance and Politics (2002), 27 Feb. 2008, http://hemi.nyu.edu/eng/seminar/peru/call/workgroups/latheatrelfiet.shtml.

18. Disidentification is conceptualized by José Esteban Muñoz as a way for minority subjects to negotiate their existence in the majority culture that excludes them. It is a disidentification, as he explains, because it does not mean total assimilation or exclusion, but rather a transformation by these subjects of the majority culture for their own artistic purposes (José Esteban Muñoz, *Disidentifications, Queers of Color and the Performance of Politics* [Minneapolis: University of Minnesota Press 1999]).

19. Muñoz, *Disidentifications*, 5.

20. Nathaniel Mackey, *Discrepant Engagement, Dissonance, Cross-Culturality, and Experimental Writing* (Tuscaloosa: University of Alabama Press, 2000), 163.

21. Glissant, *Caribbean*, 139

22. *Ibid.*, xxxv.

23. Ochoa, *Aurality*, 9.

24. Fragment of *Circular Poem for Vieques*, 2000.

25. Román was referring to the memory of the sound of a hurricane, which is a fairly common trope in the Puerto Rican imaginary. However, Hurricane María in 2017 has taken the concept of hurricane and loss into completely uncharted territory. We still don't know

what the performer's response to this devastation will be. As I write (October 2017), the island is in full survival mode (October 2017).
  26. Glissant, *Caribbean Discourse*, xxvii.
  27. Before Hurricanes Irma and María in September 2017, Puerto Rico had an exceeding debt of $70 billion and was being forcibly administrated by a fiscal control board (La Junta de Control Fiscal) that was recommending extreme austerity for already impoverished social groups such as the elderly and government workers, as well as for cultural programs.
  28. Beliza Torres Narvaez "Entrevista con la performera puertorriqueña Ivette Román." http://hemi.nyu.edu/eng/seminar/peru/call/studentprojects/btnarvaez2.shtml. 2002. Web.
  29. Luis Ahmed Irizarry is a composer-musician, and music producer, and Freddie Mercado is a renowned performance artist in Puerto Rico known for his cross-gender costume creation, among many other visually stunning transformations of his body. See Larry Lafontaine, "Puerto Rican Rasanblaj Freddie Mercado's Gender Disruption," 2017.
  30. Torres Navarez, "Entrevista," 2002.
  31. Viveca Vázquez forms part of the generation of dancers, like Sally Silvers and Merián Soto of Pepatián, to name two who are all still active in the production of dance and movement and pursue the incessant questioning of human and social practices that reveal patterns of inequality in the process of othering, among other formal and aesthetic concerns in their recent works. In 1988 Viveca Vázquez created an iconic piece in New York, titled *Mascando Inglés* (Chewing English), referring to language relations in Puerto Rican migration to the United States and the way language relations also contribute greatly to oppression of the "other" non-native English speaker (http://www.nytimes.com/1988/10/26/arts/reviews-dance-salute-to-hispanic-arts.html).
  32. The criminalization of brown and black bodies in Puerto Rico is studied by Raquel Rivera in the book *Reggeatón*, dedicated to this musical explosion of the 1990s.
  33. Zaire Dinzey-Flores, *Locked-In, Locked-Out: Gated Communities in a Puerto Rican City* (Philadelphia: University of Pennsylvania Press, 2013).
  34. The *quinqué* is a gas lamp that was in many iconic images of Puerto Rican popular culture as a reminder of pre-industrial poverty. This symbol has taken on meaning even beyond what Hernández intended in industrialized Puerto Rico with the island's electricity woes, due both to corruption and to the 2017 hurricanes.
  35. José Luis González, *El país de cuatro pisos* (San Juan: Ediciones Huracán, 1980). Web.
  36. See Juan Flores in *Cortijo's Vengeance* (1992).

BIBLIOGRAPHY

Banes, Sally. *Writing Dancing in the Age of Postmodernism*. Hanover, NH: Wesleyan University Press, 1994. 53–69. Print.
Benítez Rojo, Antonio. *The Repeating Island*. Barcelona: Editorial Casiopea, 1998. Print.
Bernabé, Jean, Chamosieau, Patrick, and Cofiant, Rapahel. *Eloge de la Creolite*. Paris: Gallimard, 1993.
Brathwaite, Kamau. *History of the Voice: The Development of Nation Language in Anglophone Caribbean Poetry*. London: New Beacon Books, 1984. Print.
Dinzey-Flores, Zaire. *Locked-In, Locked-Out: Gated Communities in a Puerto Rican City*. Philadelphia: University of Pennsylvania Press, 2013.
Fanon, Frantz. *Black Skins, White Mask*. New York: Grove Press. 1967.
Fiet, Lowell, "Dramaturgias para el nuevo milenio." *El teatro puertorriqueño reimaginado: Notas sobre la creación dramática y el performance*. San Juan: Ediciones Callejón, 2004. 319–365. Print.
_____. "Isabella diserta." *Segundo Simposio de Caribe 2000: Hablar, Nombrar, Pertenecer*. Ed. Lowell Fiet and Janette Becerra. Río Piedras: Caribe 2000, Universidad de Puerto Rico Facultad de Humanidades, 1998. 18–21.
_____. "Teresa Hernández: la acción y escritura teatral puertorriqueño." *Hemispheric Institute of Performance and Politics* (2002). 27 Feb. 2008. http://hemi.nyu.edu/eng/seminar/peru/call/workgroups/latheatrelfiet.shtml. Web.
Flores, Juan. "Cortijo's Vengeance: New Mappings of Puerto Rican Culture." *Divided Borders*. Houston: Arte Público, 1993.

Gelpí, Juan. *Literatura y Paternalismo en Puerto Rico*. San Juan: Editorial Universidad de Puerto Rico, 1993.
Glissant, Édouard. *Caribbean Discourse*. Charlottesville: University of Virginia Press, 1989. Print.
―――. *Poetics of Relation*. Trans. Betsy Wing. Michigan: University of Michigan Press, 1991. Print.
González, José Luis. *El país de cuatro pisos*. San Juan: Ediciones Huracán, 1980.
Harris, Wilson, "Creoleness: The Crossroads of a Civilization?" *Selected Essays of Wilson Harris: The Unfinished Genesis of the Imagination*. London: Routledge, 1999. 237–247. Print.
Hulme, Peter. *Colonial Encounters: Europe and the Native Caribbean 1492–1797*. New York: Routledge, 1992.
Lafontaine, Lawrence. "Puerto Rican *Rasanblaj*: Freddie Mercado's Gender Disruption." *Caribbean Rasanblaj*. emisferica 12.1. 2015. http://hemisphericinstitute.org/hemi/en/emisferica-121-caribbean-rasanblaj/lafountain. Web.
Mackey, Nathaniel. *Discrepant Engagement, Dissonance, Cross-Culturality, and Experimental Writing*. Tuscaloosa: University of Alabama Press, 2000. Print.
Muñoz, José Esteban. *Disidentifications: Queers of Color and the Performance of Politics*. Minneapolis: University of Minnesota Press, 1999. Print.
Ochoa, Ana María. *Aurality: Listening and Knowledge in Nineteenth-Century Colombia (Sign, Storage, Transmission)* Durham: Duke University Press. 2014.
Pratt, Mary-Louise. *Imperial Eyes, Travel Writing and Trasnculturation*. New York: Routledge, 1992.
Rivera, Raquel Z. "Policing Morality, *Mano Dura Stylee*: The Case of Underground Rap and Reggae in Puerto Rico in the Mid–1990s." *Reggaeton: Refiguring American Music*. Eds. Wayne Marshall, Deborah Paccini Hernández, Raquel Z. Rivera. Durham: Duke University Press, 2010.
Spivak, Gayatri Chakravorty. "Can the Subaltern Speak?" *Marxism and the Interpretation of Culture*. Eds. Cary Nelson and Lawrence Grossberg. Board of Trustees of the University of Illinois, 1988, 271–317.
Torres Narvaez, Beliza. "Entrevista con la performera puertorriqueña Ivette Román." http://hemi.nyu.edu/eng/seminar/peru/call/studentprojects/btnarvaez2.shtml. 2002.

PART 3
Staging the Other

There, abject and abjection are my safeguards. The primers of my culture.
—Julia Kristeva

Always Already
The Jewish Body as Victim and Victimizer

Rebecca K. Pappas

In his seminal work, *The Jew's Body*, Sander Gilman writes that visibility is the ultimate threat for assimilated Jews: "It is being visible in 'the body that betrays' that the Jew is most uncomfortable."[1] When I first read this in 2008, as I prepared to take my own choreographic foray into Jewishness and history, I felt the discomfort of recognition. What was the inheritance of being Jewish in the world? I wanted to know. How had it marked my body? Could I reject this marking or was it inevitable? What was my relationship to the legacies I carried and what was the burden of carrying them?

In this essay I will embark on a close reading of my 2009 choreographic work *Monster*, alternating between present-tense descriptions of the work, and analyses of how the work staged Jewishness. I will unpack the memorializing tools utilized in the piece, and in particular the Holocaust Effect, which is art historian Ernst Van Alphen's categorization of the distinct visual language of the Holocaust.

I have broken this essay into five sections: I first set the stage, examining the changing conventions of "Holocaust Art." In the next three sections, I unpack the bodily strategies of the work, looking at how *Monster* alternately constructs the Jewish body as victim, cartoon, and warrior. I conclude this essay by bringing *Monster* into the future, considering how my conceptions of Jewishness have changed in the years since choreographing the piece. Nestled among these sections are short snapshots in time and place—glimpses of how the work and my own Jewish body have been received in production, presentation, and practice in particular locations at particular moments.

Monster is grounded in the idea that to be a body is to be many things at once. Throughout the dance, bodies multiply and disintegrate as the dancers inhabit the contested Jewish body. They are the Holocaust body, whereby Jews are always already dead: the *ostjuden*, or nineteenth-century European Jew, unattractive, feminized, greedy, shrewd and sexually deviant, or the *Sabra*, a virile Zionist settler, strong and unyielding. They are victims and victimizers, caught within a cycle that impels their disfigurement.

But, to recognize history can be to change it. *Monster* was an attempt to choreograph my own relationship to Jewish history, to understand where my body was sited in history and how that history was sited on my form. Yet its creation opened space for motion in an exchange that felt stuck and disfiguring. Nine years later I have found a new relationship to Jewish history, a freer and more joyful duet with this ancient legacy.

The dance begins in a darkened theater with my voice coming out of the loudspeakers:

> *I used to have this dream where I was giving birth. I was in bed pushing really hard and then this baby would come out of my vagina. But it wouldn't be a normal baby, it would be this tiny, bloody, thing, the size of like a thumbnail. This tiny, bloody, thumbnail baby would come out of my vagina and I would see it and my chest and eyes would feel all burny and all of a sudden I would just lose it. In the sheets or under the bed.... And I would shake out the bedding and I would bend down and search. But I would know that it was gone, it had just disappeared, and I would feel sad, but also kind of relieved. Is that fucked up?*

The performance space is a long corridor with the audience along two sides, facing off like a tennis match. The dancers smash themselves against the walls of the theater and the lights come up slowly—they are lit dimly from behind. Their heads and chests press into the wall as their hips rise and fall, their hands arcing toward their genitals like an arrow. They perform the gesture of bringing in the Sabbath, waving their hands toward their eyes gently and then smacking them against their face—as if they just remembered why they should not be doing that.

There are four dancers: Nguyen, a Vietnamese dancer from Southern California in his early 30s; Harmony, a petite Mormon academic, a lesbian born and raised in Utah; Genevieve, a tall blonde gentile from Alaska; and Arletta, the only Jew, a half-Jew from Monterey, CA. These dancers allow their ethnically and racially diverse bodies to become the site on which the complexities of my American Jewish identity formation are enacted. Tonight they are being the Jewish body, they are being my Jewish body. They are living out the conflicts and contestations I have experienced walking around with a Jewish nose—a marked face.

Rewriting a Legacy

In 1949, Theodor Adorno famously declared that "poetry after Auschwitz is barbaric." Later critics asserted that the Holocaust could indeed be faced in art if the intention was to render its horror in accurate completeness. According to this construction, making art about the Holocaust is an almost mystical journey because it is a journey to the center of an unknowable black hole—a pilgrimage towards the darkest and most barbaric aspects of human existence. To use Sidra Ezrahi's words, "It is the unsayable, that which swallows up all the words, all the colors, and even the instruments that would measure the damage."[2] In 1988, Terence Des Pres, an influential writer and Holocaust survivor, decreed these rules for its representation:

> 1. The Holocaust shall be represented, in its totality as a unique event, as a special case and kingdom of its own, above or below or apart from history.
> 2. Representations of the Holocaust shall be accurate and faithful as possible to the facts and conditions of the event, without change or manipulation for any reason—artistic reasons included.
> 3. The Holocaust shall be approached as a solemn or even sacred event, with a seriousness admitting no response that might obscure its enormity or dishonor its dead.[3]

These rules serve to privilege the documentary evidence, the photo, and the first-hand narrative. Art about the Holocaust, which has become central to American Jewish identity, is art that tells the story as completely as possible. Narratives like *The Diary of Anne Frank* and *Night*, films like *Shoah* enforce the mandate to "never forget." Delimited by rules and strictures self-imposed by the Jewish community, Holocaust Art has had to fulfill an imperative to speak creatively about an experience at the core of Jewish cultural identity as accurately as possible.

In 1998 the Annual Survey of American Jewish Opinion by the American Jewish Committee established that "remembrance of the Holocaust" was the activity that American Jews felt was most important to their Jewish identity—more important than celebrating Jewish holidays, more important than traveling to Israel, or attending synagogue.[4] As author Dora Apel puts it, "The evocation of the Holocaust serves as a kind of unifying historical reminder of the inescapability of Jewishness."[5]

Yet, for some artists, the call is strong to break these rules, and to create work that interrogates the experience of inheriting this memory. These works often reenact, recreate or reconfigure our understanding of the Holocaust. Most often these artists are generations removed from the Holocaust, trying to understand a defining historic event that has come to them only through artistic productions and historical retellings.

Appel calls this experience secondary witnessing, writing in 2004 that contemporary artists making work about the Holocaust are, in fact, writing about the memory effects of the event.[6] Marianne Hirsch refers to it as postmemory: "'Postmemory' describes the relationship that the 'generation after' bears to the personal, collective, and cultural trauma of those who came before—to experiences they 'remember' only by means of the stories, images, and behaviors among which they grew up. But these experiences were transmitted to them so deeply and effectively as to seem to constitute memories in their own right. Postmemory's connection to the past is thus actually mediated not by recall but by imaginative investment, projection, and creation."[7] *Monster* is part of this legacy, seeking to interrogate and reject inherited notions of Jewishness and Jewish history.

Memorial Bodies

Monster's prelude ends and the dancers dress. They are wearing *zantai*, full bodysuits in flesh tones, covered by period costumes that evoke World War II–era refugees. The suit jackets and blouses have cut outs that expose a breast, or a side. One dancer is in a blouse where only the neck closure is visible and the entire front is missing. These costumes are our first nod to the historical record, a redaction or a gap that conjures the incomplete memories we inherit.

The stage is lit with five silver floor lamps that the dancers flick on and off throughout the piece. These lamps along with a single chair and a large book constitute the set, each with a referent in memorializing language: the bright lights of interrogation and work camps, the chair of power where an administrator or storyteller sits, and the book, whose pages hold memory as it is passed down. These objects accumulate to create what art historian Ernst van Alphen calls the "Holocaust Effect." He argues that visual referents like "interrogation lamps," lists of names, and hollowed out portraits are a visual language that summon the Holocaust for American and European audiences even before they know the context of what they are seeing.[8]

Van Alphen coined the term in writing about Christian Boltanski, a multimedia artist who creates large-scale photo walls of grainy black and white portraits. Initially these portraits were of Holocaust victims, but later he began to apply this memorial language to others, discovering what it was to "holocaustize"[9] the average person. Boltanski's enlarged, grainy portraits render his subjects as victims. The blown-up, hollowed eye-sockets and mouths create a sense of blackness and void rather than individuality and animation.[10] The bright lamps that shine down above them erase their features and eliminate their personhood.

The Holocaust Effect is, in part, a means of drawing attention to absence. Van Alphen argues that this effective void is achieved through a number of visual strategies, particularly replication and expansion. The sheer volume of photos, shoes, and names associated with Holocaust memorials speak to the many who are absent rather than the individual who is present. Replication transforms subjects into objects.[11] Over and over we are faced with magnitude not individuality.

As *Monster* continues the dancers enter the space and quickly begin to perform "the list." This is a series of 25–50 gestures strung together, each representing a person from their lives. These energized poses will become a touchstone, taking on a constant double meaning. Each time the dancers perform these poses, the dancer calls into the room a person who is important to them, while at the same time wiping away this person's individuality as they become part of the list. The performer takes on their person, filling with animation and life and then just as quickly letting it all drop away. The bodily legacy of the dancer's own history, and the constant memorializing efforts of Jewish history are both at work.

The dancers fall into a pile, another hallmark of the Holocaust Effect. This pile shifts slowly, inching its way across the floor. For me this is one of the most arresting images in the piece: The tangle of limbs and the weight of their piled bodies as they inch across the floor speak to an inherited Holocaust history and its immobilizing effects.

Limmud, winter 2009

I am at a Limmud Conference, an organization that dedicates itself to "Jewish Learning in all its varieties." Mary Pinkerson, an Orthodox Jewish woman who is the Community Affairs Coordinator at the UCLA Center for Jewish Studies, has arranged my visit. The organization has paid for my hotel and sponsored this performance of Monster: Portrait 0 (Thumbnail). The piece begins:

I used to have this dream where I was giving birth……..

Mary has come to see the presentation along with a half-dozen others. It turns out that modern dance is not among the most popular options at a conference with offerings such as challah baking for beginners and sacred songs for you and your children.

Mary believes in my work or she believes in the idea of my work, and of supporting UCLA students exploring Jewish culture. She has brought me here, but she is horrified. "I don't understand," she says as we sit in the hotel conference room. "The Jews are the monsters? I just assumed the Nazis were the monsters. How could the Jews be the monsters?" I don't know what to say, can't Jews be monsters too?

Later Mary and I will stay in touch. Over repeated coffees we will try to speak to each other about our differing connections to Judaism. Her daughter is a settler in the West Bank. Mary asks, "but how do you account for God's gift of Israel to Abraham?" As a cultural Jew, raised Unitarian, who is agnostic I don't know how to account for it. If God gifted me with something then he never told me........ But Mary's investment in me and my work is one thread that pulls me into a new history.

Monster returns again and again to the Holocaust Effect, activating the body as a tool for memorial. The dancers rise and smash together, becoming a portrait of a missing family killed in the Holocaust. They turn slowly away from the audience, clutching each other for dear life, but also beginning to disfigure, clawing at one another's faces, and slowly revealing themselves. The family portrait, like the list of names, is a familiar sight at Holocaust museums and in documentaries. The audience recognizes the trope.

"Family Photo" (photograph courtesy Andree Andreev).

As the dancers step away from the group, the names of the "dead" begin to fill the theater. Their presence becomes absence as the list returns, this time accompanied by the names of our memorialized, some of whom are sitting in the audience. Each name evokes a brief but specific gesture, a lilting walk or a bouncing basketball. "Brooke Webber, Angela Marino, Marcie Robbins, Lee Ciyavaughn..." While we are accustomed to hearing Jewish names in this format, a Vietnamese name catches our ears off guard. This is not simply the memorializing process of the Holocaust dead, this is memorializing that questions who is a victim, who could be?

In his 1990 work *The Reserve of the Dead*, Christian Boltanski takes a collection of Swiss citizens and renders them in the language of victimhood, asking viewers to examine why a wall of black and white portraiture equals dead Jews, and provoking the questions: Who else could appear on a wall as victim? How does the viewer's relationship to these images change when the subjects are less exotic? Boltanski takes "normal" subjects and "holocaustizes" them.

In discussing the piece he has said: "Before, I did pieces with dead Jews but 'dead' and 'Jew' go too well together. There is nothing more normal than the Swiss. There is no reason for them to die, so they are more terrifying in a way. They are us."[12] It is the "usness" that I and my dancers are exploring.

As the list continues Arletta reads names from the red book, getting louder and more forceful. The faster the names, the faster the dancers must move to perform each gesture. Family groups start to emerge: "Katherine Stockton, Caucasian Female……..Scott Carson, John Carson, Juan Pho, Asian Male." As the names come faster, Arletta drops her voice at the end of certain names, indicating that they are dead—the dancers allow the life to drain from their bodies and their faces to slack open. They've been "holocaustized."

The names come faster and faster with the dancers struggling to keep pace, letting their bodies be vehicles for person after person. Each one must be embodied and accounted for. Finally Nguyen, the male dancer, falls to the ground, the power of bearing all these histories, all these memories becomes too much and he cannot take its weight. The dancers silently look on as he utters "Noni."

From here the piece bifurcates: Harmony and Arletta head into the corner, they are rabid dogs biting and tearing at one another. Genevieve, in a severe white nurse's dress, comes around to Nguyen's head and drops into his arms, falling forward as they slide across the slick wooden floor. They approach the chair and the book. Genevieve cannot bear the pain. She holds her arms up as if to push it away and shakes her hand with urgency. She falls to the side and is caught by Nguyen who tries to sneak past her into the chair from which narratives and power emanate. She lifts a lamp and forces him out using its beam. She sits and lifts the book, reading Nguyen's story and forcing him through a ritual retelling. There is a sound of rewinding tape:

> The life of.... Chapter 1. At the time of....birth....still belonged to......And was home to nearly 120,000..... Seeking to improve the chances for a better life for her children, Mrs.....moved the family shortly before the war from....to the bustling port city of.... Chapter 2. When the....conquered....the 350,000....living in the country found themselves the targets of ever-growing....persecution. The....divided, occupying all of the north allowing....collaborators to rule most of the Southern Zone where....lived. On....mother was arrested by....collaborators and was deported to the.... camp where she was subjected to horrific medical experiments.

Resting one's identity on this recapitulation of violent legacy is an aspect of Jewish memory I am calling into question. What is gained by cleaving so tightly to an insurmountable loss? Genevieve's script is taken from the website of the Museum of Tolerance in Los Angeles and is the story of Jacques Benguigui, a child killed in the Holocaust. Redacted, like the costumes, it is missing all of the proper names, people, and places. In this way it becomes the story of anybody and everybody. We are all the Holocaust victim, we are all every victim. But what do these equivalencies gain and lose for us? Are they a justification for further violence?

Cartoon Bodies

Berkshire Fringe Festival, Summer 2010

Whenever we performed *Monster*, we would follow the piece up with an audience Q and A session. Originally performed with the audience divided in two, facing off against each other, it was meant to implicate them, to draw them into a debate. We are invited to perform in the Berkshire Fringe Fest at Bard College at Simon's Rock. The setting is bucolic but the dark work about the pain of Holocaust memory is sharing billing with a clown troupe from San Francisco. Two middle-aged women who appear Jewish have sat attentively through the performance and they want to know about the monologue. When my voice says:

> So we, I mean Jews, moved to Israel and formed kibbutz and swam, and trained in the military, and lived off the land, and became this totally gentile idea of what strong, virile people look like. And we kept being like, "never again will we be weak, never again will someone herd us to the slaughter. But it was also like never again will we act with compassion. Never again we will let our brains be in charge of our fists, and now it's just, now it's just all so fucked up, it's just a bloody mess, and I think a lot of people wish it would just go away."

One of them looks at Arletta, the only Jewish dancer in the group. "You have a tattoo on your wrist," she says. "Yes," Arletta responds. "It says, *emmet*, truth." Arletta nods. Arletta, who goes to *Shabbat* dinner each week and had a *Bat Mitzvah*. Arletta who has been dancing versions of this work for three years and has been with me on my Jewish journey since the beginning. "I just want you to know a great untruth has been spoken here tonight," the woman spits. We nod, we don't know what to say.

In the next section of *Monster*, the stage darkens, the music cuts out and we hear a cough. Suddenly Nguyen and Genevieve reemerge. They are wearing prosthetic noses, and extra penises. The dancers' faces are covered with stockings, blobs with features squashed to the side and lips curled beneath the nylon. The music is 1930s jazz and their movement is culled from Nazi era propaganda cartoons depicting the *ostjuden*. These dark Eastern European Jews were seen as hunched, deformed, sexually deviant, physically weak, and degenerate by Germans and Western Jewry. This section of the dance is Jewish minstrelsy as the dancers exaggerate these stereotypes. The motion is all rhythmically slumped shoulders, poking necks, and worrying hands that wrap around and around one another. Harmony and Arletta stand at attention, using the lamps to corral the Jewish "freaks" into the corner. The lamps expose each prosthetic part and the curse of visibility is at work.

The substitutions of "play bodies" for real ones is another strategy used by artists making work about the Holocaust; this frees up artists and their viewers to move more easily in the rigid and tragic narrative of the Holocaust. Through play they can identify with a variety of actors and situations rather than feeling locked into one role and one storyline. It can be argued that this freedom enables a more complicated and nuanced engagement with history.

Maus, one of the most famous examples of this type of play, was published by Art Spiegelman in 1972. It is a comic book based on interviews Spiegelman conducted with his father, Vladek, a Holocaust survivor, over the course of more than a decade. It uses the discrete frames of the comic to

"Nose and Lamp" (photograph courtesy Andree Andreev).

address issues of remembrance, suicide, legacy, and victimization, playing them out in Vladek's body. Born and raised in New York City, Spiegelman is part of a large group of second-generation survivors—inheritors of postmemory—who feel that the Holocaust has cast a shadow of depravity, violence and despair over their own lives. As one second-generation survivor put it: "The most important event of my life happened before I was born."[13]

In the book Jews are represented by mice, Germans by cats, Poles by pigs, and Americans by dogs. Within these categories of substitution we see also the trope of "cat and mouse." Through the mice, Spiegelman exaggerates and references Jewish noses[14] and moreover, reproduces and reclaims the strategies of Nazi propaganda that represented Jews as rodents or vermin.[15,16]

The racial stereotypes on which *Maus* and *Monster* play have deep historical roots. In *The Jew's Body*, Sander Gilman exhaustively excavates Jewish stereotypes, documenting their occurrences in anti–Semitic literatures of the eighteenth, nineteenth, and twentieth centuries produced by both Jews and gentiles. He traces the ways that Jews participated in and internalized imaginings of themselves as ugly and contaminated, and the efforts they've made to pass. For Gilman, visibility is the ultimate fear because it means being seen, not as an individual, but as another one of the "ugly" race. This is a bodily betrayal.[17] It is this inheritance of ugly history which *Monster* is trying to understand.

In the dance, Harmony's character embodies this shame. She sheds her clothes and begins to writhe under Genevieve's chair, birthing herself into the play space of the stage. She pulses through Genevieve's legs while the other dancers react in horror. The multiple bodies of the piece take hold and each dancer becomes someone else. Harmony is the bloody thumbnail, a monstrous creature, oozing, barely contained. This monster is not comedic, it is grotesque. The others look on, guards trying to subdue and shame her, the cartoonish bodies of a myth writ large.

They gasp and pull away, shielding their faces from the sight of Harmony. They pick up lamps and begin to corral her, forcing her this way and that as they perform a simple repetitive folk step, encircling her. The apparent simplicity of movement is underscored with a threat of bullying. In their "Introduction to 'Folk' as Keyword Perspectives from Dance," Lisa Doolittle, Penny Farfan, Ann Flynn and MJ Thompson discuss the inherent tension in the word folk, including its expression of self versus other, us versus them, tradition versus innovation, and individual versus communal. Here the dancers are playing out a classical monster tale of insider and outsider.[18]

Harmony quavers on *relevé* and flings her body forward and backward, sending her teetering off balance. Her steps are intricate and classical but with frayed edges, as if something deadly and depraved is bubbling up and spilling from her cool technique. It is the need to contain this bubbling that drives the others to subdue her. A monstrous body that overflows and con-

"Birth" (photograph courtesy Andree Andreev).

taminates is a traditional conceit of whether it is the blob that leaves a trail of slime, Dracula who pilfers fluids from his victims, or the Jewish monster who threatens the stability of society.

At their heart these myths are about contamination. David D. Gilmore writes, "Most authors agree that imaginary monsters provide a convenient pictorial metaphor for human qualities that have to be repudiated, externalized, and defeated."[19] Harmony is many monsters at once: she is the Jewish monster, recurring throughout Western history since the Middle Ages; she is the monstrous pain of carrying the brutalization of the Holocaust forward generation after generation, using it to justify the violence of Israel; she is my own shameful desire not to look Jewish, not to feel Jewish, not to be seen as Jewish by others. Harmony lets her body be occupied by these shames, becoming the monster from which we are all trying to hide.

For Spiegelman, substitution is a direct and nuanced way to get at the horrific and highly personal experience of the Holocaust. He says, "[Using humans] would have come out as some kind of odd plea for sympathy or 'Remember the Six Million,' and that wasn't my point exactly.... To use these ciphers, the cats and mice, is actually a way to allow you past the cipher to the people who are experiencing it. So it's really a much more direct way of dealing with the material."[20]

In *Monster* I use allegories, play, substitution, and simplification to help audiences access this complex material. Taking a page from *Maus*, I simplify and exaggerate imagery, trusting these archetypes to present a *more* complex and *more* nuanced narrative than might be possible with naturalistic performances. When bodies are simplified and distanced from ourselves, we can see more clearly the complex situations and emotions they face.

Virile Bodies

The last section of the dance asks how we hold together the contradictions of memory, history, and legacy within the containers of our body. To make this section, we bifurcated our bodies, creating a vocabulary where one side dragged behind as the other side flailed forward. We created minute-long dance phrases, then cut them apart into tiny pieces, rearranging them to create a kaleidoscopic effect, as movements overlapped and disappeared, occurring and diminishing in a dizzying pattern. How can one hold so much pain together? How does one move within complex, confusing, and interlocking histories? Each moment from this section is meant to reach a physical breaking point. At the climax of this section Harmony walks in a tight circle, falling over and over again until she can no longer stand. It is maximalist and punishing.

Nguyen repeats Harmony's choreography of collapse, landing repeatedly on the ground. He rights himself and the other dancers bear witness, watching as he pulls himself to standing and begins to shed nylon face mask and prosthetic nose. In his book *Muscular Judaism: The Jewish Body and the Politics of Regeneration*, Dr. Todd Presner carefully traces how the ravagement of the Holocaust and the persecution of Jews in Eastern Europe led to a new "muscular Judaism." The strong Zionist body of Israel was meant to assert cleanliness, health, and domination. Nguyen is in search of a purer unsullied version of himself, a body before the Holocaust. In this section of *Monster* it is not the body of the Holocaust victim, but rather its inverse, the Zionist body, that is visible. After suffering so much, how could Jews end up here as the unyielding body of the oppressor?

Nguyen begins to perform a folk dance, tearing across the stage with physicality and force. This dance is a repeat of the folk steps used to entrap the thumbnail, but here it is triumphant. It is the Israeli Jew arriving in the promised land and carving out a new identity of *machismo* and pastoral beauty. The steps and music are celebratory but in this ravaged space where so much violence has occurred, they are deflated. Scholar Nina Spiegel has written of how Israeli folk dances were part of a self-conscious effort to cement a national identity of toughness and togetherness for Israel. Invented beginning in the 1920s, they were a lynchpin in a manufactured narrative where Israel was the inevitable conclusion of Jewish history.[21]

"Maximalism" (photograph courtesy Andree Andreev).

The empty promise of Israeli *machismo* and violence is also a topic taken on by photographer Adi Ness. His photos of beefed up models posing as Israeli soldiers, oiled, and hyper-masculine, deconstruct and interrogate this myth of Israeli dominance. It is the *ostjuden* stereotypes that Zionists were trying to outrun in establishing Israel and fashioning for themselves a new "muscular Jewish" body, drawing a straight line between themselves and heroic biblical imaginings of the Jewish body, erasing from their histories any trace of this old, weak, colonized Jew.[22]

At the side of the stage the female dancers repeat sections from the list, calling up the ghosts whose deaths justified this violence. Nguyen forms a cylinder with his fist and thrusts downward as if planting a flag in the ground. The music grows darker and he crouches behind the chair, the site of Harmony's birth, the seat of power from which Genevieve removes him early in the dance. He flips it over and the book—the record of death—falls to the floor. Genevieve retrieves it and Nguyen begins to plow, using the chair as a barrier as he races through the space. Is he conquering anything that stands in his way? Is he plowing the desert? Is he lashing out unable to see, only to destroy? This final push is extreme for Nguyen and he races across the stage for a full minute, knocking over lamps and touching every part of the stage before he falls, exhausted, in front of the women. The effort required to maintain this kind of false strength and bravado fails him.

Jewish Questions

In 2008 when I was creating *Monster* I would meet regularly with Dr. David Myers, then the director of the Center for Jewish Studies at UCLA.[23,24] I was taking a class with him on Varieties of Jewish Nationalism. He would ask me how it was going with my Jewishness. I would tell him that the more we learned, the more critical I became of Israel. "Well that's very Jewish," he said. "Questioning is the most Jewish thing you can do."

At an Asylum artists' retreat in 2010 in Garrison, New York, Rebecca Guber, director of the Six Points Fellowship for Jewish artists, said the same thing. If you engage in your artwork by questioning, she tells us, that is an inherently Jewish process. I find myself comfortable with this definition. I am attracted to a version of Jewishness that is defined by its questions, by the talmudic practice of approaching one text with a series of surrounding questions, and then enwrapping those questions with more questions.

In making *Monster* I was reacting against negative definitions of Jewishness imposed from without. I did not want to think of my body as weak or infected, or of the prominence of my nose. I did not want to own the violence of the Israeli state. *Monster* was an attempt to insert myself into Jewish history, or to extricate myself from it. Through its creation I was launched into a series of communities that helped me redefine Jewishness in ways other than religion, Zionism, or the memory of the Holocaust. These communities offered me a new vantage on my own history. For them Jewishness was a shared attempt to understand the world via questions and the body; a celebration of shared heritage that encompassed a shrewd intellectualism and a knowledge of self-scarcity. Our endurance was valuable. There was an ancientness in these rooms that I could sink into.

There were resources I could embrace, too. *Monster* was initially funded with a $20,000 grant from the UCLA/Mellon Initiative for Research on the Holocaust in American and World Culture. After the Mellon award I received numerous opportunities designated for Jewish artists, Jewish scholars, or those working on issues of Jewish dance from institutions including Limmud, Asylum Arts, Tablet Magazine, and the Contemporary Jewish Museum. With all of these opportunities beckoning it was hard not to embrace my Jewishness.

There is something mercenary here: in the world of contemporary dance-making where many of us can go years without funding, presentations, or invitations, all of this is consequential. Each opportunity, each dollar, each invitation into community slowly redefined who I was as an artist and my position within an inherited history. I made *Monster* in order to push away certain legacies that I saw as being foisted upon me. In pushing away Jewish community I found myself enveloped by it.

But the sense of belonging was genuine. In these communities I did indeed feel welcomed, seen, and normalized in a way I had not experienced before in my life. I met many other artists who felt conflicted about their relationship to Jewishness. Especially those who, like myself, were half Jewish, or who had been raised in a different faith tradition, or with no particular faith tradition at all. Everyone was searching for a way to understand their cultural legacy and to identify its roots in their artistic practice.

> **Conney, Spring 2014**
>
> I am in Los Angeles, on the campus of USC. I spent eight years living in the city as an Angeleno but am eight months into a transplant to the Midwest. It is not going well. Being a Jew in the Midwest is to be an ornate and exotic weed in a sunflower field. Everyone is sturdy and forward moving and seemingly without doubt. I am twisty, and restless, and searching for a friction that I cannot find in this place where stolidness rules.
>
> Judith Brin Ingber, a long time Midwestern Jew, is here. She has been the convener for three years of the Dance Lab at the Conney Conference on Jewish Arts founded by Doug Rosenberg. Its purpose is to give Jewish dancers a site to dance alongside the conference—a site for embodied research of conference ideas and community building via the body. This is the second time we have gotten together and there are always questions. Who will lead, who will follow? What does it mean to make a Jewish Dance?

In this community Judith represents an older generation tied uncomplicatedly to Jewishness via faith. They go to synagogue and for them Jewishness is an unbroken line. For the younger generation of scholars and choreographers it is more tangled. Many of us are not religious. Some of us have confusing and troubled relationships to our Judaism. Judith brings sacred *tallit* for us to improvise with and demonstrates how we can wrap them over our heads or twine them about our waists. We are horrified. We don't want to defame someone else's sacred objects. *Tallit* are not ours to use as choreographic material.

Participation in this group brings these moments of deep alienation but it also brings glimpses of overwhelming connection and belonging. At one point I am walking in the street with two dancers and I think, "I feel beautiful. We are all beautiful."

The Visible Body

Since making *Monster* in 2009 I have accepted an academic position and moved from Los Angeles to the Midwest. James Friedman is also a Jewish Midwesterner. Working out of Columbus, OH, he became well-known in 1982 for his photo series *Self-Portrait with Jewish Nose Wandering in a Gentile World*. Purportedly undertaken after Friedman was denied tenure at Ohio State University due to anti–Semitism,[25] this work directly inserts Friedman's body into the legacy of the Holocaust.

In the series Friedman photographs himself walking through the streets of 1980s Columbus wearing a striped work camp uniform. He poses with passers-by at the supermarket and the drug store as they stare into the camera. His Jewishness and its legacy is strikingly on display and those he encounters appear dazed, as if they are gob-smacked by history.

Friedman believes that he was outed and ousted by Ohio State because of his Jewishness. In the title of these portraits he is laying bare his difference, bringing attention to his nose, and the ostracism that nose represents. The photo series itself is more than just the work camp uniform photos. It begins with photos of him as a child in 1955 wearing a coonskin hat and continues forward. In one picture he is a teenager; he holds a gun to his own chin while an older man who appears to be a relative stands on his side, holding a finger poised in a gun position to Friedman's temple. While there is a humor, there is also a constant, implied threat. In another, he is in blurry profile in the foreground staring at two white children and their father who wears a Pittsburgh Steelers jersey and looks down at Friedman with disdain.[26] There is a constant isolation that is the plight of the artist and the Jew.

For myself, making *Monster* has been my coming out project. It was my experiment with what it is to own this identity and to speak about it publicly. Drawing attention to my nose and foregrounding the history it represents has afforded me entry into a variety of communities and opportunities to which I would not have had access otherwise. It began with a history that trapped me and has afforded me the freedom to recreate my own history, to re-understand it, to embrace it, and to dance with it a little.

Notes

1. Sander Gilman, *The Jew's Body* (New York: Routledge, 1991), 193.
2. Sidra DeKoven Ezrahi, "Racism and Ethics: Constructing Alternative History," in *Impossible Images*, ed. Honrstein, Levitt, and Silberstein (New York: New York University Press, 2003), 119–20.
3. Terence Des Pres in Ernst van Alphen, "Holocaust Toys: Pedagogy of Remembrance through Play," in *Impossible Images*, ed. Hornstein, Levitt, and Silberstein (New York: New York University Press, 2003), p. 167.
4. Peter Novick in Dora Apel, *Memory Effects: The Holocaust and the Art of Secondary Witnessing* (New Brunswick: Rutgers University Press, 2002), 17.
5. Dora Apel, *Memory Effects: The Holocaust and the Art of Secondary Witnessing* (New Brunswick: Rutgers University Press, 2002), 17.
6. Dora Apel, *Memory Effects: The Holocaust and the Art of Secondary Witnessing* (New Brunswick: Rutgers University Press, 2002), 20.
7. Marianne Hirsch, "Home," *postmemory.net*, last modified 2017, http://www.postmemory.net/.
8. Ernst van Alphen, "Deadly Historians: Boltanski's Intervention in Holocaust Art," in *Visual Culture and The Holocaust*, ed. Barbie Zelizer (New Brunswick: Rutgers University Press, 2001), 49–50.
9. In working on *Monster: Portrait 5*, a dance piece largely inspired by van Alphen's theories of "Holocaust-effect," my dancers and I began to use the word "Holocaustization" as a verb. It referred to a physical attempt to wipe out the energy and *animae* behind a motion and leave behind an empty shell.
10. Ernst van Alphen, "Deadly Historians: Boltanski's Intervention in Holocaust Art," in *Visual Culture and The Holocaust*, ed. Barbie Zelizer (New Brunswick: Rutgers University Press, 2001), 55.
11. Ernst van Alphen, "Deadly Historians: Boltanski's Intervention in Holocaust Art," in *Visual Culture and The Holocaust*, ed. Barbie Zelizer (New Brunswick: Rutgers University Press, 2001), 55.
12. Christian Boltanski as cited in Georgia Marsh, "The White and the Black," 36; as cited Lynn Gumpert, *Christian Boltanski*, 128; as cited in Ernst van Alphen, "Deadly Historians: Boltanski's Intervention in Holocaust Art," in *Visual Culture and The Holocaust*, ed. Barbie Zelizer (New Brunswick: Rutgers University Press, 2001), 55.
13. Dora Appel, *Memory Effects: The Holocaust and the Art of Secondary Witnessing*. (New Brunswick: Rutgers University Press, 2002), p. 9.
14. Michael Rothbert, "We were talking Jewish: Art Spiegelman's 'Maus' as 'Holocaust Production," *Contemporary Literature* 35, no. 4 (Winter 1994): 675.
15. Andrew Huyssen, "Of Mice and Mimesis: Reading Spiegelman with Adorno," in *Visual Culture and The Holocaust*, ed. Barbie Zelizer (New Brunswick: Rutgers University Press, 2001), 33–34.
16. In fact, on the copyright page of each volume of *Maus* Spiegelman references this Nazi propaganda—giving credit where credit is due
17. Sander Gilman, *The Jew's Body* (New York: Routledge, 1991), 193.
18. Lisa Doolittle, Penny Farfan, Ann Flynn and MJ Thompson, "Introduction 'Folk' as Keyword Perspectives from Dance " from *Discourse in Dance* 5, no. 2 (2013): 12.
19. David D. Gilmore, *Monsters: Evil Beings, Mythical Beasts and All Manner of Imaginary Terrors* (Philadelphia: University of Pennsylvania Press, 2003), 4.
20. Andrew Huyssen, "Of Mice and Mimesis: Reading Spiegelman with Adorno," in *Visual Culture and The Holocaust*, ed. Barbie Zelizer (New Brunswick: Rutgers University Press, 2001), 34.

21. Nina Spiegel, "Dance in Israel: A Historical Perspective" (presentation, UCLA Center for Jewish Studies, April 22, 2008).
22. Todd Presner, *Muscular Judaism: Jewish The Jewish Body and the Politics of Regeneration* (Abingdon: Routledge, 2007).
23. Because I have no written record of these conversations I am not using quotations marks, but rather summarizing them according to my memory of these exchanges.
24. David Myers, personal conversation, 2008, University of California, Los Angeles.
25. Dora Appel, *Memory Effects: The Holocaust and the Art of Secondary Witnessing*. (New Brunswick: Rutgers University Press, 2002), 129.
26. James Friedman, "Self-Portraits with Jewish Nose Wandering in a Gentile World," JamesFriedmanPhotographer.com, last modified 2017, http://www.jamesfriedmanphotographer.com.

Bibliography

Apel, Dora. *Memory Effects: The Holocaust and the Art of Secondary Witnessing*. New Brunswick: Rutgers University Press, 2002.

DeKoven Ezrahi, Sidra. "Racism and Ethics: Constructing Alternative History," in *Impossible Images: Contemporary Art After the Holocaust*, edited by Shelley Hornstein, Laura Levitt, and Laurence J. Silberstein, 118–128. New York: New York University Press, 2003.

Doolittle, Lisa, Penny Farfan, Ann Flynn, and MJ Thompson. "Introduction 'Folk' as Keyword Perspectives from Dance." *Discourse in Dance* 5, no. 2 (2013): 3–13.

Friedman, James. "Self-Portraits with Jewish Nose Wandering in a Gentile World." James FriedmanPhotographer.com, last modified 2017, http://www.jamesfriedmanphotographer.com.

Gilman, Sander. *The Jew's Body*. New York: Routledge, 1991.

Gilmore, David D, *Monsters: Evil Beings, Mythical Beasts and all Manner of Imaginary Terrors*. Philadelphia: University of Pennsylvania Press, 2003.

Hirsch, Marianne. "Home." postmemory.net, last modified 2017, http://www.postmemory.net/.

Huyssen, Andrew. "Of Mice and Mimesis: Reading Spiegelman with Adorno," in *Visual Culture and The Holocaust* edited by Barbie Zelizer, 28–44. New Brunswick: Rutgers University Press (2001), 34.

Presner, Todd. *Muscular Judaism: The Jewish Body and the Politics of Regeneration*. Abingdon: Routledge, 2007.

Rothbert, Michael. "We were talking Jewish: Art Spiegelman's 'Maus' as 'Holocaust Production,'" in *Contemporary Literature* 35, no. 4 (Winter 1994): 661–687.

Spiegel, Nina. "Dance in Israel: A Historical Perspective." Presentation, UCLA Center for Jewish Studies, Los Angeles, April 22, 2008.

Van Alphen, Ernst. "Deadly Historians: Boltanski's Intervention in Holocaust Art," in *Visual Culture and The Holocaust*, edited by Barbie Zelizer, 45–73. New Brunswick: Rutgers University Press, 2001.

Van Alphen, Ernst. "Holocaust Toys: Pedagogy of Remembrance through Play," in *Impossible Images: Contemporary Art After the Holocaust*, edited by Shelley Hornstein, Laura Levitt, and Laurence J. Silberstein, 157–178. New York: New York University Press, 2003.

Israel Galván's Aesthetic Anarchism
An Ethics Instantiated in Motion

NINOTCHKA D. BENNAHUM

Mozarabic Tracings

Israel Galván de los Reyes, widely considered one of the most innovative choreographers in Spain today, was born in 1973, two years before General Francísco Franco's death in 1975. His surname, Galván, a word of Latin derivation, has no real Spanish point of origin, except that its ancestry is found in other parts of Europe during the Spanish Inquisition, 1478–1848. This is to say that its made-upness—its etymological creation—belongs to exiled peoples with Sephardic origins and conquered peoples of foreign origins; each forced to assimilate in namesake (placing an accent above the second "a" to the emerging Hispanic world of "New Spain") into Spain—an invented geography of pre–Islamic Hispañola. Hence, Galván's paternal last name carries with it the weight of history: diaspora, dislocation, and forced assimilation into an invented Castilian world whose ruling class shed all vestiges of mercantile Hispano-Arab Andalucía. *Galván* also commemorates the last Middle Eastern dynastic line to confront the militaristic command of Ferdinand and Isabella whose glutinous desire for gold and global domination led them to marry Spain's colonial ambitions to the Transatlantic Slave Trade. Undergirded by the Spanish Inquisition's catastrophic shedding of eight hundred years of Islamic, Sephardic and Catholic co-existence, or *convivencia*, Spain's early modern history would forever be tied to genocidal activity and xenophobic, capitalist ambition.

This refugee crisis brought about by the violence of post-colonial authoritarianism is a story about which Galván is profoundly aware as his namesake carries the two central principles of modern dance: Time and Space: Temporality (where one lives in transhistorical and/or rhythmic time), and Spatiality (the physical, geographic and psychic space negotiated by contemporary performers negotiating historical iterations of Southeast Asian, African and European embodiment).

The son of two dancers, his mother, baildora Eugenia de los Reyes, a *Gitana*, his father, José Galván, a beloved and respected Sevillian maestro of the flamenco and Spanish Classical traditions, Galván grew up in his parents' dancing academy, a room of about forty feet long and twenty feet wide, its closets stuffed to the brim with costumes, shoes and instruments. He graduated from high school into an Andalucían environment whose arts, commerce and civic life was waking up from forty years of governmental censorship. This was an exciting world of possibility, one that his parents had not been so lucky to confront. These two decades of democracy and socio-economic growth would give Israel and his muse, his sister Pastora, a path, first to university and then to global stardom.

Influenced by his great friend, French philosopher and art historian Georges Didi-Huberman, Galván has moved beyond a spiritual break with flamenco—previously described as cubist with its visual and performing artists' focus on the surface materiality of the body—to a humanist aesthetic approach, a choreographic strain driven by ethical impulse instantiated in motion. Severing ties with flamenco's twentieth-century modernist ethos whose choreographic methodology is driven by a gendered and architectural physiognomy, Galván gravitated toward Hispano-Arab and Sephardic historical source material.[1] Returning flamenco to its Early Modern Andalucían rootedness in which the poetic narrator renders ideas and events visible to his/her/their audience, Galván was able to construct an entire environment in which both movement vocabulary and its protagonist—the Gitana/o Other—were mobilized as instruments of historical consciousness.[2] Galván's elliptical phrasing and Kafkaesque marriage of Spanish thematic material, both driven by his radical breaks with flamenco compás—Spanish dance's chief thematic thread—expound politically (vs. technically) upon Spain's long history of Fascism. Galván's choreographic narratives raise awareness of Spanish history's egregious and historic alliance with authoritarianism by shedding light on ethical questions thereby engaging Early Modern Spain's humanist traditions. These are medieval Hispano-Arab and Sephardic religious and philosophical traditions concerned not with divine or human power, but rather with life as a source of vital aliveness, of being-ness: a route to the scholarly and aesthetic capacity of people to contemplate of love, soul, and God.[3] Flamenco-theater is thus for Galván an efficacious, socially progressive tool.[4]

I want to distinguish Galván's resistant, anti-modernist choreographic aesthetic from his mentors' and read this watershed break not as a rebellion

Israel Galván, *La Curva*, Skirball Center for the Performing Arts, New York University. New York Flamenco Festival, 2010 (photograph by Kevin Yatarola).

but as a spiritual reassessment of flamenco's historical undercurrents; in the words of Henri Bergson, its *élan vital*, or vital aliveness.

Galván sought to bring flamenco art in line with interrogations of Spain's ugly Fascistic past. A culturally-superior form, flamenco fell prey to Western, central and Eastern European modernist influences that, while crucial to its break from the form's music-hall past, did not serve it well. Seeking an organicity of present fused to a Spanish mystical consciousness, Galván has developed an anarchic artform. Previously defined solely by great artists' rhythmic, improvisational and stylistic interpretation of traditional musical line, Galván has opened not only the *compás*, but also the textual, sung (cante) verse to notions of affect and embodiment. His is a postmodern aesthetic that evolves from two organic processes. The first is his understanding of and conscious use of Anna Halprin and Simone Forti's creative use of the environment, the natural world in the service of storytelling. The second is

Israel Galván, *La Curva*, Skirball Center for the Performing Arts, New York University. New York Flamenco Festival, 2010 (photograph by Kevin Yatarola).

Halprin's use of the poetic text—in the case of flamenco cante, or sung verse—as animate, as possessing organicity and, thus, ethical pulse. Cante historically connects the *bailor/a/x*, or dancer, to spoken word and sung verse, to rhythmic play of instruments and bodies in space and time. Within Galván's work, corporeal affect and expressiveness takes on the internal beingfulness of the *cante*, fusing the interior self—one's consciousness—to the surrounding world. The dance, like the song, evolves from the very processes of its creation, thus asking the audience to perceive both body and sound differently.

Galván furthers his spiritual interrogation of traditional flamenco by assessing the functionality of postmodern choreographic strategies. He harnesses identifiable, utilitarian objects—human and material—as historical phenomena. A broken piano, white sand, pieces of shards of glass, chainlink fence, suspenders, his own bow-like flamenco body, serve as semiotic signifiers of Gitana/o/x persecution and survival. More nuanced historical referents include parts of danced phrases severed anatomically and spatially from the complete musical and choreographic phrase, enabling Galván's reconceptualization of the flamenco genre as an archeological phenomenon. Lastly, Galván works in episodic experience. Each moment in time is not driven solely by rhythmic phrase, as is usual in flamenco, but rather by a sense that the past is best situated in the immediate present.

Israel Galván, *La Curva*, Skirball Center for the Performing Arts, New York University. New York Flamenco Festival, 2010 (photograph by Kevin Yatarola).

In conversation with Didi-Huberman, Galván's fairly substantial *oeuvre* retains a transhistorical corporeal expression that enables viewers to enter into the image of history and possess its élan vital, or vital aliveness. We can study Galván's unique psycho-physical tension (or leitmotif) by staring at skin, hands, arms, hyperextended shoulders and upper torso—the break in the flamenco bodyline—and the inner impulse or soul of each movement that seems to emerge out of Galván's own sense of being. This corporeal collision of inner with outer emotional and physical expression produces a complex visual imaginary that forces the viewer into an engaged relationship with history. We see this in *La Curva* (2010), where the shedding of a ritualistic machismo is replaced with absurdist, hallucinatory off-kilter walks and non-sequitur gestures (flicks of the hands) executed to the serialist sounds of a dissonant guitar with electronic score that produces an odd intergenerational encounter of a postmodern flamenco with its café cantante-era past.

In this attempt to "get away from all of flamenco's clichés," to shed nostalgia and replace it with an austere, jagged-edged, stripped down realization of the body as an object of self-mockery, repetitive, Cagean atemporality meets flamenco's historic through-line through anguished performance art.[5] *La Curva*, a footnoting of Escudero's 1924 La Courbe Parisian performance, may indeed be read as not so much as a commentary on the hyper-masculinist Escudero (which must be seen within Fascistic-era Spain) but rather on the anarchist Frenchman Jean-Claude Gallota, whose surrealist bow-like breaking of the *bailaor* into two anatomical parts—female/male, Spanish/French,

Israel Galván, *La Curva*, Skirball Center for the Performing Arts, New York University. New York Flamenco Festival, 2010 (photograph by Kevin Yatarola).

masculine and androgynous, matador/bull, murderer/victim—set the stage for a postmodern interrogation of Spanish dance.[6]

In *The Legend of Don Juan*, which premiered as a curatorial centerpiece of the 1992 Lyon Biennale de la Danse, Galotta offered an interesting iconographic/historical critique of the bailaor-cum-matador, introducing long moments of stillness and pose into his body that appeared more like a haunting than a matador going in for the kill. His matador grew in size through scenic media, offering a dematerialized facsimile of a matador, a one-hundred-foot digital double whose mediated presence overtook any human allusion. Galván has absorbed Gallotta's display of the power of liveness, mediated presence and stillness, a physical re-configuration of the relationship of the embodiment to space that arose in the feminist pedestrianism of 60s radical dance artists, Anna Halprin, Simone Forti and Yvonne Rainer.

Galván, like Gallotta, Angel Margarit and a range of Catalan dance artists, also draws inspiration from Pina Bausch's neorealist dance interrogation of Spanish dance as the basis of a post–Fascist Spanish dance. Containing a visual imaginary that extends back in time to the Spanish Inquisition, Galván's physical, kinesthetic fragmentation lends a neorealist, post–Francoist ambiguity to flamenco tradition, thus destabilizing the form and opening it to deeper reflection.

And like Galván's global predecessors—the Russian-born architect of American ballet George Balanchine, who shed aristocratic gentility, marrying Russian classical ballet to Africanist dance traditions to create a global classical idiom, and British-born Antony Tudor, the inventor of psychological realism—Israel uses *La Curva* and *Lo Real* as choreo-history. He establishes a series of physical moments, *extremeses*, that can be read not only as interrogations of the flamenco form and confrontations of history itself, but as probing a series of questions: Where is the Gitano in this history of dance? Where is the Jew in this history of dance? Where is the refugee in this history of dance? How can flamenco be harnessed in the service of choreographies of social justice? What is its capacity to evacuate evil from time and space?

Galván's broken figurines, hyperextended limbs, detachments from his corps, oddly androgynous and, at times, distended use of torso, head, hands, even organs produce a Benjaminian visual imaginary that seeks to connect present to past historical tragedies through the medium of his body. At times a fetishization of Holocaust history, *Lo Real*'s image phantoms enable Galván to embody and display a textural—physically proprioceptive, iconographic archive that seeks to close the divide between then and now, between a series of steps and a series of shared experiences.

While Galván has spent nearly three decades mining the capacity of his own body to reconfigure itself culturally, politically and aesthetically, it is his unapologetic interrogation or reconceptualization of flamenco as having an intrinsic historical value separate from its dynastic stylization—its utopianism—that emerges most profoundly in *Lo Real*, Galván's commissioned meditation on the Nazi genocide against the Gitano/Gitana populations of the Eastern Mediterranean and Northern, Eastern, Southern and Western European worlds. A brutal work of Spartan design filled with a unique dramaturgy woven together skillfully in episodic moments that surveil the cruelty of SS officer Heinrich Himmler's barbarous cruelty in his extermination of 600,000 Gypsies between 1944 and 1955 at Auschwitz. A work of activist theater that owes a debt to Augusto Boal's *Theatre of the Oppressed*, *Lo Real* forces the audience into a complicit relationship with history by virtue of its gazing; we are made to feel morally reprehensibly responsible for history's silence during and after the events we witness onstage.

In the end, one might argue Galván's greatest virtue is his modesty. "I want to be seen as a piece of rubbish onstage," Galván told Victoria Looseleaf of the *LA Times* in 2003. In the words of Argentinian anthropologist Gastón

Lo Real. Israel Galván performing the commissioned work *Lo Real* at the Teatro Real de Madrid (photograph by Javier del Real).

Gordillo, in his book *Rubble. The Afterlife of Destruction*, that rubble is like layered sediment, the rings of a tree, buried time: constellations of objects (or in the case of Galván, movements) understood in relationship to historical processes.[7]

His work is a clearly-stated allegiance to the 1960s' concept of the ordinary body as land, an ecological terrain that, like a tree, decays, erodes, and renews in geologic, rather than human time.[8] Galván's disassembling of flamenco's materiality (its decorative silhouettes and heteronormative architecture), into a rhetorical tool of moral confrontation, absorbs the humility that ecology teaches us, enabling him to embed an ethical code into the dystopian worlds he forges. "I am a pathological nonconformist. I would rather generate extreme responses than indifference, but this is not rooted in anger or defiance, rather a desire to experiment with looking at what happens if the things we have always aesthetically admired are allowed to be questioned."[9] I take this a step further and say that the intergenerational conversation of one dancer to another, Galván, for instance with Escudero, became a transhistorical conversation of Galván in 2017 and a Galván in 1508.

Conclusion

Like the twelfth-century Spanish epic *Poema El Cid* that the exiled Spanish historian Americo Castró described as "based on five centuries of shared existence with Islam … evidence of Spanish integralism," Galván's anarchist flamenco is more than just a stylistic break with Gitano-flamenco's aesthetic past. Each resistant work, a rebellious universe unto itself, opens a path to inner existence leading to "a consciousness of self" from which to embrace the exterior world.[10] Galván's autonomous and illusory work offers in the words of Theodor Adorno, "the social antithesis of society," an anarchist spirit that enters each dance through Galván's repetitive struggle with flamenco and Spanish history herself.[11] Translated physically through broken body parts—limbs stretched to their corporeal limit—Galván's anti–Fascist philosophy reconfigures flamenco, away from art and toward an historical process of *Gitano(a)* experience which he rescues from its burial ground.

In his sonic-visual imaginaries, Galván exumes historical truths fusing their wings to present time. Bearing Cervantean roots, Galván's physical theater emboldened by a psycho-political rhetoric, stretches the boundaries of art and memory, breaking flamenco history in two.[12] His early works' iconographic tracings bear the mark of the hyper-modernist Escudero (1892–1980) whose masculinist aesthetic was crafted under the repressive surveillance of a fascist state. Escudero's self-reflexive, some might say paranoid burial of any circular bodyline or sensuous gesture that could be read as feminine, became comedic to those born during the Civil Rights protests that swept the globe.

Alongside Escudero came Galván's next artistic father, Mario Maya (1937–2008), in whose company he came of age between 1993 and 1998. Maya's Lorca-infused 1960s *cuadros* with their 1960s ethic of communitarian revolt signaled the arrival of a Gitanoist aesthetic, one that would soon be rid of its repressive leader.[13] However, unlike Maya whose flamenco cuadros are choreographed

Israel Galván dancing in the Teatro de la Maestranza de Sevilla, 2008 (photograph by Luis Castilla).

"as a single body, a rightly-synced group that carves space and sound ... a reincarnation of the Gypsy family," Galván's characters bear no utopian groupings; their hellish environments more Kafkaesque than Judson Dance Theater.[14]

But it is Galván's political father Antonio Gades (1936–2004) who helps us to understand the existential nature of Galván's aesthetic revolt. "My political position is clear.... I want to defend the rights of men and the rights of workers. I want a more just society.... These political positions have repercussions in the life my company."[15] Driven no longer by Escudero's engagement with the surface materiality of corporeal form, Galván's aesthetic fragmentation becomes an exercise in disbelief; disbelief in the catastrophe of history as shared tradition. The spiritual nature of Galván's anti-capitalist flamenco philosophy points toward an ontology of being.

NOTES

1. Americo Castro. "Deism in Fourteenth-Century Castile," *The Structure of Spanish Society*. Princeton: Princeton University Press, 1957, 673.
2. Americo Castro. "Deism in Fourteenth-Century Castile," *The Structure of Spanish History*. Princeton: Princeton University Press, 1954, 672–673.
3. *Ibid.*, 674.
4. *Ibid.*, 81.
5. Contemporary American composer John Cage (1912–1992) was the student of Arnold

Schoenberg, the Jewish-Viennese refugee artist and the inventor of atonal "serial music." Cage's avant-garde compositions absorbed the Zen-inspired belief that music is an ongoing natural process akin to a wave crashing on a beach or the sound of air rustling through trees. Musical lines cannot thus be broken up neatly but must rather follow the natural ebb and flow of life itself. Time—that which governs music—cannot be manipulated, broken up, or even composed as it is an ongoing process vs. an artform.

6. I attended the 1992 Biennale de la Danse de Lyon *Pasión de Espana* Spanish Dance Festival as a *Dance Magazine* correspondent and, thus, write from my own critical perspective having witnessed this festival. The Festival's curator, Guy Darmet, curated a substantial performance book centered on the complex historiography of Spanish dance and Gitana/o flamenco. The book is available through the Bibliothèque Nationale de France: call #: FRBNF35764549 and through WorldCat.

7. Gaston Gordillo. *Rubble: The Afterlife of Destruction*. Durham: Duke University Press, 2014.

8. For more on the ordinary body as political act of resistance, please read: Ninotchka Bennahum. "Anna Halprin's Radical Body in Motion," *Radical Bodies: Anna Halprin, Simone Forti & Yvonne Rainer in California and New York, 1955–1972*. Berkeley: University of California Press, 2017.

9. Israel Galván quoted by Valerie Gladstone. "Flamenco Like Something Out of Kafka," *The New York Times*. 25 January 2004.

10. Americo Castro. *The Structure of Spanish History*. Princeton: Princeton University Press, 1954, 344.

11. Theodor Adorno. *Aesthetic Theory*. Minneapolis: University of Minnesota Press, 1970, 8; 227.

12. For a magnificent assessment of Galván's use of flamenco as social protest, please read: Michelle Heffner Hayes, "Choreographing Contemporaneity: Cultural Legacy and Experimental Imperative," *Flamenco on the Global Stage: Historical, Critical and Theoretical Perspectives*. Jefferson, NC: McFarland, 2015, 280–291.

13. Ninotchka Bennahum and K. Meira Goldberg. "A Modern Masculinity," *100 Years of Flamenco on the New York Stage*. New York: The New York Public Library for the Performing Arts, Lincoln Center, 2013, 112–115.

14. *Ibid*. 112.

15. http://www.antoniogades.com.

BIBLIOGRAPHY

Adorno, Theodor. *Aesthetic Theory*. Minneapolis: University of Minnesota, 1970.
Benjamin, Walter. *Illuminations. Essays and Reflections*. New York: Schocken Books, 2007.
Bennahum, Ninotchka. *Carmen, a Gypsy Geography*. Middletown, CT: Wesleyan University Press, 2013.
Bennahum, Ninotchka, Goldberg, Meira K., and Heffner-Hayes, Michelle. *Flamenco on the Global Stage: Historical, Critical and Theoretical Perspectives*. Jefferson, NC: McFarland, 2015.
Bennahum, Ninotchka, and Goldberg, K. Meira. *100 Years of Flamenco on the New York Stage*. New York: The New York Public Library for the Performing Arts, Lincoln Center, 2013.
Bennahum, Ninotchka, Perron, Wendy, and Robertson, Bruce. *Radical Bodies: Anna Halprin, Simone Forti & Yvonne Rainer in California and New York, 1955–1972*. Berkeley: University of California Press, 2017.
Bergson, Henri. *The Two Sources of Morality and Religion*. New York. Henry Holt, 1935.
Blitzer, Jonathan. "The Gypsies' Dance." *New York Times*. 26 December 2012.
Castro, Americo. *The Structure of Spanish Society*. Princeton: Princeton University Press, 1957.
Diénes, Jean-Claude. *Pasión de España*. Lyon Biennale de la Danse Program, 1992
Dunning, Jennifer. "Beyond Classical: Exuberance as the Mother of Invention." *New York Times*. 3 February 2004.
Gladstone, Valerie. "Flamenco Like Something Out of Kafka." *New York Times*. 25 January 2004.

Gordillo, Gaston. *Rubble: The Afterlife of Destruction*. Durham: Duke University Press, 2014.
Kisselgoff, Anna. "Distilling Flamenco to Its Fiery Base." *New York Times*. 30 September 1993.
_____. "Flamenco Dares the Unpredictable." *New York Times*. 29 January 2002.
_____. "Maria Maya, Interpreter of Flamenco Style, Is Dead at 71." *New York Times*. 3 October 2008.
Kourlas, Gia. "One-Man Flamenco, Without Instruments." *New York Times*. 19 June 2008.
La Rocca, Claudia. "Flamenco Kept Simple, but Fun." *New York Times*. 23 February 2010.
Macaulay, Alastair. "Showing His Mastery Just by Standing Still." *New York Times*. 22 September 2011.
Mackrell, Judith. "Israel Galván: FLA.CO.MEN review—'It's an evening that screams avant garde!'" *The Guardian*. 17 February 2017.
Seibert, Brian. "Devilishly Slashing Narrow Notions of Flamenco." *New York Times*. 14 March 2014.

Brown and Black

Performing Transmission in Trisha Brown's Locus *and Hosoe Eikoh and Hijikata Tatsumi's* Kamaitachi

MICHAEL SAKAMOTO *and*
CHRISTOPHER-RASHEEM MCMILLAN

A dance conversation by Michael Sakamoto and Christopher-Rasheem McMillan, presented at the Dance Studies Association International Conference, October 2017

Brown and Black is an interdisciplinary performative presentation that examines the philosophical and embodied tensions between Western postmodern dance, performance forms rooted in marginalized communities, and vernacular modes of expression. The work utilizes conventions of intellectual discourse, scholar-artist conversation, concert dance, and performance art.

In this essay, the left column contains our text, including dialogue and choreographic direction, and the right column contains discussion of the theoretical and cultural underpinnings explored in the creative process for Sakamoto's performative photo essay, MuNK, McMillan's installation performance, *Black Lōkəs*, and *Brown and Black* itself.

We as interdisciplinary practitioners do not want to offer only one side of this exploration, but rather to lead with the text from embodiment, and in some ways put theory in the margins.

Michael (left) and Chris performing *Brown and Black* (2017).

Michael and Chris sit on opposite sides near the audience.

Michael:
   Hey, Chris.

Chris:
   Hi, Michael. How's it goin'?

Michael:
   It's goin'. Whatchu been up to?

Chris:
   Livin'.

Michael:
   Cool. What does that entail?

Chris:
   Makin' dances. Savin' lives.

Michael:
   So making dances for you means saving lives?

Chris:
   Yes, and saving my life, in particular.

Michael: Why does it matter to save lives with dance? Can one?

Chris: Artists often have a little of "I want to save the world" in them. I do as well, but it's okay if I only save one person, and it's okay if that person is me. Surely, if black men loving black men is a revolutionary act, then loving myself is a divine one. (See Riggs 1989.)

Michael:
   Huh. Can you show me what that looks like?

Chris:
   Yeah.

Chris plays a video, and he and Michael watch a projection of Chris preparing to dance in an 8' × 8' metal frame in a white gallery space. After smiling to the camera, Chris begins dancing elements of Trisha Brown's Locus.

6' × 6' is one of a number of standard jail cell dimensions. 8' × 8' is slightly larger in order to accommodate Chris's larger dancing frame and the broad, tumbling reach of Trisha's early postmodern dance style.

Michael:
   What is this?

Chris:
   It's *Locus*. *Locus* is a work Trisha Brown made in 1975 using numbers and letters as points in space as a score.

Brown's *Locus* followed a diagrammatic score: an imaginary cube organized around numerical points, each correlating to one of the twenty-six letters of the alphabet. Located at the center of the cube, a twenty-seventh point designated the space between letters.

Michael:
   You find that interesting?

Chris:
   I do.

Michael:
   Because?

Chris quoting Trisha Brown's *Locus* in his performance of *Black Lōkəs* at Legion Arts, Cedar Rapids, Iowa, April 2017.

Chris:
Because it's task based, non-virtuosic movement that highlights the era. It's iconic because it's a part of everything Trisha Brown made since then. It's like Yvonne Rainer said. No to virtuosity. No to spectacle. Just no. (Bryan-Wilson 2012: 64.)

Michael:
And that matters to you?

Chris:
Yeah, but mostly it matters to white women in spandex.

Michael:
And you identify with white women in spandex?

Chris:
I do.

Michael:
But you're black.

Chris:
I am.

Michael:
Do you think blackness matters to white women in spandex?

Chris:
In that historical moment, I don't think it's something they considered. I think postmodern dancers at that time weren't thinking about meaning, but just bodies in space.

Michael:
You mean white, heterosexual, middle class, women's bodies?

Chris:
Yeah, because historically, in terms of who had access in the late Sixties and Seventies, while white people already had access to non-virtuosic, heady dancing, black people were just being invited to the party.

Michael:
Do you think any of that mattered to early postmodern dancers?

Chris:
No, because going off on a tangent in dance is performing difference, but you need privilege to

Trisha left some of herself in *Locus*. How could she not have? The very construction of it began with the biographical information, "Trisha Brown Was…," and through the enactments and reconstructions of *Locus* one might see that "Trisha Brown Was" can be transformed into "Trisha Brown Is." See Susan Rosenberg, *Trisha Brown: Choreography as Visual Art* (151–182, 2017).

Chris: Can you name me two black early postmodern dance makers?… I will wait.

do that, and the machine has to allow you the space to make such decisions, which at that time, only white bodies had access and time to do that, the privilege to break away from the modernist structures of dance.

Michael:
    So there's no blackness in early postmodern dance?

Chris:
    No, there is blackness there, but it's unacknowledged and quiet. Or rather, it's cool.

Michael:
    It's cool?

Chris:
    Yeah.

Michael:
    It's cool like we're cool?

Chris:
    No, it's cool like I'm cool.

Michael claiming agency as an artist of color speaking partially through African American Vernacular English that he and his friends adopted as children in a mixed Mexican-American and Asian-American neighborhood of Los Angeles in the Seventies.

Chris performing *Black Lōkəs* (2017).

Michael:
 You?

Chris:
 Yeah, I'm cool cuz postmodern is cool. It's in the simultaneous specificity and not caring, or rather, in the not caring too much. Or as Trisha would say, "The doing of it and getting off of it" (Senter 2016).

Michael:
 So you don't care much about blackness?

Chris:
 Oh, no. I care too much about blackness. I care enough to go back and use the tools of the postmodern era to gaze on it with black eyes.

Michael:
 Alright, can you show me what you're talking about?

Chris:
 (pointing to the video) Yeah, right there.

In the video, Chris has begun dancing a different sequence of Locus*-like movements.*

Michael:
 It's different now? What is this?

Chris:
 It's *Black Lōkəs*.

Michael:
 Black Lōkəs?

Chris:
 Black Lōkəs uses the structure of *Locus*, but instead of Trisha's number system, I replace it with the names of people who've been recently killed by police and others in the last ten years. I also use biographical information, such as the chronological order of when they were shot and the number of times they were shot, ranging from Trayvon Martin to Freddie Grey.

Michael:
 That's cool. Yo, so you're not just pointing to numbers in space…

Chris:
 I'm pointing to black people in space, using whiteness to make blackness visible.

Chris: A black body reconstructing this material is always somehow something other. I do, however, read coolness and in some ways blackness onto Trisha's relaxed form. That has been my entry point into the archive. That is where *Black Lōkəs* starts. I sought to highlight this sense of cool and bring it out in a full way in *Black Lōkəs*. See Rebecca Walker (ed.), *Black Cool: One Thousand Streams of Blackness* (XI, 2012).

Chris: I want to use the master's tools, not to dismantle his house, but to evacuate hegemony from the premises and move in myself (Lorde 2007: 112).

Chris is resurrecting both early white postmodern dancers and murdered black bodies. The names of people killed by police violence are never mentioned by name, they're only present through Chris's bodily enactment of them, making them physical.

Chris: Regarding legacy, there are current company members (TBDC) who have received "Trisha" in the same manner, although not in the same intensity of purpose, as I have. They are submersed in her—I was just visiting. They have received the essence of "Trisha" from an embodied archive, an archive that does not only depend on videos of her work, but the bodies of the keepers of her work. I hope that *Black Lōkəs* is an alternative legacy, an offshoot, a connection that disrupts as well as affirms the archive. To put it simply,

Michael:
   Using the tools of people who didn't care about what it meant to be black.

Chris:
   Except they wanted to be, but they just didn't know it.

Michael:
   Yeah?

Chris:
   Well, yeah, you know. The relaxedness of the form. The do it and get off of it. Like you can't mean to do it. You just have to do it. Like swagger.

Michael:
   Do you swagger?

Chris:
   Yeah. What, you can't tell when I dance?

Michael:
   Sometimes.

Chris:
   Maybe that's because postmodern dance is always performing itself, but the thing about swagger is that if you try and have it, you don't. If you try to be cool, you're not cool. The thing about *Locus* is that it's cool because it's task-based, because the approach within the structure is cool. *Locus* has swagger, and *Black Lōkəs* highlights it and brings it out more.

Michael:
   Okay. So what you're saying is that you're a black man who loves white women early postmodern choreographers who don't care about race but who actually want to be black but don't know it.

Chris:
   Yeah.

Michael:
   Cool.

Chris:
   Cool. What about you? Where does your dance come from?

Michael:
   I'm a butoh artist.

through *Black Lōkəs*, through blackness.... Trisha also remains.

Michael: I'm not sure about Chris's strategy. How can you fully and effectively wield a tool without acceding to its fundamental value structure? Like Audre Lorde emphasizing that a theoretical approach is not enough: "survival is not an academic skill" (2007: 112).

Chris: Black people are making do all the time. Like chitlins. We took this idea of what's left over and created something that's actually good. In the same way, even if the master's house is problematic, blackness is able to recuperate and regenerate.

In the article, "Whose culture has capital? A critical race theory discussion of community cultural wealth" (2005), theorist Tara Yosso articulates and carefully mines the questions of whose culture has wealth and who can create and disseminate knowledge. Yosso challenges the traditional Bourdieuean cultural capital theory by arguing that communities of color have other kinds of cultural wealth that may go unnoticed as potential sources of wealth and information from which to draw: "The dominant groups within society are able to maintain power because access is limited to acquiring and learning strategies to use these forms of capital for social mobility" (2005: 76). Blackness has a wealth that is futuristic and ancestral.

Chris:
What's butoh?

Michael:
It's a dance and performance art practice developed by a dancer named Hijikata Tatsumi in Japan in the 1950s and 1960s, after World War Two and the American Occupation, during a time of rapid Westernization, and especially Americanization, of the culture and a deeply rebellious attitude in the Tokyo avant-garde against the imposition of late capitalism on the socioeconomic arena. It's the Japanese identity crisis of that era, or as Hijikata put it, butoh is the body in crisis.

Michael betrays his bias here of focusing largely on Hijikata's concept of the body in crisis for his ultimate scholarly-artistic goal of extrapolating it interculturally and transnationally to other bodies. There were other core aspects and influences as well, however, on butoh's development, such as the predominance of subjective expressivities, a focus on the carnal body (*nikutai*) and Japaneseness (*nihonjiron*) in Japan's post–World War II era, and an ongoing dialogue with Western avant-garde aesthetics and strategies from the early twentieth century onward. See, for example, John W. Dower (1999), Miryam Sas (2011), and Miriam Silverberg (2009).

Chris:
Great, but what does that have to do with your body?

Michael:
Well, it's my body in crisis.

Chris:
Is your body in crisis?

Michael:
Aren't all bodies in crisis?

Chris:
Amen to that. And while we talking about it, can you show me some?

Michael:
Yeah.

In the projection, Michael brings up photos from Hijikata and Hosoe Eikoh's Kamaitachi, *showing Hijikata performing on the streets of Tokyo and the Tohoku countryside.*

Chris:
What is this?

Michael:
Kamaitachi (Hosoe 2009). A photo essay collaboration in the 1960s between Hijikata and photographer Hosoe Eikoh shot partially in Tokyo, where they both lived and worked, but mostly in their home region of Tohoku in northern Japan. Hosoe was trying to capture the feeling of his childhood memories through images, and Hijikata was doing the same through his body. They thought of *Kamaitachi* as a subjective documentary.

Chris:
Kamai-what?

Michael:
Kamaitachi. Sickle weasel. A mythical creature from Tohoku that hides in wind flurries on deserted roads, sneaks up on unsuspecting passersby, and, you know, cuts their skin open, inserts poison into them, seals them back up, then disappears into the night to let the poison eat away at the person's soul for years to come.

Chris reacts with disgust and fear.

Michael:
Yeah. That's how it is in butoh.

Chris:
So, is this butoh or is this pictures of butoh?

Michael:
Yes and no. It's a bit like Zen. There's no separation between body and image in the mind. It's the essence of butoh.

Chris:
Is it really?

Michael:
For me, it is.

Chris:
So for you, photos are butoh.

Michael:
They have agency as knowledge of butoh. For many of us, butoh photos were part of how we first knew very well what we were doing. *Kamaitachi* is a foundational text. The photos are kinda like a bible.

Chris:
But you know all bibles are individually interpreted.

Michael:
Right. You have to interpret your own life to create your own butoh, because all butoh is personal.

Chris:
So these are personal for you.

Michael:
They're in me.

Michael's butoh movement and Zen contemplative practices are fairly intertwined. For the latter, he draws primarily from essays and koans of Soto Zen lineage founder Eihei Dogen (1200–1253), especially the foundational texts, *Zazengi* and *Genjokoan*, and the koan collection, *Shobogenzo*, both in translations by Kazuaki Tanahashi (2000, 2005).

Chris:
How's that? Could you show me?

Michael nods, and switches the projection to a new set of autobiographical photos, similar in style to Kamaitachi.

Chris:
What's this?

Michael:
MuNK.

Chris:
MuNK

Michael:
Yeah. *Kamaitachi* is where and how Hosoe's vision and Hijikata's body come from. MuNK is where my eyes and body come from. In addition to locations associated with my own autobiography, I also photographed in Tokyo and the village of Tashiro, the same autobiographical places where those men created their images, except it's my kind of awkwardly intellectual, transnational performing self. It's basically me in them, and vice versa.

Japan and Hijikata are not here, but we glimpse them through Michael's dancing body and photo-images. For both Hijikata and Michael, dancing in Tashiro is looking to the past to perform and redefine the present for the purpose of imagining the future.

Chris:
So place matters to you?

Michael:
Butoh is where your body comes from. It's all about place.

Michael's practice adapts the placeness of butoh/Tashiro/Tohoku into the space of American postmodern performance. He finds space within place. Chris turns space into place. Body is the locus of both space and place.

Chris:
Is this butoh?

Michael:
It's me.

Chris:
Are you butoh?

Michael:
Yes and no.

Chris:
(snide look askance at the audience) If this is and is not butoh, and you are and are not butoh, where does butoh end and you begin?

Michael: *(after Chris)* Maybe that's because butoh, like Zen practice, is always performing itself, but the thing about Zen is that if you try and have it, you don't. If you try to be butoh, you're not butoh. The thing about *Kamaitachi* is that it's butoh because it's self-based, because the approach within the structure is butoh. *Kamaitachi* is butoh, and MuNK highlights *Kamaitachi*'s effects in me and brings them out more.

Michael:
I never really looked at that question before, but now I suppose it starts with me as a boy playing at being Japanese.

From the series MuNK (2017).

Chris:
   Are you Japanese?

Michael:
   That's what they say.

Chris:
   What do you say?

Michael:
   I can't be. I'm fourth generation American. I grew up in East L.A. waking up every night to police sirens and eating Mexican food. I've lived my whole life in America. I don't feel any need or desire to be Japanese.

Chris:
   Okay. So what you're saying is that you're a Japanese-American man who loves Japanese early postmodern male dancers who resisted modern Japaneseness by idealizing pre-modern Japaneseness and rejecting Americanness by using the tools of their American masters.

Michael:
   Yes.

Chris:
   Cool.

Is Michael a hypocrite, or does he slip the trap of consistency by embracing, trickster-like, the contradiction of butoh's socialized-intuitive body in crisis? Hijikata attempted to be both what he was and was not as a self-fashioned, social entity culturalized by both Eastern and Western influences, and Michael traces this strategy through generations of butoh artists since the 1960s to the present, including through his own body.

Brown and Black (Sakamoto and McMillan) 213

Michael:
　Cool.

Chris:
　Cool. So if you don't want to be Japanese anymore, what do you want to be?

Michael:
　Black.

Chris:
　(laughs incredulously) Yeah?

Michael:
　Yeah, I mean, you know, when I was a kid. It was the Seventies. Who else was I gonna look up to?

Chris:
　And you danced to black music?

Michael:
　Well, in a way.

Chris:
　Meaning?

Michael:
　Disco.

Chris:
　Disco? You mean, like, black disco?

Michael:
　The Bee Gees.

Chris:
　Michael, the Bee Gees is not black music, by the way.

Michael:
　Yeah, I know it's problematic, but it got me where I needed to go.

Chris:
　Where's that?

Michael:
　Hip-hop.

Chris:
　(laughs) So you're b-boying now?

Michael:
　My first dance form was popping.

Chris:
　Same thing.

George Clinton: "To me, disco was like fucking with one stroke. You could phone that shit in. Disco itself was funk. But all they did was take one funk beat and sanitize it to no end. It's irritating. I loved Donna Summer's records. But too much of it.... The slogan behind '(Not Just) Knee Deep' was, 'Let's rescue dance music from the blahs.'" (Clinton 1990)

Between July 1976 and November 1978, the Bee Gees charted four times in the Billboard R&B Top 10. Michael was not aware of this crossover status at the time, however, as he did not begin listening to black radio stations until late 1979, when members of his Japanese-American, little league basketball team began singing "Rapper's Delight" during practice.

214 Part 3: Staging the Other

Michael:
   Chris, b-boying and popping are not the same thing. Don't you know anything about hip-hop?

Chris:
   (interrupting) Yeah, yeah, I know, I know.

M stares back questioningly/judgmentally at Chris.

Chris:
   Yeah, I know, how black am I? This is where my queerness meets the road.

Michael:
   So your dance is complicated.

Chris:
   No, but my body is. My body is always displaced, displaced from Africa because of slavery, displaced from blackness because of queerness, and displaced from postmodern dance because of blackness. My body resides in exile.

Michael:
   Sounds like your dance is a place, not just a space.

Chris:
   No, it comes from space too. That cube I was dancing in? That's the space of the dance.

Michael:
   But it's just numbers and points.

Chris:
   No, not just. Let me show you.

Chris plays an archival film of Locus *performed in studio by Lisa Krause.*

Michael:
   OK, to me, this isn't numbers and points in space. This is a specific person in a specific time and place.

Chris:
   No, it's a space.

Michael:
   A white space.

Chris:
   Not when I dance it.

Often referred to in mainstream culture as *breaking* or *break-dancing*, *b-boying* is generally considered the original and/or primary form of street dance out of the amalgam of vernacular cultural practices that came to be known as Hip-Hop by the early 1980s. *Popping* is an umbrella term for many funk-based, street dance techniques originating in California in the 1970s, such as popping, waving, ticking, roboting, and tutting, among others.

Chris: Whenever I tell people that I am a dancer they say, "Oh, so you do hip-hop"? The fact that I am a black postmodern dancer is juxtaposed to the stereotype that black men in dance must only dance their own cultural forms.

Black Lōkəs is, at least methodologically speaking, another way of engaging with Trisha Brown's legacy. It is an afterlife of *Locus*. To cite André Lepecki, "One re-enacts not to fix a work in its singular (originating) possibilization but to unlock, release, and actualize a work's many (virtual) com- and incompossibilities, which the originating instantiation of the work kept in reserve, virtually" (2010: 31). Black bodies were always in there, and now through the very structures of the choreographic, they are invited to be made visible, to be made present, to be accounted for, to matter. See Lepecki, *The Body as Archive: Will to Re-Enact and the Afterlives of Dances* (31).

Michael sees primarily through the lens of corporeal subjectivity and historical moments. Chris identifies how the stage and other performance spaces become mirrors and sites for reproducing oppression and domination in real time. Our strategies are varied and sometimes at odds, but we share the goal of creating effective space for difference in contemporary dance.

We're reimagining an American dance body and then reclaiming it through our practices as artists of color because our experience is always a question and is always questioned.

Brown and Black (Sakamoto and McMillan)

Gradually, the soundtrack from Black Lōkəs *fades up.*

Michael:
  Then what is it?

Chris:
  A black cage.

Michael:
  A what?

Chris: White cube = black cage. It's about the death of the police shooting victims, but also about making them alive again, through my black body in a cube.

Chris:
  Yeah. A black body in a box means it can't be anything other than a cage. Whether it's imagined or not.

Michael:
  Does it always feel that way?

Chris:
  I guess.

Michael:
  You're not sure.

Chris:
  Well, when I was a young dance student, I wasn't sure.

Michael:
  About the cage?

Chris:
  About my body.

Chris begins dancing tentatively in a mix of various stock postmodern and contemporary movements.

As the Black Lōkəs *soundtrack fades down, a voiceover begins of Chris's memories.*

Chris (VO):
  I was nineteen. I was sitting in UMass Fine Arts Centre, first row. I was there to see the Trisha Brown Dance Company do *Set and Reset*. I was glued, immersed in watching the slick, steady, bodies pouring in and off the stage. It was strong and pretty at the same time. It was then when I thought, "This is it."... I enrolled in modern dance the next day. I can recall feeling like a big black tree swaying in the background... and then while rolling around on the floor I thought, "This is not what they were doing on the stage last night," ... but this was not my first dance experience.
  At my oldest cousin's wedding, my grandma

walks up to me sitting alone at one of the tables, she says, "Chile, come learn this," and starts to do the electric slide just as the electric boogie comes on… she says, "Naw you gotta feel it, do it like this and then do it like you do it."

There were five male cousins at this wedding and gun violence has claimed three of them.

At first, it's movement from modern/postmodern class, then gradually transitioning into Trisha movement. At the very end, foreshadowing Set and Reset.

As sound transitions back to Black Lōkəs *soundtrack…*

Michael:
So that's how it all started?

Chris:
Yeah. How did it begin for you?

Michael:
I don't remember exactly. Just, one day in high school, dance was suddenly in my life. I feel like it's always been a part of everything I do. And ever since then, I've been workin' at it, day and night, just tryin' to keep a hold on it, a minute at a time…

Michael ponders his body and tentatively starts butoh popping to the Black Lōkəs *soundtrack.*

Michael and Chris do one sequence of the Electric Slide.

Michael learned popping as a teenager, and then butoh movement ten years later. Butoh is in his spirit, but popping is in his bones.

At the kick-out, Chris does a slow motion citation of Set and Reset-Wall One, *and Michael steps downstage into a dance evoking chaotic memories of police and gang violence in his childhood in East Los Angeles.*

As Michael ends in a hands-up freeze, facing the audience, the Set and Reset *score begins playing.*

Chris dances Set and Reset-*inspired movement around Michael and begins a live Trisha sermon.*

Set and Reset (1985), a work choreographed by Trisha Brown, is billed as a collaboration between herself, the painter Robert Rauschenberg, and the composer and performance artist Laurie Anderson. The lighting was by Beverly Emmons.

See American Bible Society, Psalms 23, King James Version (1990) for original wording.

Chris:
Trisha's structure is my shepherd I shall not want, her form leadeth me beside still waters. *Set and Reset* restoreth my soul. Yea though I walk through the valley of shitty dance, I will fear no spectacle for she is with me.

Her rod and her staff, they comfort me. She Preparest a table before me full of luscious movement possibilities. She anointest my head with oil;

my cup runneth over. Surely good dance making shall follow me all the days of my life; and I will dwell in the house of the *Locus* for ever. *(shouldhavedroveahonda)*

Chris is citing Glossolalia (speaking in tongues), which is a part of black church expressionism.

Michael's body gradually breaks down into butoh popping.

Chris sermons about his dad.

Chris and Michael end up in slow unison steps, foreshadowing the Electric Slide.

As music crossfades into the Bee Gees' More Than a Woman, *Michael and Chris begin dancing the Electric Slide full tempo.*

Chris uses the 22-step sequence, which is how he learned it from his grandma at a family wedding. Michael uses the eighteen-step version, which he learned at a nightclub in Los Angeles.

One sequence unison—no emotion.

One sequence unison—real feeling.

One sequence–Michael solo and Chris slide.

One sequence–Michael slide and Chris solo.

One sequence unison—full party mode.

One sequence—full emotional solos.

One sequence—no emotion (with music fade).

Music fades out, and Michael and Chris sit down.

Michael:
    So you are a postmodern dancer.

Chris:
    Yeah. I guess so. You black?

Michael:
    Nah, I ain't.

END.

Bibliography

Anzaldúa, Gloria. *Borderlands: la frontera*, vol. 3. San Francisco: Aunt Lute, 1987.
Bryan-Wilson, Julia. "Practicing *Trio A*." OCTOBER 140 (Spring 2012): 54–74.
Dogen, Eihei, *Enlightenment Unfolds: The Essential Teachings of Zen Master Dogen*. Ed. and trans. Kazuaki Tanahashi. Boston: Shambhala, 2000.
_____. *The True Dharma Eye: Zen Master Dogen's Three Hundred Koans*. Trans. Kazuaki Tanahashi and John Daido Loori. Boston: Shambhala, 2005.
Dower, John W. *Embracing Defeat: Japan in the Wake of World War II*. New York, NY: W. W. Norton/The New Press, 1999.
Foster, S. L. *Dances That Describe Themselves: The Improvised Choreography of Richard Bull*. Middletown, CT: Wesleyan University Press, 2002.

Fricke, David. "George Clinton: The Rolling Stone Interview." *Rolling Stone*, September 20, 1990. http://www.rollingstone.com/music/features/george-clinton-19900920.
Hosoe, Eikoh. *Kamaitachi*. New York: Aperture Foundation, 2009.
Lepecki, André. "The body as Archive: Will to Re-Enact and the Afterlives of Dances." *Dance Research Journal* 42, no. 2 (2010): 28–48.
Lorde, Audre. *Sister Outsider*. New York: Crossing Press, 2007.
Tongues Untied. Directed by Marlon Riggs. Performed by Michael Bell, Kerrigan Black. Signifyin' Works, 1989. VHS.
Rosenberg, Susan. *Trisha Brown: Choreography as Visual Art*. Cambridge: MIT Press, 2012.
Sas, Miryam. *Experimental Arts in Postwar Japan: Moments of Encounter, Engagement, and Imagined Return*. Cambridge: Harvard University Press, 2011.
Schneider, Rebecca. 2001. "Performance Remains." *Performance Research* 6, no. 2 (2001): 100–108.
Senter, Shelley. 2017. Conversation with the author, New York, New York, February 13, 2016.
Silverberg, Miriam. *Erotic Grotesque Nonsense: The Mass Culture of Japanese Modern Times*. Berkeley: University of California Press, 2009.
Taylor, Diana. *The Archive and the Repertoire: Performing Cultural Memory in the Americas*. Durham: Duke University Press, 2003.
Walker, Rebecca. *Black Cool: One Thousand Streams of Blackness*. Berkeley: Soft Skull Press, 2012.

The Bustle, the Body and Stillness
Re-Centering Modernities through the Broadway Musical

GWYNETH SHANKS

In 2008 Steven Sondheim and James Lapine's Tony-award winning musical, *Sunday in the Park with George*, was revived on Broadway. The musical, originally created by the two as the hundredth anniversary of Georges Seurat's painting neared, largely lacks a narrative, focused instead on what it means to create art. Themes of artistic genius and creation unite the two acts, set a hundred years apart from each other and each focalized around a central artist named George. Act I, the longer of the two acts (and indeed originally imagined as a stand-alone work), is set in 1880s Paris as Seurat completes *A Sunday Afternoon on the Island of La Grande Jatte*; the act offers glimpses into the lives of the various figures that populate Seurat's famous painting. The second act jumps some 100 years forward to a modern art museum's opening gala, and focuses on a contemporary artist whose art and creative processes have seemingly fully capitulated to art market forces.

The revival, directed by Sam Buntrock, was lauded for its immersive projections, designed by Tim Bird; the three walls of the stage became projection screens for an interactive animation that illustrated the incremental creation of Seurat's painting. Focused on this revival and on Act I, I question how the musical refracts narratives of artistic genius and creation, canonization and art historical legacies through a distinctly gendered understanding of visual cultures and modernity. While I likewise examine *Sunday*'s book, composition, and the original 1984 Broadway production, the 2008 revival's staging importantly foregrounds the musical's fraught relationship to class and to gender, the latter of which I particularly focus on here.[1] Here I address the ways in which the first act stages the historic gendering of urban space at the *fin de siècle* and the concurrent ways gender, class, and race—*vis-à-vis* whiteness—are unspooled alongside an emerging consumerism.

Particularly emerging in the early 1980s, art historians theorized and historicized the Impressionists and Post-Impressionists in relationship to a burgeoning modernity.[2] Much of this scholarship has focused on the important role these artists played in documenting the changing landscape of urban centers, the rise of consumer goods, and for prefiguring the emergence of modernist art practices in the early twentieth century. For example, leading art historian T.J. Clark argued in his 1984 book on Manet, that modernity indicates "a set of social relations that…[take] place within the new spaces of capitalism."[3] While he importantly placed modernity within a broader context, his frame of critical reference remained almost wholly focused on questions of capitalism. A contingent understanding of social relations—and thus modernity—as dependent upon gender and race alongside class was subsumed, in his and others work, by a critique of capitalism. Challenging such scholarship, however, feminist art historians and literary theorists, like Griselda Pollock, Roselyn Deutsche, Aruna D'Souza, Linda Nochlin, and Janet Wolff, framed Euro-American cultural production from the late nineteenth century through considerations of gender.[4] Such work often focused on the division between a public and a private realm; this framework, though, as scholars like Pollock and Deutsche have argued, overly determined how we might understand the relationship among cultural production, gender, urban life, and modernity, too easily castigating women's roles and labor into private sphere.[5] It is in relationship to this body of art historical and literary scholarship that this essay enters, thinking through considerations of gender and modernity through theatrical address.

Why is it useful to pose such questions in relationship to a twentieth-century Broadway musical, and further, its twenty-first-century Broadway revival? As scholars like Sean Metzger remind us, theatrical conventions—in his case costuming—provide a means to think anew about gender, race, and modernity precisely because such forms of theatrical address engage a politics of embodiment through "contest[ing] linear progressive histories."[6] Engaging questions of modernity through *Sunday*, as opposed to the markers of modernity legible in *fin de siècle* French literature and art, opens up the concept to its plural; no longer modernity, we instead are grappling with a set of *modernities*. Indeed, the stage becomes an ideal place to question such temporal complexities, as, within it, we see reflected a succession of representational frames and eras. Gender in *Sunday*, for example, is nuanced in relationship to nineteenth-century Parisian histories, a history of the American musical, the evolution of the musical in the late 1970s and into the early '80s, and, finally, in 2008.

Here, then, I engage *Sunday* as a historiographic site and one with the ability to re-map, if subtly, the ways we come to understand these histories.[7] While my aim here is to expand upon the art historical treatments of Seurat's work, my argument also returns to an intellectual and cultural moment (the turn of the nineteenth century into the twentieth in France) that has become key to the development of contemporary critical theory, and to fields like performance and dance studies. Engaging a set of ideas around movement, urban space, gender, and modernity, this essay reveals how theatrical practice offers a way to reconsider such concerns.

Applying a feminist visual and historiographic analysis to the production, I explore how a close reading of the costumes and staging in *Sunday* offer a way to re-read constructions of gender in late nineteenth-century Paris. I focus on Dot, George's scorned lover, the mother of his unclaimed child, and the woman with the pronounced bustle in the foreground of Seurat's painting. *Sunday* might take its name from George, but the musical makes clear the two's imbricated relationship. There is, of course, a key distinction between them: George's role within *Sunday* is always framed through his painting and artistic genius. The comparable act of creation given Dot is pregnancy and childbirth. My aim here is to offer a feminist counter-reading of the relationship between George and Dot, recuperating her as central to

Georges Seurat (1859–1891), *A Sunday on the Island of La Grande Jatte*, 1884/86, oil on canvas, 81¾ × 121¼ in. (207.5 × 308.1 cm), Helen Birch Bartlett Memorial Collection, 1926 (Art Institute of Chicago/Art Resource, New York).

the musical's dramaturgy and to an expanded understanding of Euro-American modernity.

In this piece, I examine Dot's pronounced bustle and closely analyze the ways her garment is discussed (or rather sung about) and how instances of prolonged stillness—Dot maintaining a single pose or posture—reify certain notions of gender and class within the musical. By drawing together *fin de siècle* feminine visual cultures, legible in female-authored artworks and French fashion magazines of the period, I locate a feminine visuality in Dot's bustle, challenging the masculinist one forwarded by Seurat's painting as well as Sondheim and Lapine's musical. With regards to the use of stillness, *Sunday*'s staging is built around the use of *tableaux vivant*, a device maintained even in the far more technologically sophisticated 2008 revival.[8] While I parse these two conventions at time separately, the character's costume and her staging are thoroughly entangled, revealing a theatrical representation (and continuation) of emerging modern sensibilities regarding gender norms, urban space, and agency. Dot, as a costumed and blocked character, articulates a feminist counter-dramaturgy that centers not only feminine visual culture within the musical, but also shifts *Sunday*'s thematic thrust from masculine artistic genius to distinctly feminized labor and agency.

The 2008 revival of *Sunday* opens with an empty stage. As the lights come up, the audience sees the three white walls of the simple set. While a few set pieces and props are downstage left and right, the overwhelming visual impact of this initial moment is a bare stage, evoking the blank page or canvas waiting to be filled. This opening tableau dissolves as Daniel Evens, the actor playing George, enters. As he speaks, "White. A blank page or canvas," a black line snakes its way across the stage's three walls as though sketched on a page. As Evens continues speaking—"Through design"—the line resolves into an approximation of the shoreline between park and lake that bisects the horizontal plane of Seurat's painting.

Dot enters, played by Jenna Russell in the revival, as the animation springs into life behind George—images of sailboats lazily coast across the three walls, and figures walk through the park.[9] Soon after her entrance the opening song of the musical begins; sung by Dot, "Sunday in the Park with George" establishes not only the tenor of the two's relationship but also sets in motion the musical's themes of "creative genius … and … the creation of a timeless, transcendent artifact."[10] Annoyed by George's admonishment to stay still, Dot sings of the boredom and fatigue of posing.[11]

As the musical theme for "Sunday in the Park with George" swells, George and Dot are still speaking. In these opening moments of the music, Russell, like the woman in the painting, is costumed in a skirt with a prominent bustle, and she balances a small indigo parasol on her shoulder. Over the music, Dot begins to speak:

DOT: I feel foolish.
GEORGE: Why?
DOT: I hate this thing. (*Indicating bustle*)
GEORGE: Then why wear it?
DOT: Why wear it? Everyone is wearing them!
GEORGE: Everyone...
DOT: You know they are.
GEORGE: Stand still, please.
DOT: I read they're even wearing them in America.
GEORGE: They are fighting Indians in America—and you cannot read.
...
George: Could you drop your head a little, please. (*She drops her head completely*)
Dot! (*She looks up, giggling*)
If you wish to be a good model you must learn to concentrate. Hold the pose. Look out at the water.[12]

This exchange sets up the still pose—or the *tableau vivant*—as a site of conflict and a bodily posture to which Dot and others are constantly admonished to return to by George. Stillness, and related terms like fixed, are spoken and sung throughout the first act. George chides Dot—as well as his other sitters—to "stand still," while Dot describes George as "Fixed. Cold."[13] The question of stillness—and its inherent failure—animates a discussion of the gendered politics of artistic creation and the interface between the "still" painting and live performance, creating, in the fractious dialogue between the visual and the aural, a provocation about what "artistic creation" looks like.

While George does respond to Dot's comments, each of his rejoinders is patronizing. His initial question, "Then why wear it?" implies a critique of (feminized) fashion trends—the idiocy of wearing uncomfortable or constricting clothing simply because others are. Dot's response, nevertheless, affirms the necessity of dressing like everyone else, linking her choices to those of women in America. While George's questions reveal a contrary individualism (in line with the character's actions throughout the musical), Dot is invested in being *en vogue*, which, for her, allows for a sense of communal, transatlantic, belonging linking her to a community of affluent (or at least upwardly mobile or working class) women across Western Europe and the United States. Such belonging, while parsed as frivolous by George, might more productively be imagined as the mapping of geopolitical and cultural affiliations onto consumer goods and their circulation.

Theater scholar Joseph Roach argues costumes and props function as synecdoche and metonyms on the theatrical stage; he writes, "when a crown arrives onstage as a visual metonym, it substitutes for many of the things its wearer would otherwise have to say and do to introduce ... herself convincingly to the audience."[14] Following Roach, we might image the bustle, then, as a visual metonym for fashion trends, which travel from Paris to the United States. Dot's justification for wearing the uncomfortable bustle, after all, is that it is also being worn in America, placing her and her bustle within a transatlantic circulation of goods and ideas.[15] While the bustle signifies such transatlantic circulations, on stage it also serves "to identify or locate the wearer," not only for an audience, but for George.[16] The bustle specifically, "stand[s] in for…[Dot]…as a whole," as the character's costume, more broadly, stands in for Dot within the scope of the musical.[17]

The two's banter about fashion is quickly foreclosed by George; he ends continued discussion of Dot's bustle by stating, "Stand still, please," and finally, "Hold the pose." This interlude reveals the patronizing and gendered logic of their relationship: those that pose are subordinate to the masculinized act of artistic creation. Dot's discomfort with her clothing, her inability to read, or the U.S. government's policies ("They are fighting Indians in American") are subsumed by the totalizing concentration required for George to create. While his admonishment that she stand still renders legible the gendered dimension of the stance, the topic of their conversation—the bustle—frames the male-dominated and valorized field of 'high art' through the preoccupations of nineteenth-century feminine visual culture.[18]

Dot's allusion to American women wearing bustles indicates she, like many Parisian women in the nineteenth century, is a consumer of fashion magazines. In nineteenth-century Europe, there were few avenues open to women to create images and few genres of images created for a female viewership. One notable and prolific exception was fashion plates, which circulated in the pages of fashion magazines to which Dot's off-handed comment alludes.[19] Art historian Anne Higonnet argues that fashion plates drew upon the iconography of women's private drawings and watercolors to advertise and sell the latest styles. Such drawings were a popular activity amongst primarily middle and upper class, white women. Most usually such images—self-portraits, portraits of children, home interiors—were collected into a single album, allowing a woman to flip through a self-authored visual compilation of her life. Higonnet's study reveals the interconnectedness of a certain class of women's private artistic creations and the far more visible fashion images.[20] This link establishes the larger cultural importance of such albums through the widespread adoption of their aesthetic and eventual transatlantic circulation within the pages of fashion magazines. While Higonnet's study aims to elevate the status of women's private albums in the nineteenth century, women almost exclusively created fashion plates and the editorial content in such magazines. *La Mode Illustrée*'s editor and the team in charge of its plate production were all women; similarly, *Le Moniteur de la Mode* and *La Mode Artistique* were both edited by women. Fashion magazines generated a distinctly feminine aesthetic and were also a space in which women could work, producing widely circulated content. Indeed, such magazines' reach was on par with, if not greater than, contemporaneous political magazines.

In 1866, *La Mode Illustrée* had 58,00 subscribers, which was "more than ten times the circulation of its July Monarchy counterparts and equal to that of.... *Le Figaro*," a popular political magazine.[21] By 1890, their subscriber base was 100,000. The "political" magazine *L'Illustration*, justifying its creation of a journal specifically dedicated to fashion, argued, "'Fashion today is everywhere. The fashion journal has become as compelling a necessity for the wife as the political journal is for the husband.'"[22] While *L'Illustration* maintained a gendered ideological divide, which underwrote much nineteenth-century discourse around fashion, the magazine's opening salvo—"fashion today is everywhere"—is a compelling argument for collapsing claims that sought to iterate divides between the political and the feminine. Fashion magazines presented, circulated, and repeated, with each new edition, a distinctly feminine visuality, centering it within French visual culture more broadly.

To return again to *Sunday*'s opening conversation, Dot, then, is referencing a female-authored and female-aimed iteration of visual culture. It is a reference that might be read as an assertion of a feminine visuality counter to the masculinist one framed by Seurat's fine art. Her allusion to reading about American women wearing bustles comes while George is viewing her, creating a subtle re-mapping of a unidirectional male, authorial gaze. She, like George, is positioned as a consumer of images of women, if to a distinctly different end. While such a shift in visual referents is, perhaps, not a large-scale change in how historians have discussed modernity in nineteenth-century France, it nevertheless decenters Impressionist and Post-Impressionist visuality as a predominate frame through which to understand modernity.

I do not mean to chart this relationship between women's private diaries and fashion magazines to uncritically champion the feminist potential of fashion magazines or Dot's allusion to them. Certainly, an important analysis of this relationship is the subsumption of the non-capitalist, intimate iconography of women's albums by the financial interests of fashion magazines. Here, however, I note that fine art, like Impressionist paintings, largely avoids such assessment. While it is imbricated in market forces, such art is imagined as harnessing rather than succumbing to the market. Dot's bustle, so clearly positioned as a luxury consumer good, renders the other object that circulates in the musical, Seurat's painting, legible as such.

It is now owned by the Art Institute of Chicago (AIC), and in the decades following its completion in 1886, was exhibited in Paris, Brussels, Minneapolis, and Boston before being bequeathed to the AIC in 1926.[23] The painting's exhibition history traces an emerging global pathway, foretelling the ubiquity of Impressionist works in museums around the world and the way in which such images have been re-purposed to adorn all manner of things. Numerous writers in the nineteenth century and contemporary art historians have noted Impressionist and Post-Impressionist painters' acumen at depicting the things increasingly available in a modernizing Paris; it is, perhaps, a fitting legacy for such works that they are some of the most reproduced images in museum gift stores, adoring umbrellas and tote bags, posters and scarves.[24]

The bustle, representing mass production, foregrounds Seurat's painting not as a unique object of artistic creation, but rather as a mass-produced, globally circulating consumer good: available for purchase, whether in a gift store or through the price of a ticket to Sondheim and Lapine's musical. Framing the first act through Dot's bustle exposes fashion and the painting as co-conspirators in a project of aligning aesthetics and consumption, codified notions of gender, race, and class, exhibition, and visuality.

The brief exchange between Dot and George reveals the parallel consumer histories of their respective objects, her bustle and his painting, while, nevertheless privileging the former. Clear in the oblique relationship between Dot's bustle and George's painting is the cultural cache afforded each object and, by extension, each character. The bustle was synonymous with frivolity and the ludicrousness of feminized commodity culture. Throughout the end of the nineteenth century, it was repeatedly satirized in print culture and in theatrical works like George Bernard Shaw's *Arms and the Man*.[25] Conversely, the very existence of *Sunday* bears out the continued and lasting legacy of Seurat and his Impressionist and Post-Impressionist peers. George's directive to Dot to "stand still," as she asserts a degree of self-reflection about notions of individuality (her conflicted response towards her bustle) reveals how the painting exerts a measure of control over and above the bodies of the people it depicts. Indeed, the slippage in my argument from the historic relationship between feminine visual cultures and Impressionism, on the one hand, and the musical's dramaturgy, on the other, renders the persistence of masculine artistic narratives and masculine articulation of modernity legible.[26]

Dot's lyrics for "Sunday in the Park with George" reveal the persistence of such narratives, describing as she does how it feels to be George's model and sometimes lover.

> DOT: George.
> Why is it you always get to sit in the shade
> while I have to stand in the sun?

> George?
> Hello, George?
> There is someone in this dress! (*Twitches*)
>
> A trickle of sweat. (*Twitch*)
>
> The back of the—(*twitch*)
> —head.
> He always does this.
>
> Now the foot is dead.
> Sunday in the park with George.
> One more Su—(*twitch*)
>
> The collar is damp,
> Beginning to pinch.
> The bustle's slipping—(*twitch*)
>
> I won't budge one inch. (*Tiny twitches*)
> …
> Artists are bizarre. Fixed. Cold.
> That's you, George, you're bizarre. Fixed. Cold.
> …
>
> (*George races over to Dot and rearranges her a bit, as if she were an object, then returns to his easel and resumes sketching.*)
>
> …
> Well, if you want bread
> And respect
> And attention
> Not to say connection,
> Modeling's no profession. (*Poses*)
>
> If you want instead,
> When you're dead,
> Some more public
> And more permanent
> Expression
> (*Poses*)
> Of affection,
> You want a painter[27]

Her lyrics reveal the corporeal conditions of posing, describing modeling as requiring concentration, physical stamina, and an ability to ignore bodily discomfort. It is only through her continual recitation of the way posing in her bustle affects her body that the task of wearing a certain garment and standing in a certain way becomes legible *as* labor. Dot's lyrics ask an audience member to understand the creation of Seurat's painting through their own bodies—imagining the phenomenological labor required to stand still. While Dot enumerates the discomforts of modeling (sweat, itches, feet that fall asleep), the verse "Well, if you want bread/And respect/And attention/Not to say connection,/Modeling's no profession," moves beyond the bodily effects of posing in her bustle to the socio-economic implications of such work. As Dot sings, modeling does not bring one economic security, respect, or connections, benefits that variously accrue to the musical's male characters. To stand still is not a neutral act, the song reveals, but rather one tied to gendered expectations of muse and artist.[28]

The constant repetition of the need to pose and stay still creates a useful and potent embodiment of the masculinist and authorial directives, which underwrote the visual hierarchies of nineteenth-century French culture. In "Gossip," sung by the ensemble, the group holds a tableau while they sing about their various relationships with each other and with work. The tableau uses the suspension of movement—after all, actors can only remain still for so long—as a theatricalized and controlled choreography of suspense. When will the song end? When will the tableau break? When will a fight break out on stage? The staging heightens the lyrics' barely contained discontent with class hierarchies. For example, the Boatman, continually cancerous throughout the show, sings:

> Overprivileged women
> Complaining,
> Silly little simpering
> Shopgirls,
> Condescending artists
> "Observing,"
> "Perceiving" …
> Well, screw them![29]

His lyrics strike out against his more affluent park companions, his ire particularly gendered, reserved for upper-class women and silly shopgirls. Yet the controlled and largely still staging of the song offers an embodied contrast to his lyrics, underlining the tense balance required to maintain class status as Paris rapidly modernized. The ensembles' *tableau vivant* frames stillness, then, as the corporeal state just before the class revolution.

Unlike the members of the ensemble, who hold their still poses throughout "Gossip," Dot leaves her still pose mid-way through "Sunday in the Park with George." Following his manhandling, the lights change, indicating a shift in how the audience is meant to interpret Dot's song. She is no longer in the same scene as George, dutifully posing for him; rather, she is singing her innermost thoughts and feelings to the audience. As she sings, "Well, if you want bread," Dot begins to move across the stage space. Hooking her parasol over George's arm, she claims the downstage, arms spread wide as she sings. Her dynamic movement is linked to a shift in the animation. A bright white silhouette of Dot appears on the wall behind her; it is this silhouette that George continues to sketch, as Dot voices her frustration with George's inability to connect. In the 1984 Broadway production, Bernadette

Peters' dress opened, allowing her to move about the stage space dressed in corset and bloomers, the padded infrastructure of her bustle clearly visible.[30] While Peters' staging was coquettish, and foreshowed the upcoming scene in which the character appears at her dressing table in her underthings, the '08 design choice eschewed this sexualized construction of femininity.[31]

The silhouette underscores the extent to which Dot is merely a shape within George's artistic process. Metzger writes, "the silhouette connotes a ... form but also a surface without substance." and, etymologically, indicates "something done cheaply."[32] Her animated silhouette can be contextualized in relationship to notions of feminized frivolity that attached to fashion but also to its shifting economic structure, in which goods could be made ever more cheaply for sale in large department stores. Within histories of visual culture, the silhouette is also particularly gendered and sexualized. A tantalizing mystery, a woman's silhouette reveals just the shape of a body—or garment—evacuating the individual. Dot's bustle identifies her character; it also, as the animated silhouette makes clear, substitutes Dot for her garment. The animated outline, in other words, masks the woman herself. While the silhouette trades on the (sexualized) form of the women's body, clothed or otherwise, Metzger's analysis ends arguing that the silhouette makes visible a transformational process.[33] The silhouette appears in theater and film, he argues, during those moments in which a turning point is in store for a character; the outline of the character's form indicated that what might fill in that outline is not yet clear, not yet decided. Disaggregating the character Dot from her form, the projected silhouette allows her to leave her still pose and sing out to the audience her inner most feelings. The appearance of her silhouette marks a moment of transformation for Dot, or at least the beginning of the character's narrative arch, which concludes with her leaving George and immigrating to America.

Dot and her bustle offer a way of re-framing the discourse that surrounds *Sunday*, valorizing and privileging not Seurat's artistic practice but rather considerations of feminine visual cultures and the labor linked to such visuality. If the musical seeks to complexify the discourse surrounding Seurat's legacy (the artist a stand-in for both a larger canon of nineteenth-century European artists and the more abstracted notion of artistic genius and obsession) it, nevertheless, privileges the iconic figure. Clear in the oblique relationship between Dot's bustle and George's painting is the cultural cache afforded each object and, by extension, each character. The bustle was synonymous with frivolity and the ludicrousness of feminized commodity culture. Conversely, the very existence of *Sunday* bears out the continued and lasting legacy of Seurat and his Impressionist and Post-Impressionist peers.

Dot's song ends with "But most of all/ I love your painting...."[34] Musically, the melody that supports her lines is made up of a plethora of eighth notes, resembling "the flickering light effects that Seurat and his contemporaries would achieve with their dots."[35] The musical composition of her song reveals Sondheim's desire to demonstrate the compositional similarities between his music and Seurat's painterly techniques.[36] As the last notes of "Sunday," fade, Dot returns to her initial still pose. The last few lines of the song are semi-spoken,

> DOT: The tip of a stay.
>   Right under the tit.
>   No, don't give in, just
>   Life the arm a bit...
> GEORGE: Don't lift the arm, please.
> DOT: Sunday in the park with George...
> GEORGE: The bustle high, please.[37]

While the visual placement of the notes on the page and their verbal performance calls back to Seurat, Dot's lines add a gendered dimension to this compositional allusion to Seurat's painterly technique. Her lyric directly follows her pronouncement of how much she loves George's paintings is "I think I'm fainting." Dot speaks the line as she returns to her position, and the white animated silhouette on the stage wall disappears. As Russell walks back across the stage, though, her body cast a long dark shadow across the walls of the stage, a dark twin to the white animated silhouette still projected on the wall. Her shadow adds layers of visual depth to the animation: an embodied indexical mark, lingering across the stage's walls. Russell's shadow functions as an indexical rejoinder to the animation, marking the actress's embodied presence on stage, and challenging the gendered layers of stillness—the painting, the animated silhouette, George's directs to 'stay still'—to which Russell's character was subject.

Dot represents a distillation of Seurat's painting. The pun on the character's name, after all, indicates that she stands in as an embodied placeholder for Seurat's particular painterly technique.[38] If her name interpolates Dot as the material for Seurat's painting, she is also meant to function as an embodiment of Sondheim's score, the eighth notes in "Sunday in the Park with George" a transposition of Seurat's dots into music.[39] She offers a re-imagining, then, of Seurat as well as of Sondheim. On a meta-theatrical level, the character functions as not only an often-critical observer of George, with whom she shares the stage, but also, I would argue, as a critical observer of Sondheim. Dot's characterization, and my reading of it here, parallels the development of certain strains of feminist art history published in the late 1970s and into the 1980s. Scholars like Griselda Pollack and Linda Nochlin, most famously in the U.S. academy, championed a re-evaluation of the role of (albeit primarily white) women artists in art historical canons, and, in particular, drew attention to the ways white, Euro-American women's labor and visibility was occluded in previous narratives of *fin de siècle* artistic production. Analyzing Dot through her bustle and the extended network of feminine visuality and labor it surfaces, challenges what is meant by "artistic genius" in all its gendered dimensions, asking a viewer of the musical or listener of the score to reflect upon Seurat and Sondheim's role. Dot renders legible, in other words, the sexist visual politic of not only late nineteenth-century France but also the musical's core dramaturgy.

The acts of embodied stillness throughout *Sunday* frame a notion of modernity indebted not to movement—Walter Benjamin's influential notion of the *flâneur*—but rather the stasis of the held pose.[40] The *tableaux vivant*, as an apparatus of theatrical address, functions, therefore, as a "controlling mechanism for regulating *the ways in which women became* [and become] *visible in public*" and on the stage.[41] The *flâneur* represents a (masculine) figure of urban conquest understood through the sites and spaces of an emerged consumerism. In other words, if the *flâneur* has long stood as an archetype for modernity, then *Sunday*'s staging offers a way to re-write such histories, drawing our attention instead to what it means to be still and stilled within certain spaces and times—or at least in the theatricalization of them. Dot renders legible not this masculine figuration of urbanity, but rather a feminized one. Stillness draws our attention to the margins of urban space, in which domestic staff, shop girls, sex workers, bar maids, and married women dwell.

The still pose, thus, does two things: First, it articulates a site of control and masculine mastery, a stance that falls in line with much art historical and literary scholarship on nineteenth-century Paris and the divisions between masculinity and femininity. Second, however, it also functions as a site of potentiality, which prefigures a profound move towards greater (self-)actualization. To be still is the nascent movement before the move outwards. As a minor thread in contemporary critical theory, stillness is most often framed as a gesture of refusal or a politics of passivity. For example, in *Empire* Michael Hardt and Antonio Negri frame stillness as a revolutionary gesture in its nascent stages. They write, "[stillness or] refusal is the beginning of a liberatory politics, but it is only a beginning."[42] The still pose is the stance one strikes right before bursting into action, abandoning passivity for full engagement in the existing social order.

Dot's stillness then, as well as her bustle, mark her figure and narrative arch as a site of profound disconnect and nascent action. By standing still, after all, we are able to see her fashionable attire and attest to her figure's propinquity to the fashion plate. This relationship is one of visual similarity, but also embodied affinity. Dot, like the fashion plates to which her held pose renders her so similar and to which she alludes in her comment to George, circulates. She, like those plates, travels across the Atlantic, leaves George and Paris behind. It is in these moments of stillness that we see a kind of refusal; it is a refusal of George's lack of attention and control, of being treated as an object and reduced to her geometrically pleasing parts. These still poses, then, prefigure Dot's abandonment of a highly gendered notion of passivity—the artist's model—and see the character propelling herself into a most modern of trajectories: transatlantic immigration. Fashion and stillness within *fin de siècle* Paris function as sites/sights of, paradoxically, motility and mobility for a female viewership, indicating not an individual mastery of urban space, but rather a means of looking elsewhere—at the plates of a fashion magazine or the sweat dripping down Dot's spine, at female-authored texts or the periphery characters that populate the stage.

NOTES

1. I use the abbreviation *Sunday* throughout to reference the musical's complete title.
2. See among other works: T.J. Clark, *The Painting of Modern Life: Paris in the art of Manet and His Followers* (New York: Knopf, 1984); Theodore Reff, *Manet and Modern Paris* (Chicago: University of Chicago Press, 1982); Anna Gruetzner Robins, *A Fragile Modernism: Whistler and his Impressionist Followers* (New Haven: Yale University Press, Paul Mellon Centre for Studies in British Art, 2007).
3. Aruna D'Souza and Tom McDonough, eds., *The Invisible Flâneuse? Gender, Public Space, and Visual Cultures in Nineteenth-Century Paris* (Manchester: Manchester University Press, 2006), 5.
4. See among other words: Roselyn Deutsche, *Evictions* (Cambridge: MIT Press, 1996); Aruna D'Souza, "Why the Impressionists Never Painted the Department Store," in *The Invisible Flâneuse? Gender, Public Space, and Visual Cultures in Nineteenth-Century Paris*, ed. Aruna D'Souza and Tom McDonough (Manchester: Manchester University Press, 2006), 129–147; Linda Nochlin, "Why Have There Been No Great Women Artists?" in *Women, Art, and Power: and Other Essays* (New York: Harper & Row, 1988), 21–34; Griselda Pollock, "Modernity and the Spaces of Femininity," in *Vision and Difference: Feminism, Femininity, and the Histories of Art* (London: Routledge, 2003), 50–90; Janet Wolff, "The Invisible *Flâneuse*: Women and the Literature of Modernity," *Theory, Culture & Society* 2.3 (1985): 37–46.
5. Deutsche wrote of Clark's work, "[his] repression of feminism is not ... necessitated by his interest in the category of class. Instead, it is authorized by his image of the social as a complete entity in which a single set of social relations are privileged as determinate." Deutsche, 196–198.
6. Sean Metzger, *Chinese Looks: Fashion, Performance, Race* (Bloomington: Indiana University Press, 2014).
7. For extensive work on the Broadway musical and Sondheim's work in relationship to gender, race, and class see Sandor Goodhart, "Reading Sondheim: The End of Ever After," *Reading Stephen Sondheim: A Collection of Critical Essays*, ed. Sandor Goodhart (London: Routledge, 2000); David Savran, *Highbrow/lowdown: Theater, Jazz, and the Making of the New Middle Class* (Ann Arbor: University of Michigan Press, 2009); Stacy E. Wolf, *Changed for Good: A Feminist History of the Broadway Musical* (New York: Oxford University Press, 2010); "Keeping Company with Sondheim's Women," *The Oxford Handbook of Sondheim Studies*, ed. Robert Gordon (New York: Oxford University Press, 2014), 365–383; *A Problem Like Maria: Gender and Sexuality in the American Musical* (Ann Arbor: University of Michigan Press, 2002).
8. Tableaux vivants are a theatrical convention dating back to at least the "late medieval and early renaissance periods," and performances of the *Quem Quaeritis* and similar liturgical dramas. The convention was used again in Europe, beginning in 1500s, as means of celebrating or commemorating royal visits or anniversaries. The use of tableaux vivants in theatre, though, is perhaps most associated with the nineteenth-century; theatrical troupes integrated the staging of a painting into the narrative of a play, either by, as in Victorian melodramas,

having the actors form a picture on stage or by "constructing a play out of the story implied by a painting." Jack W. McCullough, *Living Pictures on the New York Stage* (Ann Arbor: UMI Research Press, 1981), 1–10.

The French philosopher and writer Denis Diderot used the tableau vivant as a means of producing a "certain kind of subject (an Enlightened, rational one) as a certain kind of object," i.e., the *mise en scéne* of the tableau. The tableau vivant for Diderot functioned as an embodied illustration, which indexed an interface between "live" and "art." Deborah Levitt, "Living Pictures: From Tableaux Vivants to Puppets and Para-Selves," *Acting and Performance in Moving Image Culture: Bodies, Screens, Renderings*, ed. Jörg Sternagel, et al. (Bielefeld: Transcript Verlag, 2012), 179.

    9. In the 1985 Broadway production, Mandy Patinkin and Bernadette Peters originated the roles of George and Dot, respectively.

    10. Edward T. Bonahue, Jr., "Portraits of the Artist: Sunday in the Park with George as 'Postmodern' Drama," *Stephen Sondheim: A Casebook*, ed. Joanne Gordon (New York: Garland, 1997) 172.

    11. Sondheim refers to "Sunday" as an "inner monologue song." Richard Kislan, "Stephen Sondheim," *The Musical: A Look at the American Musical Theater* (Englewood Cliffs, NJ: Prentice-Hall, 1980) 161.

    12. Sondheim and Lapine 18–19.

    13. *Ibid.*, 18, 22.

    14. Joseph Roach, "The Global Parasol: Accessorizing the Four Corners of the World," in *The Global Eighteenth Century*, ed. Felicity A. Nussbaum (Baltimore: Johns Hopkins University Press, 2003), 98.

    15. Concurrent with the Haussmannization of Paris was the rise of mass production and industrialization, harbingers of the creation of the department store and ready-to-wear clothing; fashion, thus, could be enjoyed (or at least lusted after) by many segments of Parisian, European, and U.S. society. The expanding rail and postal networks meant that it was easier to ship Parisian goods to the French provinces and the broader world. Justine De Young, "Representing the Modern Woman: The Fashion Place Reconsidered (1865–75)," in *Women, Femininity and Public Space in European Visual Culture, 1789–1914*, ed. Temma Balducci and Heather Belnap Jensen (Farnham, Surrey: Ashgate, 2014), 98.

    16. Roach, "Global Parasol," 98.

    17. *Ibid.*

    18. See note four.

    19. De Young, "Representing the Modern Woman," 97; De Young, "Representing the Modern Woman," 99.

    20. Anne Higonnet, "Secluded Vision: Images of Feminine Experience in Nineteenth-Century Europe," in *The Expanding Discourse: Feminism and Art History*, ed. Norma Broude and Mary Garrard (New York: HarperCollins 1992), 171–185.

    21. De Young, 97.

    22. De Young, 99.

    23. "About This Artwork: *A Sunday on La Grande Jatte*," Art Institute of Chicago, accessed October 27, 2015, http://www.artic.edu/aic/collections/artwork/27992.

    24. One can purchase, for instance, at the AIC's store posters and matted prints of the painting as well as art books, an umbrella, or a collectable figurine of Seurat with an easel beside him display a tiny version of *A Sunday on La Grande Jatte*.

    25. George Bernard Shaw, *Arms and the Man* (New York: Dodd, Mead, 1926); Amber Butchart, "Exhibitionism: Fashion Fantasies—Fashion Plates and Fashion Satire 117–1925," *Amber Butchart: Fashion Historian*, 21 March 2011, 27 October 2015 http://amberbutchart.com/2011/03/21/exhibition-review-fashion-fantasies-fashion-plates-and-fashion-satire-1775-1925/.

    26. Historically, of course, *plein air* painting struggled to gain acceptance from the art establishment then dominated by the tradition and structures of academic painting. However, the circle of artists who advocated for and practiced what came to be known as Impressionism gained favor from the public if art critics remained skeptical of their work. In a similar fashion, we might note the musical's struggle to be considered on par with "straight" plays or tragedy.

234   Part 3: Staging the Other

27. Sondheim and Lapine 21–24.

28. Linda Nochlin, "Morisot's *Wet Nurse*: The Construction of Work and Leisure in Impressionist Painting," in *The Expanding Discourse: Feminism and Art History*, ed. Norma Broude and Mary Garrard (New York: HarperCollins 1992), 231–243.

29. Sondheim and Lapine 44.

30. The early 1980s marked, in the U.S. academy, an initial wave of feminist art historians re-writing long-established narratives of the fin de siècle, Impressionist art, and early modernists movements.

31. The duet "Color and Light," stages yet another of Seurat's paintings, *Jeune femme se poudrant*. Dot, sitting at her vanity, embodies the women in the painting. Originally, an additional painting, *Une baignade, Asniéres*, was also to have its own song, but, in the final production, it is only a short scene near the start of the show. A large frame is wheeled on stage, the painting projected into it. The song the boys in the painting were to sing, "Yoo-hoo!," survives only as underscoring. Banfield 358.

32. Metzger, 139.

33. Metzger 156, 158.

34. Sondheim and Lapine, *Sunday in the Park*, 24.

35. Stephen Banfield, *Sondheim's Broadway Musicals* (Ann Arbor: University of Michigan Press, 1993), 351.

36. This musical homage to another genre of art making, however, is not limited to the obvious allusion to Seurat. Stephen Banfield argues that throughout *Sunday*, and most overtly in Dot's opening song, Sondheim references Wagner. While Banfield is interested in the ways in which Sondheim seemingly incorporated musical forms and techniques occurring contemporaneously to Seurat's painting practice, here I am interested in the subtle allusion to the *gesamtkunstwerk*. The aural link within the song's final refrain to an artist committed to thinking across aesthetic practices—music, dance, painting—asserts a link between art practices imagined as distinct.

37. Sondheim and Lapine, *Sunday in the Park*, 24–25.

38. Thomas P. Alder, "The Sung and the Said: Literary Value and the Musical Dramas of Stephen Sondheim," in *Reading Stephen Sondheim: A Collection of Critical Essays*, ed. Sandor Goodhart (New York: Routledge, 2000), 52.

39. The character's name is attributed to Lapine.

40. Walter Benjamin, *The Arcades Project*, trans. Howard Eiland and Kevin McLaughlin (Cambridge: Belknap Press of Harvard University Press, 2002).

41. D'Souza and McDonough, 4.

42. Michael Hardt and Antonio Negri, *Empire* (Cambridge: Harvard University Press, 2000), 204.

Gradations of Presence
Armida in Nineteenth-Century Italian Dance Librettos

MELISSA MELPIGNANO

There is no such a thing as an innocent transposition.
—Gérard Genette, *Paratexts*[1]

Armida, one of the female—and most uncanny—protagonists of Torquato Tasso's Italian Renaissance poem *Jerusalem Delivered* (1575), has received great attention in the history of modern European theater.[2] Set during the First Crusade,[3] this poem portrays Armida as a beautiful Arab sorceress, who receives from the lord of Damascus the task of distracting the Christian soldiers.[4] Pretending to be an heir to the throne threatened by a tyrannical uncle, she asks the crusaders to help her defeat him; in exchange, she offers military support and subjection to their vassalage despite the religious difference. Enchanted by her beauty and magic, the soldiers give up their duties, abandon their bases, and follow her into her magical realm. The captain of the crusaders calls upon the best of his knights, Rinaldo, in order to rescue his companions. Because of his extraordinary, heroic skills, he succeeds. Informed of their liberation, through her magic powers, Armida attracts Rinaldo to her castle and makes him fall asleep so that she can kill him. But, when she sees him, she instantly falls in love. She transports them both to her enchanted island so that they can live in their love, undisturbed. Rinaldo forgets his military duties, until two crusaders find him and break Armida's love spell; they successfully persuade Rinaldo to leave the sorceress. Abandoned, an infuriated Armida destroys her castle and magical realm, promising revenge against her lover. She obtains the help of the Egyptian army and, displaying her beauty in warrior garments, finally faces Rinaldo in fight. Again about to kill him, her love once again prevents her from harming the object of her love. Defeated, she flees to the forest. As she is about to commit suicide, Rinaldo arrives and promises to love and marry her if she converts to Christianity. Armida happily accepts Rinaldo's plea to abandon her religion and love him.

Generally, theatrical stagings of literary episodes reduce the literary source to the most spectacular or dramatic moments. Dance and opera transpositions of *Jerusalem Delivered*, for instance, mainly focus on three scenes: Armida's desire of revenge because of the liberation of the soldiers; the enchanted time of their love and passion; Rinaldo's abandonment and her renewed desire to revenge manifested in the destruction of her enchanted realm. Musicologists Giovanni Morelli and Elvidio Surian claim that this episode of Armida and Rinaldo initiated a "secular religion of expressive eros" and a tradition of "sentimentalization of conflicts" in the Western theater.[5] The dance transpositions of the episode often intensify these qualities that Morelli and Surian identify, while also reifying normative gender dynamics. Along these lines, the orientalized and exoticized portrayal of Armida as the Arab seductress—nonetheless presented as blond in the poem to respond to the white, Eurocentric canon of beauty—is perpetuated in several dance librettos in order to foster a specific ideology of the female body as sexualized, and to make statements about women's moral conduct.[6]

In this essay, I offer an analysis of nineteenth-century ballet rewritings of Armida through a close reading of ten Italian dance librettos (1799–1873), in order to delineate the complex transformations of this female character throughout the nineteenth-century European ballet scene. Dance librettos propose variations of Tasso's narrative and characterization of Armida by altering, for instance, the plot or her portrayal in terms of energy and emotion. My claim is that dance librettos are able to convey the corporeal presence of dancing bodies whose memory would be lost otherwise. In order to grasp the tactility of nineteenth-century dancing bodies in these peculiar texts, I first offer a methodology for reading dance librettos. In the light of this, I will subsequently explore the statements that dance librettos pose about Armida's body, illustrating how shifts in Armida's corporeality correspond to shifts in the conceptualization of the female body as well as in her construction as the uncanny "other."

Text as Dance: A Methodology to Read a Dance Libretto

A dance libretto is a short text that narrates the plot of a dance performed on stage. In Italian, *libretto* means "small book." Its writing is characterized by the economy and fragmentation of its prose. It differs from the staged dance, an opera libretto (which contains the lines sung by the singers), a music score, and a dance notation. At the same time, its writing is informed by the choreographic intensity of the dancing bodies to which the libretto relates, so that the *body* of the text moves with a rhetorical energy that addresses the energy of the dance on stage, as I will show in the following pages.[7] Dance librettos operate *in partnership* with the bodies of the dance they narrate. The correspondence between bodies and text occurs through the variations between the danced scene and the libretto's textuality: a correspondence based on the difference between the two practices—dancing and writing.[8]

The methodological possibility of reaffirming the dancing bodies through the librettos and claiming a correspondence between text and danced scene lies in the assumption that the latter is not inevitably lost, inescapably ephemeral. To claim, as the performance studies scholar Peggy Phelan does, that performance is an ephemeral practice, invites one to question if also the performing bodies are ephemeral, and, thus, prone to be forgotten.[9] Phelan does not merely deny the presence of the bodies; on the contrary, she claims that presence can affirm itself as the immaterial repository of memory, so that the body of the remembered subject can be inscribed within the memorial horizon of the remembering subject. In this way, according to Phelan's remembering subject, memory is expressed as the performance of a loss.

One problem with this conceptualization, as performance studies scholar Annalisa Sacchi claims, is that it can easily lead to the production of nostalgia for the lost subject. This in turn can prevent us, as readers in the present, from grasping the potential of the performing bodies of the past.[10] For this reason, Sacchi proposes subverting the vertical hierarchy between the "original" performance as ephemeral, and its remains. In light of this, I subscribe to Sacchi's idea that, today, we are "delayed spectators" of a performance that, despite being concluded, we can decide to encounter in our present.[11] Thus, I consider the dance libretto as a "performance remain" of the staged dance it narrates.[12] In this way, I claim that dance librettos give us the opportunity to encounter nineteenth-century dancing bodies through their movement in the text. That is, in reading a dance libretto, I assume a co-presence of text and dance. While I consider the dance libretto as autonomous from the staged dance, I read it as a threshold for the emergence of dancing bodies that would be otherwise lost. Thus, I claim a correspondence between textuality and corporeality. By corporeality, I am referring to the way the reality of the bodies is constructed in their tactile materiality. It is the text that preserves such a tangible permanence.[13] For this reason, I look at the rhetorical strategies in motion in the libretto that cooperate in the creation of movement in the text.[14]

A dance libretto neither lists nor describes the steps nor the position of the arms. The bodies remain in the writing through textual signals, such as word choices and syntactical structures, that indicate their spatiality, temporality, and energy. Writing does not just paint or draw the dance in terms of representation but shows the author's engagement with the bodies and the ideas that preside over their organization. The dancing bodies in their choreographic organization partner with the writing of the libretto, which is comprised of short, quick, incisive structures. Punctuation works as an intervention in the rhythm of a movement or marker of expressive qualities. Thus, the libretto does not *literally* correspond to the scene, which maintains its spatio-temporal autonomy in relation to the written word, as much as the writing mode of the libretto—slender and agile—dodges a *literal* inscription of the bodies on stage. To account for this, I claim that the dance libretto constitutes an autonomous literary genre that allows us to access the dancing bodies of the past in their material presence through a writing that, referring to and informed by the dance, invites the reader to approach it corporeally, sensing and imagining the presence of the bodies—a text that wants to be read as dance, a text that dances.[15]

From the Literary Source to the Dance Stage

In the *canto* IV, when Armida appears to the Christian soldiers, Tasso describes her entrance as a luminous *crescendo*, enhanced by the brightness of her blond hair. While this illuminating vocabulary introduces the enchanted atmosphere associated to Armida, her outstanding attractiveness and sensuality is what attracts and seduces the soldiers. By portraying, what Fredi Chiappelli calls "the image of turmoil per se,"[16] Armida represents the uncanny "other" of the Christian soldier. In his poetry, Tasso expresses these qualities of her presence by intensifying the frequency of verbs denoting movement, descriptions of movement trajectories, and word-choices that convey Armida's turbulent energy.[17]

In the poem, Armida draws a highly dynamic arc—from evil deceiver, through sensual lover of the most prominent crusader, to her final conversion to Christianity. Because of the complexity of her character, which resides in the emotional process and vicissitudes that lead her to a radical transformation, Armida enjoyed great fortune in eighteenth and nineteenth-century dance, drama, and opera theater.[18] Authors writing for the theatrical stage intervened in the literary episode often by selecting specific scenes, and by modifying or adding elements—characters, actions, affects—in order to enhance its representational efficacy and conflictual elements.[19] However, it is pantomime ballet that confers an emotionally and physically articulated human presence upon Amida through dramaturgical interventions aimed at emphasizing an expressive and dramatic physicality. These textual interventions, rely on a rhetoric of passions that, in choreography, translate as modulations of energy.[20] In the following section, I will show how the dramaturgical and choreographic variations in the librettos trace Armida's gradations of presence in nineteenth-century ballet.[21]

Gradations of Presence of Armida

In relation to the emergence of a system of emotions in European ballet, dance scholar Christina Thurner states: "This privileging of the emotions should function in [ballet] by means of a system of codes, which, however, was no longer to be recognizable as such: that system, after all, was no longer based on external rules of representation, as in court ballet, but on the notion of an understanding that was 'interior' and sensual. Thus the demand for the performative signs of this understanding was that they be of a universal character, relevant to all mankind."[22] On the nineteenth-century European stage, dancing bodies faced the task of expressing emotion in an explicit manner through "performative signs," or the construction of a gestural, expressive, corporeal dramaturgy. In the dance performance on stage, as well as in the one constructed in the dance libretto, dramaturgy refers to a multi-layered system of expressivity that includes linguistic choices able to perform ideas of body, the energy of a gesture, the proximity among bodies, the temporality that regulates the sequence of actions, the virtuosic and spectacular use of bodies and technique, and the suspension of action to intensify the emotional and visual turmoil. The variations among dance librettos and the ways in which emotions and modes of expression of turmoil are conveyed, implicate different constructions of the body of Armida and different gradations through which her presence emerges from these dancing texts.

One of the most important and renowned ballet choreographers in Europe between the eighteenth and nineteenth century, Onorato Viganò (1739–1811), staged, in 1790, a ballet in three parts to represent the whole episode of Armida.[23] In the final *tableau*, the libretto announces that Armida "vividly" expresses her unhappiness to the audience. In terms of choreographic construction of the scene, the energy represented by the adverb "vividly" and the gestural force of the expression of sadness fill a stage occupied only by her body.[24] The adverb possibly refers to a frantic or vigorous movement sequence that, in that spatial configuration, highlights the solitude of the character.

Different from the original poem, Giovanni Battista Checchi's *Armida Abandoned* (1799), imagines Armida and Rinaldo exchanging rings on the shore of her enchanted island, thus getting married before her conversion.[25] The whole ensemble celebrates the event with a general dance, prioritized over the religious matter. In the middle of the dance, two fellow soldiers arrive interrupting the festive atmosphere, and convince Rinaldo, adorned with flowers and "voluptuous" garments, that love weakened him and that he needs to go back to the crusaders' camp. The moral issue concerns Rinaldo's feminization and deviation from the patriarchal norm. However, the interruption of Rinaldo's sensual discovery also arrests Armida's sexual experience. The feeling of rejection, but also of frustration, drags Armida down to the ground: "She falls fainted," as Checchi performatively writes.[26] When she wakes up, she regains consciousness and recovers her demoniac powers, so that she can perform her despair by flying away on a cart pulled by dragons. Ultimately, in this libretto's scheme, while Rinaldo goes back to the 'righteous' path of the patriarchal order, Armida represents an idea of femininity that coincides with pagan sin and un-granted redemption.

In 1817, Antonio Landini choreographs an Armida psychologically more sophisticated.[27] When she meets Rinaldo, she falls in love "*gradatamente*," "gradually." While Viganò portrayed an Armida with a prompt, immediate, energetic presence, Landini shows an Armida that progressively distills her turmoil. In the libretto, this is marked by the intensification of the commas along with the seductive languidness of verbs such as "to induce" and "to doubt" that suggest how her actions slow down and reduce their intensity.

The following year, Giacomo Piglia writes that Rinaldo and Armida fall in love "*vicendevolmente*," "reciprocally."[28] The Italian six-syllable adverb underlines the temporal intensity of the scene. Such a long adverb, within a meager text such as this dance libretto, visually performs the long lace that joins Armida and Rinaldo, who embody the two opposites *par excellence*. Later in the text, Armida's action turns abrupt and violent when Rinaldo is about to leave her island: "suddenly he is blocked" by Armida. This note indicates a dramaturgical change in the representation and use of Armida's body: from the delicate body that moves gradually in Landini, to the dominant and sudden gesture that surprises Rinaldo as well as the audience, in Piglia.

Three years after his first version, Antonio Landini re-choreographs Armida, augmenting the dramatic effect of the falling-in-love scene.[29] While before she fell in love "gradually," now, she remains "*immobile*," motionless. This suggests a different conceptualization of the body, which responds to a different economy of energy, and performs a different kinesthetic system on stage. While, in 1817, Armida's dancing body releases her energy progressively, in the 1820 libretto, her energy is suspended and held in order to create a different expressive effect.

In Bernardo Vestris's *Armida* (1844), the *pathos* further increases.[30] The libretto, here, is a constellation of adverbs and expressions of time: *sollecita-*

mente; improvvisamente; allorquando; prima; poscia; sul punto ... quando ...; frettoloso; affréttansi; per qualche momento; da un istante all'altro; dopo breve momento (quickly; suddenly; as soon as; before; then; about to ... when ...; in a hurry; hurrying up; for a little while; at any given time; after a little while). Here, Armida can affirm her presence through her body's ability to leave an instantaneous trace in the audience's memory. As Jonathan Crary affirms in his work on vision in the nineteenth century: "As observation is increasingly tied to the body in the early nineteenth century, temporality and vision become inseparable. The shifting processes of one's own subjectivity experienced in time became synonymous with the act of seeing, dissolving the Cartesian ideal of an observer completely focused on an object."[31]

This epistemic change materializes in the dance libretto through the intensification of the temporal presence of the characters. The kinesthetic empathy to which Crary refers, and that which Susan Foster has theorized as culturally informed, resides already in the writing of the libretto.[32] In Vestris, Armida's body seems to materialize and rematerialize before the readers' eyes each time an adverb or expression of time affirms her mobility and corporeality. In this libretto, Armida affirms the dancing subject's desire to move, the livelihood of her body, and her resistance to discourses that oppress her presence.

In his *Armida e Rinaldo* (1845), Antonio Cortesi (1796–1879) offers a highly contrasting interpretation of Armida's character.[33] Cortesi strongly emphasizes the moral binary at play in the narrative by opening the ballet with the figure of the "pious Goffredo," the head of the Christian army and symbol of rectitude in Tasso's poem. His presence is unusual in dance librettos based on the episode of Armida and Rinaldo. Armida is portrayed as deceptive in her beauty, and insistent in her seductive moves and gazes. The moral juxtaposition is highlighted by another unusual character in dance librettos: Armida's uncle, "the perfidious sorcerer." However, when Armida falls in love with Rinaldo, Cortesi offers a docile interpretation of Armida as "tender" and "lovely." At this point, her energetic impulse is strongly mitigated. But her docile body still allows a representation of sensuality full of malice. Such a combination is evident in the use of the verb "*folleggiare*," translatable as "to have frivolous and crazy fun" (also a marker of Violetta's morality in Giuseppe Verdi's *Traviata*). This association of Armida to frivolity and excess aims at delegitimizing her within a patriarchal society, whose bourgeois morality condemns women's sexuality while sexualizing their bodies.

Giuseppe Rota's (1822–1865) *The Loves of Armida and Rinaldo* (1853) characterizes Armida's body as prompt and nervous but less impulsive.[34] In the libretto, Armida does not linger over the beautiful image of a sleeping Rinaldo to fall in love with him. Instead, after being shot by cupid's arrow, she strategically decides with an extemporaneous gesture to have him carried to her magic garden. Rota emphasizes how Armida makes this decision "*improvvisamente*" ("suddenly"). The adverb refers not only to the quality of her movement but of her decision-making process. This Armida does not indulge: she is practical and strategic. Rota underlines this quality by describing her as "*pensosa*" ("pensive") and one who is able to act with order and cognition, as testified by the strict logical planning of her revenge when Rinaldo abandons her: "Informed of Rinaldo's escape, Armida asks Hate for a dagger, and flies away to look for the perjurer."[35] Rota's Armida is the last of these dance librettos which maintains a clear narrative connection to the original poem; like the previous abandoned Armidas, in the final scene, she displays an unprecedented self-control over the events before weeping and fainting.

Throughout the nineteenth century, the representation of emotions acquires complexity by involving the whole body in the creation of an expressive effect. For instance, while in Viganò's libretto, Armida demonstrates her anger by biting her lips, then, in Piglia's libretto her whole body "trembles," so that movement is conceived of as the result of information transmitted through the psyche. Armida's fainting episodes are also the signs of an uncontainable energy that literally cannot be held within her body. Even her desire of revenge is the consequence of her desiring body longing for Rinaldo. His departure—which symbolizes his suppression of his own desire––corresponds to an abrupt interruption of Armida's sensual life too. Armida exhibits the consequences of such a loss on her own body, not just on the environment, but also in the poem and in the eighteenth-century librettos. In the mid-nineteenth century, Armida's body becomes the site for the manifestation of her emotional state. Her hysterical portrayal resides in her fear of the subtraction of her own bodily vitality. Anticipating psychoanalytical discourses, dance presents the female body not just as an object of desire, but also as a desiring agent. In fact, Armida does not only lament Rinaldo's loss but so too mourns the forced denial of her sexuality by the patriarchal order.

Armida represents a female body whose expressivity reacts to a patriarchal system that diminishes her body's political potential by inscribing her in the regime of the uncanny, namely the paradigm of the seductive temptress. A choreographic and cultural analysis of dance librettos helps unveil the patriarchy-informed conception of hysteria as situated in the woman by displaying the patriarchal system of representation and construction of the narrative. Such a patriarchal system is supported by and works together with the colonial mindset that orientalizes Armida's body as uncanny. What remains of Armida's "Arabness" is her exoticized sexuality and eroticism, which is progressively eroded, towards the end of the nineteenth century, in order to discipline the female body and fulfill the normative demands of the European bourgeois theater.[36]

Gradations of Presence (Melpignano) 243

Dance histories written in the West and lacking a cultural studies perspective have treated orientalism and exoticization as a mere matter of fashion or taste,[37] excluding the fundamental role played by the modern imperialist and colonialist mindset that informed an eighteenth and nineteenth-century ballet aesthetic and culture. In *Orientalism*, Edward Said examines European literary classics in order to show how the European, binary construction of "the East" as an inversion of "the West" serves as a legitimizing tool for the affirmation of a Western hegemony. For instance, in an analysis of Gustave Flaubert's accounts of his travels to "the Orient" from the 1850s, where the French writer portrays the Egyptian dancer Kuchuck Hanem, Said observes:

> The Oriental woman is an occasion and an opportunity for Flaubert's musings; he is entranced by her self-sufficiency, by her emotional carelessness, and also by what, lying next to him, she allows him to think. Less a woman than a display of impressive but verbally inexpressive femininity, Kuchuk is the prototype of Flaubert's Salammbô and Salomé, as well as of all the versions of carnal female temptation to which his Saint Anthony is subject. Like the Queen of Sheba (who also danced "The Bee") she could say—were she able to speak—"*Je ne suis pas une femme, je suis un monde.*" Looked at from another angle, Kuchuk is a disturbing symbol of fecundity, peculiarly Oriental in her luxuriant and seemingly unbounded sexuality.... Woven through all of Flaubert's Oriental experiences, exciting or disappointing, is an almost uniform association between the Orient and sex. In making this association Flaubert was neither the first nor the most exaggerated instance of a remarkably persistent motif in Western attitudes to the Orient.[38]

If we consider, like Said, Flaubert's orientalism as epitomizing European mid-nineteenth-century orientalism, we can immediately recognize a parallel between Kuchuck Hanem and Armida's main features. The Arab woman, and more specifically, the dancing Arab woman is the absolute Uncanny of the European man, for, on the one hand, she represents the possibility of unleashing a carnal desire that does not respond to the European, bourgeois moral and behavioral norm; on the other hand, she represents an excess that, as such, needs to be contained, thus reinforcing repressive codes.

Towards the end of the nineteenth century, the increased transparency of female costumes and the progressive shortening from *tutulette* to *tutu* to docilely serve the male gaze parallels the conceptualization of Armida's corporeality itself. In Pratesi's libretto (1872), Armida becomes a benevolent genie.[39] This dematerialization manifests the will of domesticating the character. In 1873, Filippo Senatori choreographs a Parisian Armida, whose authoritarian father wants to contain the sensual excess of his daughter.[40] In *Armida's Dream*, Armida and her friends celebrate All Saints' Day by dancing in the streets when her father drags her away, blaming her for "such libertinage."[41] Confined in her room, her girlfriends appear as the demons of sin, and secretly bring her to a ball, where she dances with the uncanny Ermando. Her father and brother, in disguise, find Armida. She is about to leave with

Ermando but her father takes his mask off, and asks Ermando "to cede him the girl."[42] Despite her apologies, the father curses her. She flees to a forest and prays. There, another man offers to marry her in order to save her. They are about to celebrate the wedding, when a storm starts to rage; everybody blames Armida for the storm and curses her. On top of that, her father's spirit materializes and curses her again. Extremely shocked, Armida collapses and dies. Unable to become disciplined, she is punished with death. Already in Pratesi, Armida was no longer an energetic heroine, but her body was disciplined by making it evanescent.

These late bodies deprived of the possibility of expressing eros parallel the dematerialization of the dance librettos' narrative. Throughout the century, the texts get increasingly shorter, and the description of steps and spatial configurations substitute the narrative. Ultimately, this indicates how the narrative itself becomes less relevant and technical virtuosity takes over expressivity in late nineteenth-century ballet.

Armida as the Other/Armida's Other

At the end of the eighteenth century, in Viganò's libretto, Armida's vulnerable presence complies with the tradition of the *comédie larmoyante* in the final, moving reconciliation of the lovers and the "moral triumph," meaning the affirmation of the norm. Differently, Checchi's libretto shows Armida as the embodiment of evil and a woman who cannot be granted redemption. The increasing occurrence of adverbs in nineteenth-century dance librettos nuance a schematic representation of Armida, developing an expressive and choreographic complexity—as Landini and Piglia show—and realizing the Noverrian dictate. Vestris's Armida emphasizes this urgency of affirming Armida's corporeality through an extreme investment in expressions and adverbs of time—such as "promptly," "in a hurry," "immediately"—that affirm Armida's presence in the moment. The libretto extends beyond its narrative and choreographic contingent purpose, becoming the site for an explicit theoretical investigation of the dancing body.

In Cortesi's libretto, which emphasizes the bourgeois moral binary, Armida's body must be condemned as the uncanny object of male, heteronormative sexual desire. She performs resistance by rebelling against her uncle's in order to kill Rinaldo and affirm her love for the Christian soldier. Mitigating an investment in her emotions, Rota's Armida displays control, until the end when both Armida and Rinaldo cry. However, while Rinaldo's tears are the signs of remorse, Armida cries for the realization of patriarchy's ultimate triumph over her: "the image of a horrendous chaos."[43] In the end, Armida does not generically decide to "take revenge" but, more lucidly, "decides to pursue [Rinaldo]."[44] Rota's Armida is not at the mercy of the patriarchal order but wants to pursue it.

Towards the end of the century, the energy of this Armida gets depleted in two ways: first, by neglecting her corporeality on the most physical level through her de-corporealization in Pratesi's libretto; secondarily, by frustrating, denying, and punishing her body as a body of desire through proponents of patriarchy in Senatori's libretto.

Ultimately, throughout these nineteenth-century Italian dance librettos, Armida works as the "other" of a patriarchal norm that aims at denying the complexity of her corporeality and attempting to discipline her behavior by domesticating her into a heteronormative order and conservative behavioral system. Already in the second half of the eighteenth century, architect Francesco Milizia claimed that "huge public theaters of Italy promoted an erotic scopophilia that was deleterious to social mores and proper relations between the sexes."[45] The bourgeois audience's need for a sexualized object of desire had to be repressed through its progressive ephemeralization. The dematerialization of Armida's body and resistive force as a woman at the end of the nineteenth century follows her disappearance as the uncanny Arab sorceress and her transformation into the more manageable object of bourgeois moral and sexual order.

Armida's progressive de-corporealization goes together with the erasure of her ethnicity, which responds to ballet's "physical refinement and purification" as signs of Western "bodily civility."[46] This doubly violent, colonial gesture speaks about the politics embedded in—and promoted through—bourgeois morality and its *dispositifs* of representation (like ballet). Ballet is a system of knowledge that shows the power that the Western, patriarchal order has to subjugate, and also erase, bodies, in particular female bodies, more specifically non-normative bodies—meaning those bodies that do not promote and glorify Western hegemony. Nevertheless, these dance librettos, insisting on Armida's sexual power and disempowerment, work as the Western "site of the unconscious,"[47] marked by anxieties of dominion and control, which Armida's corporeal arc itself reveals and problematizes. Hence, the remains of Armida's energetic clan and the presence of the female dancing body's physical labor endure in the traces drawn by the writing of the dance librettos. It is up to us, readers of the present, to reactivate the political energy of those bodies through a reading that unveils the performative force and political potential of the dancing bodies of the past.

NOTES

1. Gérard Genette, *Paratexts: Thresholds of Interpretation* (Cambridge: Cambridge University Press, 1997), 294.
2. By uncanny, in the context of this essay, I refer to the particular use made in dance studies scholarship in reference to the representation of the female body as object of desire

and, simultaneously, of patriarchal control, in nineteenth-century ballet. I refer, in particular, to Susan Foster's use in *Choreography and Narrative* (Bloomington: Indiana University Press, 1996), where uncanny is the ballerina's presence (5) when considered, not merely as a dancing body on "the historical stage of vanished dancing" (5) but as the engine of ballet's "machinery of desire" (1) as seen from "the theoretical stage of choreographic strategies for representation" (5). Although Foster does not explicitly mention Freud, her use of "uncanny" connects to *Das Unheimliche* (1919). According to Freud, the uncanny is something or someone that manifests social taboos or represents the repressed impulses of the viewer, who, thus, feels fear and fascination at once. See Sigmund Freud, "The 'Uncanny,'" in *The Standard Edition of the Complete Psychological Works of Sigmund Freud*, vol. XVII (1917–1919): "An Infantile Neurosis and Other Works" (London: Vintage/Hogarth Press, 2001), 217–256. See, also, Susan Foster, "The Ballerina's Phallic Pointe," in *Corporealities*, ed. Susan Foster (London: Routledge, 1996), 1–24.

3. On the legacy of Armida and Tasso's characters in European art, see Giovanni Careri, *La Jérusalem Délivrée du Tasse: Poésie, Peinture, Musique, Ballet; Actes du Colloque Organisé au Musée du Louvre. Paris, les 13 et 14 Novembre 1996* (Paris: Klincksieck, 1999), and Giovanni Careri, *Gestes d'Amour et de Guerre: la Jérusalem Délivrée, Images et Affects, XVIe-XVIIIe siècle* (Paris: Editions de l'École des Hautes Études en Sciences Sociales, 2005).

4. The most recent English edition of Tasso's poem is Torquato Tasso, *Jerusalem Delivered*, trans. Anthony M. Esolen (Baltimore: Johns Hopkins University Press, 2000). For the original Italian, I recommend Torquato Tasso, *Gerusalemme Liberata*, ed. Fredi Chiappelli (Milano: Rusconi, 1982).

5. Giovanni Morelli and Elvidio Surian, "Contagi d'Armida," in *Torquato Tasso: Letteratura, Musica, Teatro, Arti Figurative*, ed. Andrea Buzzoni (Bologna: Nuova Alfa Editoriale, 1985), 159. My translation.

6. For a wider discourse on the intersections of colonialism, race, and gender in European theater, see, for instance, Wendy Sutherland, *Staging Blackness and Performing Whiteness in Eighteenth-Century German Drama* (New York: Routledge, 2016). For a specific study on the representation of the female body in nineteenth-century ballet, see Susan Leigh Foster, *Choreography & Narrative: Ballet's Staging of Story and Desire* (Bloomington: Indiana University Press, 1998). For a case study on Orientalism in dance librettos, see Claudia Jeschke, Gabi Vettermann, and Nicole Haitzinger, *Les Choses Espagnoles: Research into the Hispanomania of 19th Century Dance* (München: Epodium, 2009).

7. This essay does not aim to offer a genealogy of the development of the dance libretto. However, for historical clarity, note that dance librettos in their narrative form (like the ones at which I look here) mark the passage from the baroque period to the eighteenth-century dance Reform era as conceptualized by ballet masters and theoreticians such as Jean-Georges Noverre, Gasparo Angiolini, and John Weaver, with the emergence of the *ballo pantomimo*, in Italy, and the *ballet d'action*, in France. In particular, Noverre, in his *Lettres sur la Danse* (1760, re-edited in 1803), gives the dance libretto a programmatic role. He takes a distance from the *ballet de cour*, whose goal was to geometrically and graphically inscribe the dancing body in order to represent the centralization of sovereign power. For the pre-Reform ballet discourse, see Mark Franko, *Dance as Text: Ideologies of the Baroque Body* (New York: Cambridge University Press, 1993; revised edition New York: Oxford University Press, 2015). Rather than focusing on the creation of figurative patterns, Noverre stresses the importance of the expressive and emotional element, which needs a new system of dance writing able to convey the succession of events, passions, and affects. Furthermore, Noverre strongly defends the autonomy of the dance libretto as a proof of choreographic authorship and as a tool for the legitimization of ballet as an autonomous art. In this essay, I show the livelihood of Armida's character as theorized in reformed ballet, and the potential of the dance libretto as a primary source for the study of corporeality. Pioneering studies for the cultural analysis of dance librettos are Judith Chazin-Bennahum, *Dance in the Shadow of the Guillotine* (Carbondale: Southern Illinois University Press, 1988); Selma Jeanne Cohen and Katy Matheson, eds., *Dance as a Theatre Art: Source Readings in Dance History from 1581 to the Present* (Princeton: Princeton Book Company, 1992); and Foster, *Choreography & Narrative*. See also Susan Foster's genealogy of the term "choreography" in *Choreographing Empathy: Kinesthesia in Performance* (New York: Routledge, 2011), 15–72.

8. One of the most debated issues in the study of dance librettos concerns their authorship. Dance librettos are not formally signed, and the authorship of the text is not indicated as different from the function of the choreographer (also called "composer of the ballet"). My assumption is that the two functions coincided, as conventionally thought. In my opinion, debates about authorial authenticity constitute a false problem that distracts from the analysis of the libretto's textual-performative potential. For a treatise of the problem of authorship, and for a different approach to the study of dance librettos, see Edward Nye, *Mime, Music and Drama on Eighteenth-Century Stage: The Ballet d'Action* (Cambridge: Cambridge University Press, 2011).

9. For an in-depth study of Phelan's positions on the matter of ephemerality, see Peggy Phelan, *Unmarked: The Politics of Performance* (London: Routledge, 1993), and Peggy Phelan, "Trisha Brown's *Orfeo*: Two Takes on Double Endings," in *Of the Presence of the Body: Essays on Dance and Performance Theory*, ed. André Lepecki (Middletown, CT: Wesleyan University Press, 2004), 13–28. On presence and ephemerality, see also the other essays in Lepecki, *Of the Presence of the Body*.

10. See Annalisa Sacchi, "Oltre la nostalgia: spettatori ritardatari e performance storiche," in *Biennale Danza 2013. Abitare il mondo: trasmissione e pratiche* (Festival catalogue, June 29–30, 2013), ed. Stefano Tomassini, with Daniela Giuliano (Venezia: La Biennale di Venezia, 2013), 100–105.

11. Sacchi, "Oltre la nostalgia," 104. My translation.

12. I borrow the expression "performance remain" from Rebecca Schneider, *Performing Remains: Art and War in Times of Theatrical Reenactment* (Abingdon: Routledge, 2011). In a section entitled "Troubling Disappearance," Schneider writes: "In the theatre the issue of remains as material document, and the issue of performance as documentable, becomes complicated—necessarily imbricated, chiasmically, with the live body.... If we consider performance as of disappearance, of an ephemerality read as vanishment and loss, are we perhaps limiting ourselves to an understanding of performance predetermined by our cultural habituation to the logic of the archive? To the degree that performance is *not* its own document (as [Richard] Schechner, [Herbert] Blau, and [Peggy] Phelan have argued), it is, constitutively, that which *does not remain*" (97–98). Similarly, dance scholar Stefano Tomassini describes dance librettos as "relics." See "Il fiore in rivolta," in *Variazioni su Adone*, vol II (Lucca: M. Pacini Fazzi, 2009), 13. To consider the dance libretto as a relic prevents its subordination to the staged representation, and helps us release it from an arbitrary hierarchy that neutralizes its potential in the present as well as the corporeal tensions of its writing. Furthermore, following Diana Taylor's theorization, I suggest that the dance libretto works as both archive and repertoire. As an "archive," it retains the tangible memory of a performance; it also constitutes a "repertoire," because its writing directly refers to embodied acts on stage. See Diana Taylor, *The Archive and the Repertoire: Performing Cultural Memory in the Americas* (Durham: Duke University Press, 2007), 19–23. As my analysis shows, dance librettos as archives work also as a systematizing and controlling platform for the female body. At the same time, following Schneider's critique of Taylor's that points out how archives are also "houses of the theatrical slip and slide" (Schneider, *Performing Remains*, 107), dance librettos are not only repositories of dominion but sites for the re-emergence of the female dancing bodies in their potentialities.

13. The methodology I propose to read a dance libretto considers the direct relation between dance and text. This is a case in which, in my opinion, text does not subjugate or limit the body; on the contrary, it works as a bridge between us and the ballet bodies of the past. To expand on the debate about the hierarchical relationship between body and text, see, for example, Ronald J. Pelias, "Performative Writing as Scholarship: An Apology, an Argument, an Anecdote," *Cultural Studies, Critical Methodologies* 5, no. 4 (2005): 415–424; Ellen W. Goellner and Jacqueline Shea Murphy, eds., *Bodies of the Text: Dance as Theory, Literature as Dance* (New Brunswick: Rutgers University Press, 1995); Laura Colombo and Stefano Genetti, eds., *Pas de Mots. De la Littérature à la Danse* (Paris: Hermann Éditeurs, 2010); Edward Nye, ed., *Sur Quel Pied Danser? Danse et Littérature* (Amsterdam: Rodopi, 2005).

14. Susan Foster defines "corporeality" as "the study of bodies through a consideration of bodily reality, not as natural or absolute given but as a tangible and substantial category

248   Part 3: Staging the Other

of cultural experience.... From the beginning, the body is capable of being scripted, being written. In that writing, the body's movements become the source of interpretations and judgements"; see her "Introduction," in *Corporealities: Dancing Knowledge, Culture and Power*, ed. Susan Foster (London: Routledge, 1996), x.

   15. I expand on the rhetorical mechanisms and literary autonomy of the dance libretto in "Scritture della Presenza: il Libretto di Ballo come Genere Letterario" ["Writings of the Presence: the Dance Libretto as a Literary Genre"], in Stefano Tomassini and Andrea Torre, eds., *Parole su Due Piedi: Danza e Letteratura nella Cultura Italiana* (Roma: Editoria & Spettacolo, 2018, in print).

   16. Fredi Chiappelli, handwritten note (probably for a university seminar), "Center for Medieval and Renaissance Studies," University of California, Los Angeles. On exoticizing and orientalizing *dispositifs* in Renaissance literature and performance, see Dominique de Courcelles, ed., *Littérature et exotisme, XVIe-XVIIIe siècle* (Paris: École des Chartes, 1997); C. C. Barfoot and Theo D'haen, eds., *Oriental Prospects: Western Literature and the Lure of the East* (Amsterdam: Rodopi, 1998); Emily C. Bartels, *Spectacles of Strangeness: Imperialism, Alienation, and Marlowe* (Philadelphia: University of Pennsylvania Press, 1993); Kathleen M. Ashley, "'Strange and Exotic': Representing the Other in Medieval and Renaissance Performance," in *East of West: Cross-Cultural Performance and the Staging of Difference*, ed. Claire Sponsler and Xiaomei Chen (New York: Palgrave, 2000), 77–91. For specific studies in the context of dance librettos, see Jeschke, Vattermann, and Haitzinger, *Les Choses Espagnoles*, and, in particular, Foster, *Choreography and Narrative*, for the relationship among exoticism, desire, and the uncanny.

   17. While the episode of Armida and Rinaldo in Tasso's poem covers the *canti* IV, V, VII, X, XIV, XVI, XVII, XIX, and XX, the ones to which dance librettos usually refer are the *canti* IV, XIV, and XVI—the ones with the more sensual scenes and where Tasso's writing expresses Armida's physicality and energy. For instance, the following lines convey her fierce posture: "Lovely Armida, proud of her peerless form, | of the charms of her youth and womanhood, | takes up the challenge, leaves that very night, | traveling ever along the hidden road | and in her woman's habit nursing hope | to rout armed legions never yet subdued" (IV, 27, 1–6). And, in the most sensual *canto*, we can see how Tasso plays with movement: "And with that word she smiled, and ne'ertheless | Her love–toys still she used, and pleasures bold! | Her hair, that done, she twisted up in tress, | And looser locks in silken laces rolled, | Her curles garlandwise she did up–dress, | Wherein, like rich enamel laid on gold, | The twisted flowers smiled, and her white breast | The lilies there that spring with roses dressed" (XVI, 23).

   18. On the legacy of the episode of Armida and Rinaldo in theater and popular culture, see Jean Starobinski, *Enchantment: the Seductress of Opera* (New York: Columbia University Press, 2008); Jean Starobinski, *Rousseau e Tasso* (Torino: Bollati Boringhieri, 1994); and Marzio Pieri, *Tasso e l'Opera* (Parma: Edizioni Parma, 1985).

   19. Scholars have extensively acknowledged how the *Armide* by Quinault-Lully (1686) marked the beginning of the long tradition of the episode on the European theatrical scene. For instance, Quinault added the allegorical character of *Sagesse* (Wisdom), whose role is to anticipate and explicate the moral content of the episode, and focused on the conflict between passion and duty in Armida--instead of privileging this issue in Rinaldo, like in the poem. Quinault also offers a very human, psychologically defined portrait of Armida as suffering of insomnia due to anxiety, as stuck in moments of *empasse* before an important dramaturgical turn. While in the seventeenth century, the moral dilemma and the relationship between Armida and her interlocutors prevails, in the nineteenth century rewritings of the episode, passion and pleasure are the central issues at stake.

   20. Noverre, in his *Lettres sur la Danse*, categorizes passions in order to dispose them within a choreographic scheme whose organization can be recognized in the dance libretto. Noverre clearly explains the difference between "mechanical dance" and "*ballet d'action*" as follows: "La première [mechanical dance] ne parle qu'aux yeaux, et les charme par la simétrie de ses mouvemens, par le brillant des pas et la varieté des tems; par l'elévation du corps l'aplomb, la fermeté, l'élégance des attitudes, la noblesse des positions, et la bonne grace de la personne. La seconde que l'on nomme danse en action est, si j'ose m'exprimer ainsi, l'âme

de la première; ell lui donne la vie et l'expression, et en séduisant l'œil elle captive le cœur, et l'entraine aux plus vives émotions; voilà ce qui constitue l'art" (Noverre, *Lettres*, II, letter XI, 106).

21. I borrow this expression from performance studies scholar Enrico Pitozzi, who, by *gradation* indicates the "consistency" and "intensity" of presence. Pitozzi demonstrates how gradations can help us observe "a form of *passage of states of matter*" (124). I believe that his theorizations of the gradations of presence can illuminate the passage from stage to libretto, from the materiality of the written word to the tactility and corporeality of the bodies emerging from writing. See Enrico Pitozzi, "Figurazioni: uno studio sulle gradazioni di presenza," in *On Presence*, ed. Enrico Pitozzi, "Culture Teatrali. Studî, interventi e scritture sullo spettacolo," 21 Annuario 2011, 107–127.

22. Christina Thurner, "Affect, Discourse, and Dance before 1900," in *New German Dance Studies*, ed. Susan Manning and Lucia Ruprecht (Urbana: University of Illinois Press, 2012), 23.

23. The ballet was presented over two nights and, thus, with two librettos: Onorato Viganò, *Rinaldo e Armida, o sia La conquista di Sionne, ballo eroico pantomimo diviso in tre atti* [...] da rappresentarsi nel nobilissimo Teatro di San Samuele di Venezia il carnovale dell'anno 1790, and Onorato Viganò, *La conquista di Sionne, parte terza delle azioni di Rinaldo e d'Armida* [...] da rappresentarsi nel nobilissimo Teatro di San Samuele di Venezia il carnovale dell'anno 1790, in *Andromaca, dramma per musica*, da rappresentarsi nel nobilissimo Teatro di San Samuele, il carnovale dell'anno 1790, Venezia, appresso Modesto Fenzo, 47–62.

24. Viganò, *La conquista di Sionne*.

25. Giovanni Battista Checchi, *Armida abbandonata*, in *Gli amanti comici, ossia La famiglia in scompiglio, dramma giocoso per* musica, da rappresentarsi nel nobilissimo Teatro Vendramin di San Luca, nell'estate 1799, in Venezia, 1799, per il Casali, con permissione, 35–38.

26. Checchi, *Armida abbandonata*, 37.

27. Antonio Landini, *Armida e Rinaldo, ballo pantomimo* [...], in Ferdinando Paër, *La Didone, melodramma per musica*, da rappresentarsi nell'Imp. E R. Teatro di via della Pergola la primavera del 1817 sotto la protezione di S.A.I. e R. Ferdinando III Granduca di Toscana, Firenze, Stamperia Fantosini, 1817, 19–24.

28. Giacomo Piglia, *Armida e Rinaldo, ballo eroico in tre scene in tre scene* [...], in Stefano Pavesi, *Celanira, melodramma eroico in due atti*, da rappresentarsi nel Teatro Eretenio di Vicenza, l'estate dell'anno 1818, poesia di Rossi, musica di Pavesi, Vicenza, Tipografia Parise, 23–30.

29. Antonio Landini, *Rinaldo ed Armida, ovvero Il trionfo della Verità, ballo eroico in sei atti* [...], in Stefano Pavesi, *Le Danaidi romane, dramma per musica*, da rappresentarsi nel Regio Teatro di Torino nel carnovale dell'anno 1820 alla presenza delle LL. SS. RR. MM., Torino, presso Onorato Derossi Stamp. e Lib. Del R. Teatro, 1820, 41–48.

30. Bernardo Vestris, *Armida, azione mimica in sette parti* [...] da rappresentarsi nell'I. R. Teatro alla Scala il carnovale del 1844, Milano, per Gaspare Truffi, 1844.

31. Johnathan Crary, *Techniques of the Observer. On Vision and Modernity in the Nineteenth Century* (Cambridge: MIT Press, 1992), 98.

32. See Foster, *Choreographing Empathy*, and in particular 177–178 for references to eighteenth and nineteenth-century ballet.

33. Antonio Cortesi, *Armida e Rinaldo, azione mimica coreografica in tre parti* [...] da rappresentarsi nell'I. e R. Teatro in via della Pergola la primavera del 1845 sotto la protezione di S. A. I. e R. Leopoldo II Granduca di Toscana, ec. ec. ec., Firenze, Tipografia Galletti.

34. Giuseppe Rota, *Gli Amori di Armida e Rinaldo, azione mimica danzante in quattro scene* [...] da rappresentarsi nell'I. R. Teatro alla Canobbiana la primavera del 1853, Milano, coi tipi di Luigi di Giacomo Pirola.

35. Rota, *Gli Amori di Armida e Rinaldo*, n.p.

36. One of the key features of European bourgeois theater is its universalizing project: "Bourgeois theatre's association with pathos asserts self-expression and sensibility as sources of truth, while its promotion of empathy highlights what is shared by all mankind, regardless of rank" (Meg Mumford, "bourgeois theater," 2010, *The Oxford Companion to Theatre and Performance*, ed. Dennis Kennedy, accessed November 4, 2017, http://www.oxfordreference.com/view/10.1093/acref/9780199574193.001.0001/acref-9780199574193-e-522).

37. I specifically refer to Curt Sachs, who sees the dances of "oriental civilizations" as spectacle, in his *World History of the Dance* (New York: Norton, 1937); to Walter Sorell, who, in *The Dance through the Ages* (New York: Grosset & Dunlap, 1967), locates eighteenth- and nineteenth-century ballet's interest for "the Orient," culturally, in Antoine Galland's French translation of *A Thousand and One Nights*, and, economically, in the increased trade with China; and, more recently, to Carol Lee, who, in *Ballet in Western Culture: A History of its Origins and Evolution* (New York: Routledge, 2002), does not problematize "the current [late eighteenth century-] rococo rage in Europe for all things oriental" (105).

38. Edward Said, *Orientalism* (New York: Vintage Books, 1978), 187–88.

39. Ferdinando Pratesi, *Armida, ballo romantico fantastico in quattro atti e sei scene* [...] da rappresentarsi al Real Teatro Bellini nell'anno teatrale 1871-1872, Palermo, Stabilimento Tipografico Lao, 1872.

40. Filippo Senatori, *Sunto del Ballo fantastico in un Prologo e 5 Quadri intitolato Il sogno d'Armida* [...] che si rappresenta nel Teatro Cerruti in Cagliari nella Stagione d'Autunno dell'anno 1873, Cagliari, Tipografia del Commercio.

41. Senatori, *Il sogno di Armida*, n.p.

42. Ibid.

43. Cortesi, *Armida e Rinaldo*, n.p.

44. Ibid.

45. Martha Feldman, *Opera and Sovereignty: Transforming Myths in Eighteenth-Century Italy* (Chicago: University of Chicago Press, 2007), 357.

46. Foster, "The Ballerina's Phallic Point," 1.

47. Meyda Yeğenoğlu, *Colonial Fantasies: Towards a Feminist Reading of Orientalism* (Cambridge: Cambridge University Press, 1998), 26.

BIBLIOGRAPHY

Ashley, Kathleen M. "'Strange and Exotic': Representing the Other in Medieval and Renaissance Performance." In *East of West: Cross-Cultural Performance and the Staging of Difference*, edited by Claire Sponsler and Xiaomei Chen, 77–91. New York: Palgrave, 2000.

Barfoot, C. C., and Theo D'haen, eds. *Oriental Prospects: Western Literature and the Lure of the East*. Amsterdam: Rodopi, 1998.

Bartels, Emily C. *Spectacles of Strangeness: Imperialism, Alienation, and Marlowe*. Philadelphia: University of Pennsylvania Press, 1993.

Careri, Giovanni. *La Jérusalem Délivrée du Tasse: Poésie, Peinture, Musique, Ballet; Actes du Colloque Organisé au Musée du Louvre. Paris, les 13 et 14 Novembre 1996*. Paris: Klincksieck, 1999.

Careri, Giovanni. *Gestes d'Amour et de Guerre: la Jérusalem Délivrée, Images et Affects, XVIe-XVIIIe siècle*. Paris: Editions de l'École des Hautes Études en Sciences Sociales, 2005.

Chazin-Bennahum, Judith. *Dance in the shadow of the guillotine*. Carbondale: Southern Illinois University Press, 1988.

Checchi, Giovanni Battista. *Armida abbandonata*. In *Gli amanti comici, ossia La famiglia in scompiglio, dramma giocoso per musica*, da rappresentarsi nel nobilissimo Teatro Vendramin di San Luca, nell'estate 1799, in Venezia, 1799, per il Casali, con permissione: 35–38.

Cohen, Selma Jeanne, and Katy Matheson, eds. *Dance as a Theatre Art: Source Readings in Dance History from 1581 to the Present*. Princeton: Princeton Book Company, 1992.

Colombo, Laura, and Stefano Genetti, eds. *Pas de Mots. De la Littérature à la Danse*. Paris: Hermann Éditeurs, 2010.

Cortesi, Antonio. *Armida e Rinaldo, azione mimica coreografica in tre parti* [...] da rappresentarsi nell'I. e R. Teatro in via della Pergola la primavera del 1845 sotto la protezione di S. A. I. e R. Leopoldo II Granduca di Toscana, ec. ec. ec., Firenze, Tipografia Galletti.

De Courcelles, Dominique ed. *Littérature et exotisme, XVIe-XVIIIe siècle*. Paris: École des Chartes, 1997.

Feldman, Martha. *Opera and Sovereignty: Transforming Myths in Eighteenth-Century Italy*. Chicago: University of Chicago Press, 2007.

Foster, Susan Leigh. *Choreography & Narrative: Ballet's Staging of Story and Desire.* Bloomington: Indiana University Press, 1998.
Foster, Susan Leigh. *Choreographing Empathy: Kinesthesia in Performance.* New York: Routledge, 2011.
Foster, Susan Leigh, ed. *Corporealities: Dancing Knowledge, Culture and Power.* London: Routledge, 1996.
Franko, Mark. *Dance as Text: Ideologies of the Baroque Body.* New York: Cambridge University Press, 1993.
Freud, Sigmund. "The 'Uncanny.'" In *The Standard Edition of the Complete Psychological Works of Sigmund Freud*, vol. XVII (1917-1919): "An Infantile Neurosis and Other Works," 217-256. London: Vintage/Hogarth Press, 2001.
Genette, Gérard. *Paratexts: Thresholds of Interpretation.* Cambridge: Cambridge University Press, 1997.
Goellner, Ellen W., and Jacqueline Shea Murphy, eds. *Bodies of the Text: Dance as Theory, Literature as Dance.* New Brunswick: Rutgers University Press, 1995.
Jeschke, Claudia, Gabi Vettermann, and Nicole Haitzinger, *Les Choses Espagnoles: Research into the Hispanomania of 19th Century Dance.* München: Epodium, 2009.
Landini, Antonio. *Armida e Rinaldo, ballo pantomimo* [...]. In Ferdinando Paër, *La Didone, melodramma per musica*, da rappresentarsi nell'Imp. E R. Teatro di via della Pergola la primavera del 1817 sotto la protezione di S.A.I. e R. Ferdinando III Gran-duca di Toscana, Firenze, Stamperia Fantosini, 1817: 19-24.
Landini, Antonio. *Rinaldo ed Armida, ovvero Il trionfo della Verità, ballo eroico in sei atti* [...]. In Stefano Pavesi, *Le Danaidi romane, dramma per musica*, da rappresentarsi nel Regio Teatro di Torino nel carnovale dell'anno 1820 alla presenza delle LL. SS. RR. MM., Torino, presso Onorato Derossi Stamp. e Lib. Del R. Teatro, 1820: 41-48.
Lepecki, André, ed. *Of the Presence of the Body: Essays on Dance and Performance Theory.* Middletown, CT: Wesleyan University Press, 2004.
Morelli, Giovanni and Elvidio Surian, "Contagi d'Armida." In *Torquato Tasso: Letteratura, Musica, Teatro, Arti Figurative*, edited by Andrea Buzzoni, 151-65. Bologna: Nuova Alfa Editoriale, 1985.
Noverre, Jean-Georges. *Lettres sur la Danse et sur les Ballets.* Stuttgard: Aimé Delaroche, 1760. Reprint, Whitefish, MT: Kessinger Publishing, 2010.
Nye, Edward. *Mime, Music and Drama on Eighteenth-Century Stage: The Ballet d'Action.* Cambridge: Cambridge University Press, 2011.
Nye, Edward, ed. *Sur Quel Pied Danser? Danse et Littérature.* Amsterdam: Rodopi, 2005.
Pelias, Ronald J. "Performative Writing as Scholarship: An Apology, an Argument, an Anecdote." *Cultural Studies, Critical Methodologies* 5, no. 4 (2005): 415-424.
Phelan, Peggy. *Unmarked: The Politics of Performance.* London: Routledge, 1993.
Pieri, Marzio. *Tasso e l'Opera.* Parma: Edizioni Parma, 1985.
Piglia, Giacomo. *Armida e Rinaldo, ballo eroico in tre scene in tre scene* [...]. In Stefano Pavesi, *Celanira, melo-dramma eroico in due atti*, da rappresentarsi nel Teatro Eretenio di Vicenza, l'estate dell'anno 1818, poesia di Rossi, musica di Pavesi, Vicenza, Tipografia Parise: 23-30.
Pitozzi, Enrico, ed. *On Presence.* In "Culture Teatrali. Studî, interventi e scritture sullo spettacolo," n. 21, 2011.
Pratesi, Ferdinando. *Armida, ballo romantico fantastico in quattro atti e sei scene* [...] da rappresentarsi al Real Teatro Bellini nell'anno teatrale 1871-1872, Palermo, Stabilimento Tipografico Lao, 1872.
Quinault, Philippe. *Armide, tragédie lyrique en cinq actes*, représenté en 1686. In *Œuvres choisies de Quinault*, t. II, 385-430. Paris: Imprimerie de Crapelet, 1824.
Rota, Giuseppe. *Gli Amori di Armida e Rinaldo, azione mimica danzante in quattro scene* [...] da rappresentarsi nell'I. R. Teatro alla Canobbiana la primavera del 1853, Milano, coi tipi di Luigi di Giacomo Pirola.
Sacchi, Annalisa. "Oltre la nostalgia: spettatori ritardatari e performance storiche." In *Biennale Danza 2013. Abitare il mondo: trasmissione e pratiche*, Festival catalogue, June 29-30, 2013, edited by Stefano Tomassini, with Daniela Giuliano, 100-05. Venezia: La Biennale di Venezia, 2013.

Said, Edward. *Orientalism*. New York: Vintage Books, 1978.
Schneider, Rebecca. *Performing Remains: Art and War in Times of Theatrical Reenactment*. Abingdon: Routledge, 2011.
Senatori, Filippo. *Sunto del Ballo fantastico in un Prologo e 5 Quadri intitolato Il sogno d'Armida* [...] che si rappresenta nel Teatro Cerruti in Cagliari nella Stagione d'Autunno dell'anno 1873, Cagliari, Tipografia del Commercio.
Starobinski, Jean. *Rousseau e Tasso*. Torino: Bollati Boringhieri, 1994.
Starobinski, Jean. *Enchantment: the Seductress of Opera*. New York: Columbia University Press, 2008.
Sutherland, Wendy. *Staging Blackness and Performing Whiteness in Eighteenth-Century German Drama*. New York: Routledge, 2016.
Tasso, Torquato. *Jerusaleem Delivered*. Translated by Anthony M. Esolen. Baltimore: Johns Hopkins University Press, 2000.
Tasso, Torquato. *Gerusalemme Liberata*. Edited by Fredi Chiappelli. Milano: Rusconi, 1982.
Taylor, Diana. *The Archive and the Repertoire: Performing Cultural Memory in the Americas*. Durham: Duke University Press, 2007.
Thurner, Christina. "Affect, Discourse, and Dance before 1900." In *New German Dance Studies*, edited by Susan Manning and Lucia Ruprecht, 17–30. Urbana: University of Illinois Press, 2012.
Tomassini, Stefano, ed. *Variazioni su Adone, vol. II. Libretti musicali e di ballo (1614–1898)*. Lucca: M. Pacini Fazi, 2009.
Tomassini, Stefano, and Andrea Torre, eds. *Parole su Due Piedi: Danza e Letteratura nella Cultura Italiana*. Roma: Editoria & Spettacolo, 2018.
Vestris, Bernardo. *Armida, azione mimica in sette parti* [...] da rappresentarsi nell'I. R. Teatro alla Scala il carnovale del 1844, Milano, per Gaspare Truffi, 1844.
Viganò, Onorato. *Rinaldo e Armida, o sia La conquista di Sionne, ballo eroico pantomimo diviso in tre atti* [...] da rappresentarsi nel nobilissimo Teatro di San Samuele di Venezia il carnovale dell'anno 1790.
Viganò, Onorato. *La conquista di Sionne, parte terza delle azioni di Rinaldo e d'Armida* [...] da rappresentarsi nel nobilissimo Teatro di San Samuele di Venezia il carnovale dell'anno 1790. In *Andromaca, dramma per musica*, da rappresentarsi nel nobilissimo Teatro di San Samuele, il carnovale dell'anno 1790, Venezia, appresso Modesto Fenzo: 47–62.
Yeğenoğlu, Meyda. *Colonial Fantasies: Towards a Feminist Reading of Orientalism*. Cambridge: Cambridge University Press, 1998.

Choreogrammatics
About the Cover of This Book

LINDA CARREIRO

In 2011, I created ten textual veils entitled *Poetry Skins*, linking a series of poems letter by letter into a rice-paper lattice of horizontal words. The words dangled sideways, effectively coercing viewers to bend laterally in order to decipher the disoriented text. Extremely light and delicate, these works responded to every breath or approach, allowing them to almost dance with the viewer. Opening the possibilities of bodily movement while reading was an extension of the shifting reading paths required to navigate the fragmented and broken texts in my earlier artworks, here with the idea of the whole body elicited. *Poetry Skins* implicated the viewer in a performance of reading, acutely aware of their tilting body as both they and the tissue-like pieces moved. I have created subsequent works which emphasize the relationship between kinesthetics and reading, to reveal how the meaning of words can be opened and even intensified through a bodily performance of text. Employing complicated textured surfaces, curvatures and expansive lengths, the words in these pieces cannot be scanned from a distance nor in a stationary position; because most of the words are partially hidden or closed off, these works necessitate that the reader engage in a series of pronounced movements including a side-to-side, up and down, or across a space viewing. In this merging of textuality and motility, I have conceived of the term "choreogrammatics."

Choreogrammatics is a theory of meaning-making centered around the body's experience with words, particularly how the words appear to open up in conjunction with the viewer's movements. Building upon the linguistic modes of syntagmatic and paradigmatic cognition, choreogrammatics maintains that alternate understandings of words can occur through the act of moving while reading. Many language and reading theorists have established that an understanding of words are formed by both associative and connotative codes; the first influenced by personal correlations and the second embedded through cultural ascriptions. Compounding these with kinesthetic responses, choreogrammatics proposes to shake loose some of the prescribed readings, allowing for diverse interpretations. Establishing a study with participants of varying heights, ages, genders, cultural backgrounds, language learning, and degrees of mobility, I interviewed, observed and recorded the individuals as they navigated the artworks. In many cases, the readers became animated and responsive to their own gestures while reading the text, movements which started to amplify with each subsequent reading. In turn, readers began to play with the meanings of words through each engagement with the work.

All participants strongly conveyed a heightened awareness of their motions while reading, citing their own bodies as significant factors in the work. They noted 'performing' with the piece, or having the sense that they were taking part in a dance. Maxine Sheets-Johnstone asserts the primacy of movement as a significant form of knowledge, one acquired by individuals even before language.[1] In *The Primacy of Movement*, she forges the impact that movement has on knowledge: "In making kinetic sense of ourselves, we progressively attain complex conceptual understandings having to do with *containment*, with *consequential relationships*, with *weight*, with *effort*, and with myriad other bodily-anchored happening and phenomena that in turn anchor our sense of the world and *its* happening and phenomena."[2] Creating an amalgam of two seemingly unrelated activities—reading and moving—the ambulatory viewer is mobilized into the position of reader, where the intensified actions of weight, effort and space appear to challenge and heighten the relationship to the words.

Emerging from the lateral orientation involved in reading *Poetry Skins*, my piece *Relay* (2015) attempts to expand the reading act over a span of length and time. Over 30 feet of words on tissuous paper are suspended as a scroll at various heights across the center of a room, with the start of the work near the ceiling and the end hanging onto the floor (Fig. 1). A kind of relay race of language emerges, as word slips, homophones and phonemes are printed down the length of the scroll like a musical score, each a few centimeters apart. The slippage of words, where letters pass and then repeat with a small change in letterforms, form a reading pace with which viewers can keep step. While small sections can be easily viewed, the soft-bronze letters on ascending and descending waves, plus the sheer expanse of words, necessitate movement in order to fully engage with the piece (Fig. 2). Many levels of complication are provided to the reader: through the length of the work; through the distress of words on the tissuey paper; as the bronze words catch the light and

About the Cover of This Book (Carreiro)   255

force a sideways toggle; through sub-vocalization, vocalization and gesturing; and as an act of reading in pairs.

The rhythm of reading is accentuated by the shifts of the reader, not only as they traverse the scroll, but as they maneuver through word interventions. Readers navigate the text through a metered, gaited reading, walking frontward, backwards or sideways alongside the piece as they move through the lines of shifting words. As people explore the piece from start to finish, the pace of reading is echoed by their actions: they read and move at a slowed, measured gait, sometimes stopping at a trickier run of words, then resuming, creating a cadence of reading (Fig. 3). This work relies first on lulling readers into rhythm, mirroring their movements as they read, then stopping and starting throughout the length as the run of words become challenging. This proposes that rhythm and reading are merged, and the result becomes like musical chairs where the music starts and stops, and the participants follow suit. While the experience of reading aloud is familiar to many, by adding walking, stretching, sideways shifting and bending, along with the spatiality

Figure 1. *Relay* installed at Harcourt House Gallery, Edmonton, March 2015. The piece, composed of hand-printed bronze pigment on 30 feet of Asarakusui paper, undulates throughout the space, forcing viewers to move while reading (photograph by Stacey Cann).

Figure 2. A viewer interacting with *Relay* at the Nickle Galleries, Calgary, October 2016 (photograph by Mohsen Kamalzadeh).

of the gallery setting and the prolonged word game, the reading act is heightened as participatory and consciously performed.

While *Relay* generates a leisurely, gradual pace to reading, with subtle twists and turns of the body, *Gallimaufry*, elicits dynamic side-to-side and up-and-down movements where readers bend and shimmy across a wall. Positioned as swooping ribbons, the banners of words invite the reader to oscillate as they progress along the splayed sections. The honeycomb texture of the paper varies when pulled or manipulated, creating notable fissures in the words depending upon the viewer's stance. This creates an illusory effect while reading; certain letters lose their distinction when viewers are standing directly in front of them, and come into focus only when the reader repositions slightly left or right of the word (Fig. 4). Forcing the readers to sidestep until words are decipherable enables a dance-like interplay through the reading act, as viewers pivot and dodge to find or focus upon fractured words. Reading cannot happen in one scan, nor in one spot; viewers have to pause, shift sideways, then jump to different points of the text. Much like *Relay*, it is impossible for readers not to be aware of their movements while reading, since they sway and shimmy while relocating across the wall. This rhythm brings attention to the physicality of reading, instilling awareness of their own body as viewers ambulate the piece through a variety of movements. As

About the Cover of This Book (Carreiro) 257

Figure 3. The tissuous paper of *Relay* allows words to be seen through the backside, and many viewers respond by reading both sides of the piece (photograph by Mohsen Kamalzadeh).

one participant noted: "I feel like I'm part of a performance piece. I feel like my interaction is part of the work, like a dance. I become the artwork."[3] In another instance, a participant noted the lack of punctuation marks within the text, and began to perceive her movements—up, down, side to side—as standing in for the missing symbols as period (stop), exclamation point (look), and comma (pause).[4] This interaction of body and text creates possibilities for how the piece is performed, with the words and their placement acting as a loose score to be improvised. One reader fell to a crouch each time he read the phrase "My mother said to pick the very best one and you are not it" positioned at the bottom of *Gallimaufry*—a response to the words that were echoed in his movements, or inversely, a movement influencing his interpretation of the phrase as "heavy."[5] Another participant noted a compulsion to reread the work several times in different ways, since each experience with words was more "involved."[6] She crisscrossed the wall over and over, stopping at certain areas to reread or fixate on a phrase, creating connections and interpretations with other areas. The text in *Gallimaufry* shifts significantly in each reading just as the viewer shifts with the piece, fragmented and undulating. Like a choreographer, I place a set of potential navigations for readers of the artworks; readers approach each reading with a

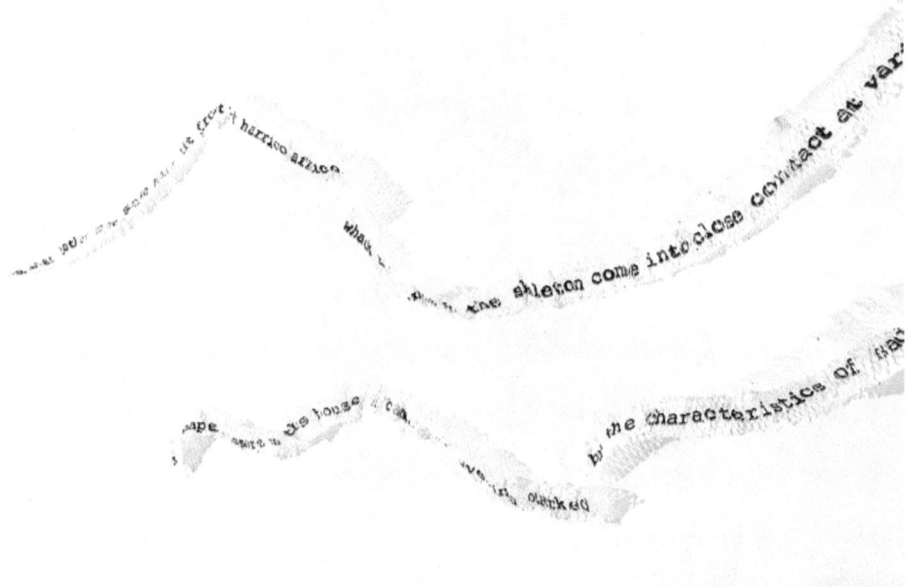

Figure 4. Detail of *Gallimaufry* sections showing the fissures from the die-cut honey-combed paper, which necessitates oscillating in order for words to come into view (photograph by Blake Chorley).

new set of movements, and every reader with their own set of movements, with a subsequent range of interpretations emerging from each physical encounter.

When text becomes "impudent"[7]—when it misbehaves—our reading habits also misbehave, permitting a thumbing at the rules. Just as placing something upside-down will initiate an echoed tilt in the curious viewer, the words within these artworks—staged unconventionally in order to compel movement—seem to unroll meanings, making the reader continuously question their presupposition with a text. As one reader kept revising her response with each performing of *Gallimaufry*, noting its increasing intensity, she commented: "We usually reread something to savor a passage, but here it's pulling my memories, but also questioning how we use our signs. You released these words onto the paper. You left them there. If I read them, then they're going to catch."[8] Sheets-Johnstone underscores that physical, dynamic processes involved in movement have the potential to generate feelings, and equally, emotions can stir our bodies.[9] With choreogrammatics, physical receptions are invoked through the interplay with words, many times urging a semantic questioning of certain phrases, creating metaphors and poetic responses, or calling up personal memories. One participant stated: "Certain words come out but there are multiple subtexts."[10] This was echoed by another who clearly conveyed how the sentences had changed from his initial reading to something more acute: "I'm going past my assumptions about the words. I have heightened awareness."[11] Pushing against the semiotics of words, which provides a set of propositions for any reader, choreogrammacy then nudges prescribed meanings into a place where assumptions about words are replaced with unfolding thoughts and responses. Movement seems to amplify and multiply the significations, the unsettling of text loosening and unravelling layers of responses. The animated body is generative here, with viewers responding to the oscillation of text, material, shadows and wall as a sensuous constituent of reading, rather than binding the words to some coded set of phrases. The curves, hand-printing and other interventions also create a sense of words as less stable or fixed, and as a result, reveal the text as less authoritative, more subjective. What Julia Kristeva refers to as "desire" in language[12] takes place through this rupture or destabilization, where the monotone/monovoice is replaced with sensuality and plurality. Building upon Mikhail Bakhtin's idea of polyphony,[13] in which multiple voices at play in any work are enabled to emerge, and intertexuality, where a multitude of texts intersect and inform each other, Kristeva foregrounds the importance of the reader to this dialogic set of responses. Referring to the space of a text as "dynamic,"[14] the subjective, associative and connotative properties of words seem to concurrently advance through choreogrammacy, so that single meanings lose adherence. This dilution of the authoritative text to an opening of multiple "subtexts" allows for diverse responses to occur, as the body works to both dislodge and re-write interpretations.

The relation of movement and reading to expanded meanings is also apparent in *Shadow Boxed* (2017), one of my recent works that is highlighted on the cover of this book. The cover conveys a detail of the larger piece, where white text on transparent white fabric sections are pinned at various heights across two gallery walls. The lack of contrast between the printed words, fabric and wall makes the text nearly invisible when standing at a distance, only becoming discernable when the viewer approaches the work. In reference to a shadow-box frame, the words, shadows, and wall create layers from the material that appears to float just above the surface. The shifts in legibility as the fabric catches the light, along with the subtle printing and oft-repeated phrases in the work, create a movement in and away from the piece as readers attempt to catch a word, then place it within the streaming, interconnected sentences. The barely-perceptible text interferes with the reader's expectations for clarity, as they bend and shuffle sideways to navigate the strings of words. Using undulations, creases, invisibility and extensive lengths of fabric, the work places numerous interventions into the reading act. In the spirit of the term "shadow boxing" as a training for some event, readers begin the process slowly, trying to piece together the subtle text. Once a spectator fully "rehearses" the text, they move back to the beginning of the piece for another performance. It is through the cast shadows that the text is most articulate, a kind of sub-vocalization or echo behind the whispered words. (Fig. 5) Not only do the shadowed words become more prominent, but a viewer's own shadows also interact, at times silhouetting their body as they perform the words; inversely, at times their own shadow impedes their view of the words. As readers vacillate between reading the faint words, reading the shadows beneath the piece, and interacting with the shimmering material, an awareness of alternating perspectives is enacted. One reader noted that while the words were imprinted and physicalized, which should "fix" or establish the text, the shadows undermined this assertiveness, creating disquiet and doubt.[15] Particularly with *Shadow Boxed*, the shadows take on more salience than the printed words, a kind of marginalia that pushes forward from the pale text at the surface. A viewer's potential to shift and change the text with each movement is analogous to the words' mutability, opening a plurality of Otherness from the sensuous layers.

The application of movement as both generating and expressing feeling contributes to the reading in these works, adding layers of complexity in response to a recognizable sign positioned in unfamiliar ways. The significations of words, both associative and connotative, formed from cultural codes, are coupled with the effects from kinetic and kinesthetic interaction with the pieces. Adding material and spatial aspects, this abundance of stim-

Figure 5. Interacting with *Shadow Boxed*, the viewer's moving body, the shadows of words and the satiny material are all significant components in reading (photograph by Aaron Moore).

ulation seeks to heighten reactions, as participants' responses suggest. Choreogrammatic reading proposes that the impulsion to move and the challenge of reading the words are synergistic: the pressing of the body seems to press against the words, and these effectively work together to squeeze the text beyond a set of signs deposited onto paper. The improvisational and interpretative possibilities enable readings beyond syntagmatic or paradigmatic approaches to one that include movement as its own form of understanding, thus forging a new relationship of textuality and corporeality.

NOTES

1. Maxine Sheets-Johnstone, *The Primacy of Movement* (Amsterdam: John Benjamins, 2011), 113.
2. Maxine Sheets-Johnstone, *The Primacy of Movement*, 118.
3. Participant, October 7, 2015.
4. Participant, October 5, 2015.
5. Participant, October 7, 2015.
6. Participant, September 15, 2015.
7. Beatrice Ward, "The Crystal Goblet, or Printing Should Be Invisible," in *Sixteen Essays on Typography* (London: Sylvan Press, 1955), 13.
8. Participant, October 5, 2015.
9. Maxine Sheets-Johnstone, "Movement and Mirror Neurons: A Challenging and Choice Conversation," *Phenomenology and the Cognitive Sciences* 11 (2012): 393.

10. Participant, September 7, 2015.
11. Participant, August 14, 2016.
12. Julia Kristeva, *Desire in Language: A Semiotic Approach* (New York: Columbia University Press, 1980).
13. Mikhail Bakhtin, *Problems of Dostoevsky's Poetics* (Minneapolis: University of Minnesota Press, 1984).
14. Julia Kristeva, "Word Dialogue Novel," in *The Kristeva Reader*, ed. Toril Moi (New York: Columbia University Press, 1986), 66.
15. Reader, December 21, 2017.

BIBLIOGRAPHY

Bakhtin, Mikhail. *Problems of Dostoevsky's Poetics*. Minneapolis: University of Minnesota Press, 1984.
Kristeva, Julia. *Desire in Language: A Semiotic Approach*. New York: Columbia University Press, 1980.
Kristeva, Julia. "Word Dialogue Novel." *The Kristeva Reader*. Edited by Toril Moi. New York: Columbia University Press, 1986.
Sheets-Johnstone, Maxine. "Movement and Mirror Neurons: A Challenging and Choice Conversation." *Phenomenology and the Cognitive Sciences* 11 (2012): 385–401.
Sheets-Johnstone, Maxine. *The Primacy of Movement*. Amsterdam: John Benjamins, 2011.
Ward, Beatrice. "The Crystal Goblet, or Printing Should Be Invisible." *Sixteen Essays on Typography*. London: Sylvan Press, 1955.

About the Contributors

Ninotchka D. **Bennahum** is a professor of theater and dance at the University of California, Santa Barbara. Her areas of research include European intellectual history and feminist historiographies of flamenco, ballet and contemporary performance.

A'Keitha **Carey** is a Bahamian artist, educator and scholar. She is a lecturer in the dance program at Miami Dade College Kendall Campus. She researches Caribbean spaces, locating movements that are indigenous, contemporary and fusion based, and investigates how Caribbean cultural performance (Bahamian Junkanoo, Trinidadian Carnival, and Jamaican Dancehall) can be viewed as praxis.

Linda E. **Carreiro** is a visual artist, associate professor and associate dean, academic, in the Faculty of Art, Ontario College of Art and Design at the University in Toronto. She pursues critical and creative research on physicalized text, resulting in solo exhibitions at Limerick Printmakers Gallery (Ireland), Center for Book Arts (New York City), Mallin Gallery (Kansas City), TRUCK Contemporary Art (Calgary), and Harcourt House (Edmonton).

K. Meira **Goldberg** is a flamenco performer, teacher, choreographer and historian. She teaches at Fashion Institute of Technology and is a visiting research scholar at the Foundation for Iberian music at the CUNY Graduate Center. Her monograph, *Sonidos Negros: On the Blackness of Flamenco*, was recently published by Oxford University Press (2018).

Michelle Heffner **Hayes** is professor and chair of dance at the University of Kansas. Her book *Flamenco: Conflicting Histories of the Dance* (McFarland, 2009) uses feminist and postcolonial theory to analyze constructions of gender, race and sexuality in representations of flamenco.

Christopher-Rasheem **McMillan** is a performance-related artist and scholar. He has a joint appointment between Dance and Gender and Women's and Sexuality Studies at the University of Iowa. McMillan's performance works have been seen at the Bates College Dance Festival, Providence International Arts Festival and the Dance Complex and Green Street Studios in Cambridge, Massachusetts.

Melissa **Melpignano** is a doctoral candidate in the Department of World Arts and Cultures/Dance at UCLA, where she works on theorizations of livability through choreography in Israel.

About the Contributors

Kiko **Mora** is a professor of the semiotics of advertising and culture industries at the University of Alicante (Spain) and of Spanish cinema for the CIEE of the same city. His present research explores the convergence of Spanish music and dance in early cinema and the early recording industry in the United States.

Rebecca K. **Pappas** is an assistant professor of dance at Ball State University and a guest artist in the master of arts in social practice at the University of Indianapolis. She makes dances that address the body as an archive for personal and social memory. Her choreography has received support from the Mellon Foundation, the Indiana Arts Commission, and Choreographers in Mentorship Exchange.

Lydia **Platón Lázaro** teaches at the University of Puerto Rico in Cayey in the Department of English. She is the author of *Defiant Itineraries: Caribbean Paradigms in American Dance and Film* (New York: Palgrave Macmillan, 2015) and numerous articles about Puerto Rican performance and dance artists.

Michael **Sakamoto** is an interdisciplinary scholar-artist and assistant professor in the Department of Dance at the University of Iowa. His work has been performed and exhibited in 14 countries at such venues as REDCAT, Vancouver International Dance Festival, Audio Art Festival-Krakow, Roulette/NYC, TACT/Fest Osaka, Jogja International Performing Arts Festival and Gøteborg Art Sounds.

Gwyneth **Shanks** is a Helena Rubinstein Curatorial Fellow in the Whitney Independent Study Program. Before that, she was the Mellon Postdoctoral Interdisciplinary Arts Fellow at the Walker Art Center. Her research focuses on curating performance in museums, and a new project explores the interface among performance, cities, and ecology.

Brynn Wein **Shiovitz** is a lecturer in the Department of Theatre at the University of California, Los Angeles, and the Department of Dance at Chapman University. Her research addresses the performance of racial masking vis-à-vis tap dance and other simultaneously aural forms of the stage, screen and sound cartoon. She has published in *Dance Chronicle*, *Dance Research Journal*, *Jazz Perspectives*, *Theatre Survey*, and *Women in Performance*.

Michelle T. **Summers** is a lecturer in the Department of Theater, Dance, and Performance Studies at UC Berkeley and adjunct faculty at the Graduate Theological Union. She holds a Ph.D. in critical dance studies from UC Riverside. Her research interests investigate the intersections of dance, religion, race, and right-wing studies. Her writing on dance can be seen in *Humanities Education Research*, *Dance Research*, and *Women and Performance Journals*, as well as in a forthcoming collection on dance and American culture.

MJ **Thompson** is a writer and teacher working on dance, performance, and visual art. She is an assistant professor of interdisciplinary studies and practices in the Department of Art Education at Concordia University, Montreal. Her articles have appeared in *Ballettanz*, *Dance Current*, *Dance Magazine*, *The Drama Review*, *Women and Performance*, and *Theatre Journal*.

Constance **Valis Hill** is a Five College professor emerita of dance at Hampshire College in Amherst, Massachusetts. Her writings have appeared in various dance magazines and edited collections devoted to dance. She is the author of *Brotherhood in Rhythm: The Jazz Tap Dancing of the Nicholas Brothers* (2000) and *Tap Dancing America, A Cultural History* (2010).

Index

Numbers in **_bold italics_** indicate pages with illustrations

abjection 4–9, 13, 19, 24, 26–27, 32, 90, 124, 146, 171; *see also* Kristeva, Julia; Shimakawa, Karen
Adorno, Theodore 174, 198
aesthetic anarchism 190 – 199
Africanist presence 9, 34, 56, 58, 68, 75–76 137–138, 140–141, 144–146, 196; *see also* cakewalk; Gottschild, Brenda Dixon
Afrofuturism 79, 119, 129; *see also* Chude-Sokei, Louis
Ahmed, Sara 4
AIDS epidemic 9, 14, 87, 89–90
Andalucía 7, 10, 41, 45–48, 51, 54, 56–58; *see also* architectural synecdoche; flamenco; fractured body; Hispano-Arab
anti-Fascist corporeality 198
appropriation 7, 56–57, 68–70, 75, 107, 114*n*7 139–140, 143
Aranda, José Otero 8, 42, 61*n*42
architectural synecdoche 7–8, 42, 45–48, 51, 56–58
Armida (fictional character) 12–13, 235– 236, 238–245; as Arab 12, 235, 242–243, 245
avant-garde 6, 56, 86, 209

baile ingles 48–49
Bakhtin, Mikhail 29, 259
ballet 49, 50–51, 56, 72, 74–76, 86–87, 95, 97–98, 104, 106, 127, 142–144, 196, 236, 238–239, 241, 243–245
Bassett, Angela 79–80
Benjamin, Walter 103, 196, 231, 261
Bergson, Henri: concept of *élan vital* 191, 192, 194
"Big Eliza" 24
black Atlantic 10, 25–26, 56, 119, 122–124, 137, 143, 153–154, 190, 214
Black Lōkəs 12, 202, ***206***–208, 214–216
blackface 2, 5, 20, 22, 29, 49; *see also* minstrelsy

body as capital 18–19, 21, 24–25, 30–32; *see also* Bourdieu, Pierre
body labor 12, 24–25, 80, 113, 140, 144, 222, 227, 229, 245; *see also* femininity
Bourdieu, Pierre 8, 18, 19, 21, 24, 26, 30–32, 34, 208
break dancing 76, 213–214, 216–217
Brown, Trisha 6, 12, 85, 97, 202–***204***, 205– 208, 211, 214–217; *see also* postmodern dance
buck dancing 1, 18, 22–27; mobile buck dance 24
bulerías 54, 136, 138, 144; *see also* flamenco
butoh 12, 208–212, 216–217

Cage, John 194
cakewalk 6, 22–23, 30, 42, 49, 53, 56, 97
Caribfunk 119, 126–131
carnival 10, 117–121, 123–124, 128–130
Castilla (place) 44–45, 48, 51, 56
Castilla, Luis **199**; *see also* flamenco
Castró, Americo 198
Catalan postmodernism 195
"choreogrammatics" 253–254, 259–261
choreographies of displacement 11, 13, 25, 122, 137, 214; *see also* displaced body; flamenco; postmodern dance
Chude-Sokei, Louis 56, 70, 79; *see also* diaspora
citizenship 74, 120–122, 126–127, 147, 178; *see also* identity
Clark, Michael 87–90
coachman's clog 22, 37*n*44
colonialism 4, 10, 42–44, 46, 69, 76, 79, 121– 127, 137, 143, 153–154, 156–160, 190, 242– 243, 245
compás (rhythmic structure) 144, 191–192
"coon" (nineteenth-century slang) 27; dancing 22; music 26 (*see also* ragtime)
corporeal erasure 42, 53, 58, 71, 97, 125, 134, 138–139, 141, 147, 176, 245

266 Index

Cunningham, Merce 6, 85, 97
La Curva (2010) 192, 193, 194, 195
cyborg body 56, 73, 78–80; see also *Strange Days*

dance manual 6, 147; see also libretto as dancing text
dancefilm 50–51, 57, 62*n*45
dancehall 117, 120–121, 129
Dauset Moreno, Carmen "Carmencita" 41, 59*n*1
Dean, Dora 53; see also cakewalk
DeFrantz, Thomas 3, 80*n*2, 81*n*7, 140, 148*n*26
desire 92–93, 158, 246*n*2; as identity 6, 24, 77, 120, 122, 183, 241, 259 (see also abjection; Kristeva, Julia); as mode of spectatorship 24, 71, 77, 158, 242–243, 245; as sexual 80, 90, 92–95, 120, 242, 244
diaspora 8–10, 12, 25, 37*n*39, 69, 74–75, 77, 81*n*2, 121–122, 125–126, 135, 138, 140, 142–145, 152, 154–155, 158, 165, 190; see also black Atlantic; migration
Didi-Huberman, Georges 191, 194
Dot 12–13
drill team 10, 105–106; see also military
Du Bois, W.E.B.: double consciousness 54, 56; talented tenth 32–33

energy 91, 117, 120, 176, 188*n*9; as empowerment 122, 130, 167, 245; as ephebism 71, 73–74, 76–77 (see also Africanist presence; Gottschild, Brenda Dixon); as erotic energy 97–98, 242, 245, 248*n*17; vis-à-vis text 236, 237–240, 242, 245, 248*n*17 (see also Armida)
Escudero, Vicente 191, 194, 198–199
essence of Virginia 1, 18, 22–28, 30, 35*n*4
Ethiopian: dances 2, *21*–23; specimens *19*, 21, *21*; see also minstrel troupes
Eurocentrism 12, 45, 49, 56–57, 69, 75, 119, 145–146, 154, 236, 242–243, 249*n*36

the feminine 7, 12, 78, 82*n*24, 89, 97, 106, 198, 222–226, 229, 231, 240, 243–244; see also performance of gender
feminism 107–108, 115*n*35, 127–128, 130, 159, 232*n*5; feminist art 6, 68–69, 79, 127–128, 123–124; feminist art history 220–222, 225, 230, 234*n*30 (see also Foster, Susan; French fashion); feminist pedestrianism (1960s) 6, 195, 197; see also postmodern dance
fetish 24–26, 29–33, *30*, *31*, 71, 90, 95, 196, 229; oral fixation 7–8, 18, 20, 27, *28*, 30–31 38*n*58; see also fractured body; the gaze
flamenco 6, 7, 8, 10, 11, 41–51, *52*, 53–58, 59*n*1, 61*n*32, 134–137, 147*n*6, 147*n*12, 148*n*25, 157, 191–199; *cuadro flamenco* 41, 53, 198; flamenco *escobilla* 48–49; *quadro flamenco* 51, *52*, 53; see also *bulerías*
flâneur 231–232
Flores, Antonio Montoya "El Farruco" 54
folkloric forms 10, 47–48, 51, 56, 57, 137–138, 139–141, 146, 165, 181, 184; in dance 10, 56, 141, 181, 184
football 105–106
Forti, Simone 6, 192, 195; see also postmodern dance
Foster, Susan 1, 3, 17, 241, 246; and "bodily writing" 3, 247*n*14; and choreography 135; and empathy 241; and the female body 246*n*2, 246*n*6, 246*n*7
fractured body: of the black body 7–8, 9, 18–21, *21*, 22–28, *28*, 28–35, 35*n*6 38*n*58, 38*n*62 68–71, 72, 74–75, 77–79, 118–120, 131; of the Gitano/a 56, 57, 61*n*38; of the Jews 172–173, 175, 180, *180*, 181–183, 184–185, 187; of the sexualized 4, 24, 47, 54, 86, 91–94, 95–96, 99, 117–119, 120, 122–123, 126, 127, 131*n*11, 173, 180, 224, 229, 236, 242, 245; see also identity
Franco, Francísco 190
Frank, Kevin 68, 71
French fashion 222–226, 229, 231–232, 233*n*15
Friedman, James 187; see also noses
Fung, Richard 70

Gades, Antonio 199
Gallimaufry 256, 258, *258*, 259
Gallota, Jean-Claude 194, 195; *The Legend of Don Juan* 195
Galván, José 191
Galván de los Reyes, Israel 190–199; concept of postmodern *flamenco* 192; see also flamenco
the gaze 71–72, 124, 207, 225, 241, 243
"Generation of '98" 42, 45–47, 57
Gilman, Sander 172, 181
Gitano/a 6, 44–45, 56, 60*n*25, 61*n*32, 148*n*25, 191–193, 196–198, 200*n*6
Glissant, Edouard 155–160, 167, 168*n*4
Gottschild, Brenda Dixon 20, 75, 140–144, 149*n*36
grotesque 18–34, *19*, *21*, *28*, 29, 37*n*36, 87, 124, 154, 163–167, 173, 180–183; see also Bakhtin, Mikhail

Halprin, Anna 6, 86, 192–193, 195; see also postmodern dance
Hernández, Teresa 11, 152–167, *165*, *166*, 168*n*2
hip hop 6, 12, 105–106, 127, 130, 143, 145, 150*n*43, 213–214
"hip wine" 10, 117–131, 131*n*11
historically black college university (HBCU) 10, 118
Hollywood (city) 33
Hollywood (industry) 71
Holocaust 196, 197, 174, 178–187, 188*n*9, 196; Art 172–173, 174–175; Effect 172–173, 175–176, 177
hybridity 6, 11, 25–26, 42, 49, 79–80, 129, 131*n*11, 140, 143, 156, 202–217; see also choreographies of displacement
hysteria 12–13, 242; see also femininity

Index 267

identity: as American 20, 22, 24–25, 32–33, 33–35, 85, 87, 90, 105–113, 127, 173–181, 206, 209, 211–214, 216, 222–232; as Afro-Caribbean 120, 124, 127–131, 141, 143; as Asian American 12, 206, 209–217; as black 1–8, 9, 10, 11, 12, 18–35, 41–58, 68–80, 106–108, 111–112, 117–131, 140, 169*n*32, 202–217; as Caribbean 10, 11, 68, 71, 79, 117–131, 138–147, 152, 153, 158–167; as Gitano 6, 42, 44–45, 56, 60*n*25, 61*n*32, 61*n*38, 148*n*25, 191–198, 200*n*6; as Hispano-Arab 190, 191; as Japanese 12, 209–217; as Jewish 11, 46, 69–70, 77, 139, 172–188, 196, 200 (*see also* ostjuden; sabra); as Mexican-American 12, 206–212; as Puerto Rican 10, 152–167, 168*n*9, 168*n*25, 169*n*29, 169*n*31, 169*n*34; as Quebecois 8–9, 68–78, 81*n*3, 82*n*30, 82*n*34, 82*n*35; as queer 8, 9, 11, 68–70, 77–80, 81*n*20, 82*n*24 86–94, 97–98 108, 169*n*29, 214; as white 20–24, 25, 34–35, 45, 57, 78, 82*n*32, 82*n*35, 104–113, 164, 207, 220, 246*n*6
Impressionism 219–232, **221**

jazz: dance 143, 145; music 6, 37*n*45, 56, 58, 138, 142, 180
Jewish body 11, 77, 172–188; "muscular Jewish" body 184–185
Jim Crow 20, 32, 111
jolly nigger bank 30–31, **31**
junkanoo 117, 124, 129

Kersands, Billy 1, 4, 7–8, 18, **19**, 19–**28**, 28–**30**, 30–35, 35*n*6, 38*n*58
Kristeva, Julia 8, 14, 20, 24, 26, 171, 259; *see also* abjection

La La La Human Steps 7, 9, 68, 69–80, 80*n*2
Lecavalier, Louise 7, 9, 68–80, 81*n*24, 82*n*30
Lepecki, André 4–5, 214
libretto 235–245, 246*n*7; as dancing text 12, 239–240, 245; as literary genre 238, 246*n*6, 247*n*8, 248*n*15
Lithgoe, Nigel 34
Little, Dick **19**, 27–28
Lo Que Queda/That Which Remains 10, 134, **142**, 142–147
Lo Real/Lo Réel/The Real (2014) 196
Lock, Édouard 9, 69–70, 73–77, 82*n*30
Locus 12
Lucas, Sam 8, 18, **19**, 21, 27, 32, 35, 35*n*5, 35*n*6
Lumière brothers 8, 41–58, 58*n*1, 59, 61*n*42, 62*n*45

Margarit, Angel 195
Masking 2, 5, 20, 22, 26, 29, 49, 56, 72, 131*n*3, 155, 184, 229, 243–244; *see also* blackface
Maya, Mario 198
McIntosh, Tom 8, 18, **21**, 32, 35, 36*n*7
Mercer, Kobena 72–75
migration 10, 25, 42, 63*n*63, 74, 122, 137, 149*n*37, 153–155, 165, 169*n*31, 190, 229, 232; *see also* diaspora
military (gesture and style) 105–107, 114*n*5, 159, 179
minstrelsy 1, 2, 3, 18–35, 49, 56, 114*n*5, 180; minstrel troupes 1–2, **19**, **21**, 22, 26, **28**, 35*n*5, 36*n*6, 36*n*7, 38*n*62; *see also* blackface; masking
modernism 5, 6, 56, 165, 191–192, 198, 206, 219–226, 231, 234*n*30
"Monster" 11, 172–187, 188*n*9
Muñoz, José Estaban 156, 168*n*18

nationalism 9, 60*n*16, 68–69, 73–74, 118, 120, 126, 165; as gendered 114*n*8; as Jewish 184–185 (*see also* Zionism); as patriotic 104–113; as Spanish 43–48, 57, 137
Nazi propaganda 180–181, 188*n*16; see also *ostjuden*
negritude 6, 56
"El Negro Meri" 8, 42, **50**, 50–58, 59*n*6, 62*n*48; *see also* Padilla, Jacinto
Nijinsky, Vaslav 9, 86, 89, 92, 97–99; see also *Rite of Spring*
1900 Paris Exhibition 7–8, 43–44, 50, 57

Orientalism 12–13, 53, 97, 236, 242–243, 246*n*6, 248*n*16, 250*n*37
Ostjuden 173, 185; *see also* femininity; grotesque
Otero Aranda, José 8, 42–43, 48–53

Padilla, Jacinto (*see also* el "Negro Meri") 8, 42, **50**, 53–55, **55**, 58, 59*n*6, 59*n*7
Padilla, Rafael "Chocolat" 54–55, **55**
palmas (hand clapping) 41, 53–54, 144–145, **145**
patriarchy 12, 105, 108, 112–113, 120, 122–123, 125, 240–245, 245*n*2
pattin' Juba 23, 35*n*6
performance of gender and sexuality 4, 9, 12, 72–73, 90–92, 98–99, 191, 219, 232, 236, 246*n*6; as queer 8, 9, 11, 68–70, 77–80, 81*n*20, 82*n*24 86–94, 97–98 108, 169*n*29, 214 (*see also* "Black Lōkəs"; Petronio, Stephen); *see also* the feminine
performance of sensuality 99, 122–123, 127, 238, 241, 259
performing reading 2, 11–12, 13, 69, 73, 78, 122, 126, 135, 140, 172, 178, 221, 225, 230, 236–237, 245, 253–255, **255**, 256, **257**, 258–261, 261
Petronio, Stephen 6, 9, 13, 85–99
photology 4; *see also* Lepecki, André
pickaninnies 26, 37*n*36
"plastic eye" 77
plurality 74, 259–260
Poetry Skins 253–254
Popick, Jeff 104–113, 114*n*12
"*postmemory*" 175, 181
postmodern dance 5–6, 11, 85–86, 135, 162,

268 Index

192–195, 202–208, 212–217; Judson Dance Theater 199
postwar Japan 209, 211–212
practice as research (PaR) 10, 141–144, 149n31
presence: as absence and presence 74, 178, 207; as embodied 6, 12, 75, 77, 79, 140–147, 148n25, 163, 230, 235–239, 240, 245, 246n2; as language and text 163, 237, 238, 239–240, 241, 244, 249n21; as mediated 33, 42, 152, 195, 241, 244
punk aesthetic 9, 69–74, 79, 87

Québécois dance 68–80, 82n34; see also La La La Human Steps

ragtime 23, 42, 56, 60n9, 140
Rainer, Yvonne 6, 85, 135, 195, 205; see also Judson Dance Theater; postmodern dance
Relay (choreography) 254, **255**, 256, **256**, **257**
religious right 6, 9, 104, 109
Remmington, Mamie 26
Requera, Anita "Anita de la Feria" **50**, 51, **52**
Reyes, Eugenia de los 191
ring shout 22
Rite of Spring (*Le Sacre du Printemps*) 9, 86, 88–95
Robinson, Bill 33
Roma 10, 42, 44, 54, 58, 134, 137–139, 148n25; see also Africanist presence; flamenco; Gitano
Román, Ivette 11, 152–159, **160**, 160–167, 168n2, 168n25

Sabra 173; see also grotesque
Sephardic 137, 139, 190–191
sevillanas 51–53, 62n51, 62n52, 138, 141; *sevillanas boleras* 41, 51–53, 62n52; see also flamenco
Shadow Boxed 260, **261**
Sheets-Johnstone, Maxine 254, 259
Shimakawa, Karen 5, 7–8
Siglo de Oro 45
So You Think You Can Dance 33–34; see also Lithgoe, Nigel; Sullivan, Melinda
Sondheim, Stephen 7, 12, 219, 222, 226, 230–231, 232n7, 233n11, 234n36; see also *Sunday in the Park with George*

soundscapes 5, 10, 56, 152–167
Spanish dance 8–9, 41–53, 58, 59n1, 135–137, 191–196, 200n6; see also flamenco; Gitano; palmas; sevillanas
Spanish Fascism 190, 191, 192
Spiegelman, Art 180–183; see also Nazi propaganda
stillness 6, 12, 99, 195, 219–223, 228–232; see also performance of gender
Strange Days 78–80
Stravinsky, Igor 9, 85–88, 90–95, 98–99
structures of whiteness 13, 34, 45, 57, 68–69, 71, 73–74, 78, 82n32, 82n35, 104–113, 164, 207, 220, 246n6; see also identity; patriarchy
Sullivan, Melinda 34–35; see also *So You Think You Can Dance*
Sunday in the Park with George (broadway show and painting) 7, 12, 219–232, 233n11, 233n23, 233n24 234n36

tango de negros 49, 148n17
tap dance 1–4, 7, 8, 18–35, 35n4, 37n27, 37n45, 149n42
Tasso, Torquato 235–238, 246n3, 246n4, 248n17
Tatsumi, Hijikata 202, 209
taxonomical approaches to the body 2–4, 13, 60n27, 153, 248n20
Taylor, Diana: and archive and repertoire 25, 247n12
Trump, Donald 6, 10, 104–113

University of Kansas 10, **142**, 142–145, **145**, 145–147, 149n29, 149n42
urban space 12, 48, 220–222, 231–232
USA Freedom Kids 10, 13, 104–113

victim 11, 86, 98, 112, 122, 172–175, 178–179, 181–184, 195, 215
victimizer 11, 172–173
Viganò, Onorato 239–240, 242–244, 249n23

Walker, Ada Overton 53, 97; see also cakewalk
writing as *différance* 4–5

Zionism 173, 184–186; see also nationalism

www.ingramcontent.com/pod-product-compliance
Lightning Source LLC
Chambersburg PA
CBHW021350300426
44114CB00012B/1159